Molecular Biology Intelligence Unit

Integrins and Development

Erik H.J. Danen, Ph.D.
Division of Cell Biology
The Netherlands Cancer Institute
Amsterdam, The Netherlands

CRC Press
Taylor & Francis Group
Boca Raton London New York

CRC Press is an imprint of the
Taylor & Francis Group, an **informa** business

INTEGRINS AND DEVELOPMENT

Molecular Biology Intelligence Unit

CRC Press
Taylor & Francis Group
6000 Broken Sound Parkway NW, Suite 300
Boca Raton, FL 33487-2742

First issued in hardback 2019

ISBN-13: 978-1-58706-293-3 (hbk)

Visit the Taylor & Francis Web site at
http://www.taylorandfrancis.com

and the CRC Press Web site at
http://www.crcpress.com

Library of Congress Cataloging-in-Publication Data

Integrins and development / [edited by] Erik H.J. Danen.
p. ; cm. -- (Molecular biology intelligence unit)
Includes bibliographical references and index.
ISBN 1-58706-293-3
1. Integrins. I. Danen, Erik H. J. II. Title. III. Series: Molecular biology intelligence unit (Unnumbered)
[DNLM: 1. Integrins--physiology. 2. Cell Adhesion--physiology. 3. Growth and Development--physiology. 4. Signal Transduction--physiology. QW 570 I585 2006]
QP552.I55I579 2006
572'.68--dc22

2006007063

CONTENTS

EDITOR

Erik H.J. Danen
Division of Cell Biology
The Netherlands Cancer Institute
Amsterdam, The Netherlands
Email: e.danen@nki.nl
Chapter 1

CONTRIBUTORS

Claudia S. Barros
The Scripps Research Institute
Department of Cell Biology
La Jolla, California, U.S.A.
Chapter 13

Cord Brakebusch
Max Planck Institute for Biochemistry
Department of Molecular Medicine
Martinsried, Germany
Email: brakebus@biochem.mpg.de
Chapter 7

Nicholas H. Brown
The Gurdon Institute and Department of Anatomy
University of Cambridge
Cambridge, U.K.
Email: n.brown@gurdon.cam.ac.uk
Chapter 6

Sara Cabodi
Department of Genetics
Biology and Biochemistry
University of Torino
Torino, Italy
Chapter 4

Alessandra Cambi
Department of Tumor Immunology
Nijmegen Centre for Molecular Life Sciences
Radboud University
Nijmegen, The Netherlands
Email: a.cambi@ncmls.ru.nl
Chapter 14

Katherine Clark
Department of Biochemistry
University of Leicester
Leicester, U.K.
Chapter 2

Paola Defilippi
Department of Genetics, Biology and Biochemistry
University of Torino
Torino, Italy
Email: paola.defilippi@unito.it
Chapter 4

Carl G. Figdor
Department of Tumor Immunology
Nijmegen Centre for Molecular Life Sciences
Radboud University
Nijmegen, The Netherlands
Email: c.figdor@ncmls.kun.nl
Chapter 14

Elisabeth Georges-Labouesse
Institut de Génétique et de Biologie Moléculaire et Cellulaire
CNRS/INSERM/ULP
Illkirch Cedex, France
Email: georges@igbmc.u-strasbg.fr
Chapter 5

Miep H. Helfrich
Department of Medicine and Therapeutics
University of Aberdeen Institute of Medical Sciences
Foresterhill, Aberdeen, U.K.
Chapter 8

Kairbaan M. Hodivala-Dilke
Cell Adhesion and Disease Group
Tumour Biology Laboratory
Cancer Research U.K. Clinical Centre
Barts and The London
Queen Mary's School of Medicine
and Dentistry
John Vane Science Centre
London, U.K.
Email:
kairbaan.hodivala-dilke@cancer.org.uk
Chapter 10

Michael A. Horton
Bone and Mineral Centre
Department of Medicine
London, U.K.
Email: m.horton@ucl.ac.uk
Chapter 8

Jordan A. Kreidberg
Department of Medicine
Children's Hospital Boston
and
Department of Pediatrics
Harvard Medical School
Boston, Massachusetts, U.S.A.
Email: jordan.kreidberg@tch.harvard.edu
Chapter 11

Michel Labouesse
Institut de Génétique et de Biologie
Moléculaire et Cellulaire
CNRS/INSERM/ULP
Illkirch Cedex, France
Email: lmichel@igbmc.u-strasbg.fr
Chapter 5

Ulrike Mayer
Wellcome Trust Centre
for Cell-Matrix Research
Faculty of Life Sciences
University of Manchester
Manchester, U.K.
Email: Ulrike.Mayer@manchester.ac.uk
Chapter 9

Ulrich Müller
The Scripps Research Institute
Department of Cell Biology
La Jolla, California, U.S.A.
Email: umueller@scripps.edu
Chapter 13

Maithreyi Narasimha
The Gurdon Institute
and Department of Anatomy
University of Cambridge
Cambridge, U.K.
Chapter 6

Matthew J. Naylor
Developmental Biology Program
Victor Chang Cardiac Research Institute
Darlinghurst, Australia
Chapter 12

Roumen Pankov
Craniofacial Developmental Biology
and Regeneration Branch
National Institute of Dental
and Craniofacial Research
National Institutes of Health
Bethesda, Maryland, U.S.A.
Email: rpankov@biofac.uni-sofia.bg
Chapter 2

Andrew R. Reynolds
Cell Adhesion and Disease Group
Tumour Biology Laboratory
Cancer Research U.K. Clinical Centre
Barts and The London
Queen Mary's School of Medicine
and Dentistry
John Vane Science Centre
London, U.K.
Chapter 10

Lisa K. Rogers
Wellcome Trust Centre
for Cell-Matrix Research
Faculty of Life Sciences
University of Manchester
Manchester, U.K.
Chapter 9

Kate A. Scott
Cell Adhesion and Disease Group
Tumour Biology Laboratory
Cancer Research U.K. Clinical Centre
Barts and The London
Queen Mary's School of Medicine
and Dentistry
John Vane Science Centre
London, U.K.
Chapter 10

Charles H. Streuli
Faculty of Life Sciences
University of Manchester
Manchester, U.K.
Email: cstreuli@manchester.ac.uk
Chapter 12

Suzanne F.G. van Helden
Department of Tumor Immunology
Nijmegen Centre for Molecular
Life Sciences
Radboud University
Nijmegen, The Netherlands
Email: s.vanhelden@ncmls.ru.nl
Chapter 14

Bernhard Wehrle-Haller
Department of Cellular Physiology
and Metabolism
Centre Medical Universitaire
Geneva, Switzerland
Email: Bernhard.Wehrle-Haller@
medecine.unige.ch
Chapter 3

PREFACE

Cell adhesion is essential for the organization of multicellular organisms. Indeed, various types of cell adhesion receptors, including cadherins and integrins are present in animals ranging from nematodes and insects to vertebrates. In this book, we focus on the integrin family, which is shared among all metazoans but has expanded considerably with vertebrate evolution. Since the cloning of the first integrin subunit, some twenty years ago, integrin biology has been—and still is—a topic of intense study. Integrin-mediated adhesion is a regulated process that, in turn, regulates the organization of the actin cytoskeleton. Moreover, it has become clear from in vitro analyses that integrin-mediated adhesion can affect virtually all aspects of cellular behavior—including polarity, motility, proliferation, survival, and differentiation. How important are these functions in the context of a whole organism? To address this question genetic studies in nematodes, flies, and mice over the past ten years have tested the roles of integrins and associated proteins in animal development, physiology, and pathology. This book aims to provide an extensive overview of the current knowledge about the regulation of developmental processes as well as the maintenance of proper tissue function, by integrin-mediated adhesion. In addition, key aspects of integrin cell biology are discussed.

Chapter 1 of this book is meant as an introduction in integrin biology and is followed by a more in-depth discussion of the roles that integrins play in extracellular matrix assembly, in cell migration, and in the regulation of intracellular signaling cascades (Chapters 2-4). Subsequently, Chapters 5 and 6 discuss what has been learned about the role of integrins and associated proteins in animal development from genetic analysis of two invertebrates—the flatworm, *C. elegans* and the fruit fly, *D. melanogaster*. The relatively limited number of genes encoding adhesion-related proteins and the relative ease and speed with which genetic experiments can be performed in these animals, have allowed researchers to study the basic principles of integrin biology in vivo. Finally, Chapters 7-14 discuss how integrin-mediated adhesion regulates the development and functionality of the different mammalian organ systems, based to a large extent on (conditional) gene knockout studies in mice and on studies in human patients.

The authors of each of the chapters are all currently working at the forefront of the different aspects of integrin research. We think this book will be of use as a book of reference for "integrin aficionados" but certainly also as an introduction to the field for those less familiar with this intriguing class of proteins.

On behalf of the authors,
Erik H.J. Danen, Ph.D.

Chapter 1

Integrins:
An Overview of Structural and Functional Aspects

Erik H.J. Danen*

Introduction

Integrins are heterodimeric transmembrane receptors that mediate cell-adhesion.[1] With their extracellular head region, most integrins bind extracellular matrix (ECM) glycoproteins such as laminins and collagens in basement membranes or connective tissue components like fibronectin. Others bind counterreceptors on neighboring cells, bacterial polysaccharides, or viral coat proteins. Through all these interactions integrins mediate stable adhesion to basement membranes, the formation of extracellular matrices and migration on such matrices, the formation of platelet aggregates, the establishment of intercellular junctions in the immune system, and bacterial and viral entry during infectious diseases. Furthermore, integrin-mediated adhesion modulates signaling cascades in control of cell motility, survival, proliferation, and differentiation. Here, a brief overview of basic structure/function aspects of integrins is given with the aim of providing background information for the following chapters in this book.

The History of Integrins

Intercellular cohesion is critical for the structural organization and intercellular communication within multicellular organisms. The origin of cell adhesion precedes multicellularity: genes encoding cadherin and C-type lectin domains were discovered in the genome of choanoflagellates, a group of metazoa-related protozoa.[2] The integrin family emerged in metazoa, which appeared during evolution over 600 million years ago. In the most ancient living animals - the sponges, no cadherins have been detected and sponge cell aggregation is mediated by an extracellular proteoglycan matrix. The receptor that binds to this matrix also contains the typical integrin-binding motif, RGD (see below). Indeed, α and β integrin subunits have been cloned from sponges and RGD peptides block sponge cell aggregation.[3] Some early metazoan must have evolved a laminin-binding integrin and an RGD-binding integrin, both of which have been preserved throughout metazoan evolution.[1] In the nematode *Caenorhabditis elegans*, one β subunit (termed βpat3) and two α subunits (termed αina1 and αpat2) form two integrins (Table 1). In the fruit fly *Drosophila melanogaster*, five integrins are formed through combination of one β subunit (termed βPS) with five α subunits (termed αPS1 through 5). With vertebrate evolution, the integrin family has expanded considerably. Many different laminin and RGD-binding integrins evolved and a new group of integrins appeared that have an extra inserted I-domain (see below) in the extracellular region and that contain the receptors for fibrillar collagens and the leukocyte-specific β2 integrins. Ultimately, sequencing of the human genome has identified as many as 18 α and 8 β subunits, from which 24 different functional integrins are currently known to be formed in humans.[4,5]

*Erik H.J. Danen—Division of Cell Biology, The Netherlands Cancer Institute, Plesmanlaan 121, 1066CX Amsterdam, The Netherlands. Email: e.danen@nki.nl

Integrins and Development, edited by Erik H.J. Danen.

Table 1. Integrins and their ligands

C. elegans Integrin	Ligand	*D. melanogaster* Integrin	Ligand	*H. sapiens* Integrin	Ligand	
αina1βpat3	LN	αPS1βPS	LN	α3β1	LN; TSP	**LAMININ**
				α6β1	LN; *ADAMs*	**RECEPTORS**
				α7β1	LN	
				α6β4	LN	
				α1β1	CO; LN	**FIBRILLAR**
				α2β1	CO; LN; MMP1	**COLLAGEN**
				α10β1	CO	**RECEPTORS**
				α11β1	CO	
αpat2βpat3	?	αPS2βPS	TG	α5β1	FN; *ADAMs*	**RGD**
				α8β1	FN; NN; TN; LAP	**RECEPTORS**
				αvβ1	FN; VN; LAP	
				αvβ3	FN; VN; VWF; OP; TN; BSP; TSP; *CD31*; *ADAMs*; MMP2	
				αIIbβ3	FN; VN; FG; VWF; CD40L; prothrombin	
				αvβ5	VN; BSP; LAP	
				αvβ6	FN; TN; LAP	
				αvβ8	FN; CO; LN; LAP	
		αPS3βPS	?	α4β1	FN; OP; *VCAM*	
		αPS4βPS	?	α9β1	TN; OP; LN; CO; *VCAM*; *ADAMs*	
		αPS5βPS	?			
				α4β7	FN; *VCAM*; *MadCAM*	**LEUKOCYTE-**
				αEβ7	*E-cadherin*	**SPECIFIC**
				αDβ2	*ICAM3*; *VCAM*	**INTEGRINS**
				αLβ2	*ICAM1-5*	
				αMβ2	*ICAM1*; *VCAM*; FG; iC3b	
				αXβ2	FG; iC3b	

Integrin heterodimers and their ligands in flatworms, flies, and mammals are shown.

ADAM: a disintegrin and metalloprotease; BSP: bone sialic protein; CO: collagen(s); FG: fibrinogen; FN: fibronectin; iC3b: inactivated complement component 3b; ICAM: intercellular adhesion molecule; LAP: TGFb latency associated protein; LN: laminin(s); MadCAM: mucosal addressin adhesion molecule; MMP: matrix metalloprotease; NN: nephronectin; OP: osteopontin; TG: tiggrin; TN: tenascin; TSP: thrombospondin; VCAM: vascular cell adhesion molecule; VN: vitronectin; VWF: von Willebrand Factor. Lines connect orthologues. Interactions involved in intercellular adhesion are indicated in italics.

Structural Aspects of Integrins

Integrins are formed through noncovalent association of two type I transmembrane glycoproteins, the α- and the β subunit. The extracellular parts are approximately 700 amino acids for α - and 1000 amino acids for β subunits and form elongated stalks and a globular ligand-binding head region[6] (Fig. 1). The part of the α subunit that forms the globular head region and participates in ligand binding consists of seven repeats of about 60 amino acids that fold into a seven-bladed β propeller structure similar to the trimeric G protein β subunit.[7] About half of the integrin α subunits has an insertion termed the I-domain within the β-propeller,

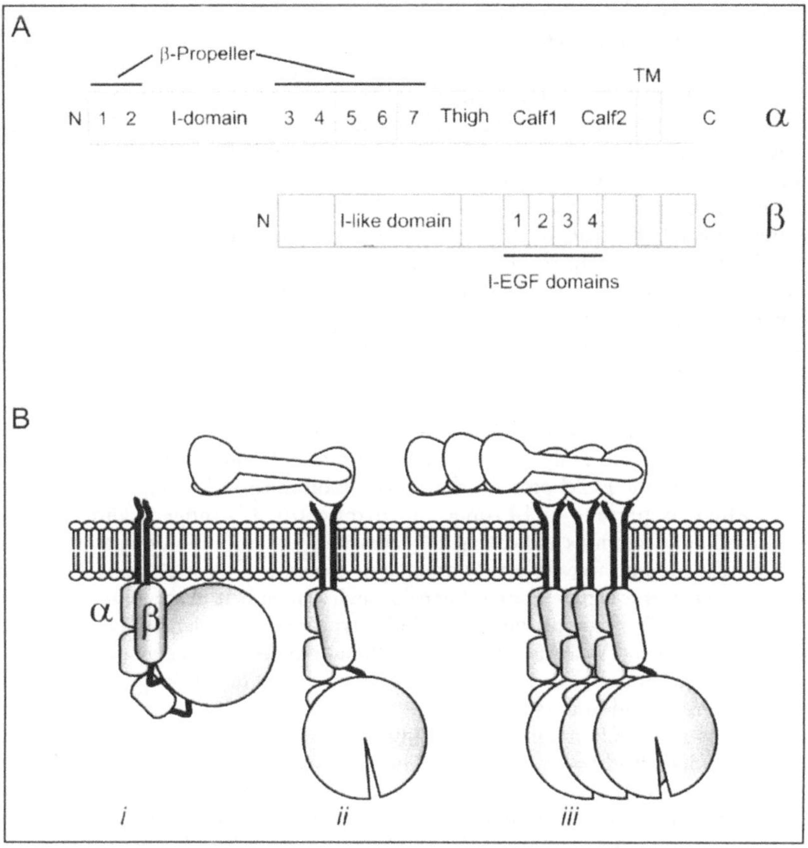

Figure 1. Structural aspects of integrins. A) Schematic drawing of integrin α and β subunits indicating basic structural units discussed in the text. B) Cartoon of inactive (i), activated (ii), and clustered integrins (iii). The exact nature of the conformatial change induced by talin-binding to the cytoplasmic region (ii) remains controversial. See text for details.

which directly participates in ligand binding and appears closely related to the trimeric G protein α subunit and small G proteins.[8] All integrin β subunits contain an "I-like domain" in the globular head region that also participates in ligand binding, perhaps most importantly in integrins lacking α I-domains. I domains contain a "metal ion-dependent adhesion site" (MIDAS) and I-like domains contain a structurally similar metal-binding motif. In analogy to small G proteins, where GTP hydrolysis leads to altered coordination of a Mg^{2+} ion causing conformational changes and loss of effector protein interaction, integrin ligand binding alters the coordination of the metal ion and shifts the I-domains from a closed to an open conformation. This process is thought to increase the affinity for ligand and to cause adhesion strengthening.[8]

The integrin cytoplasmic tails are less than 75 amino acids in length (the β4 tail forms an exception with its length of approximately 1000 amino acids containing 4 fibronectin type III repeats). There is a striking homology between the β cytoplasmic tails but the α cytoplasmic tails are highly divergent, except for a conserved GFFKR motif next to the transmembrane region that is important for association with the β tail.[9] A large number of cytoskeletal and signaling proteins has been reported to bind to β cytoplasmic tails and some have been found to interact with specific α tails. Most β tails contain an NPxY motif, which can potentially bind a wide variety of cytoskeletal and signaling proteins that contain a PTB-domain.[10] Furthermore, tyrosine phosphorylation in the NPxY motif may represent a mode of regulation of integrin interactions at the cytoplasmic face. By recruiting several proteins that bind actin filaments, the cytoplasmic tails mediate the integrin-cytoskeletal connection, which is essential for most, if not all, integrin mediated functions (see below for further details).

Finally, alternative mRNA splicing in the extracellular as well as in the intracellular regions gives rise to variant forms of several α and β subunits.[11] There is evidence that these variants can differ in ligand specificity or in their effect on signal transduction pathways. The selection for certain splice variants in specific cell types at specific stages of differentiation points to important functional differences. For instance, the switch to α7A and β1D variants that accompanies terminal differentiation of skeletal muscle cells probably reflects the need for a stronger interaction with laminin and a stronger connection to the cytoskeleton to be able to support muscle contraction.[12]

Extracellular Ligands and Lateral Associations

Most integrins bind to components of the ECM such as laminins, collagens, and fibronectin (Table 1). Many of the ECM proteins share a common integrin-binding motif, Arg-Gly-Asp (RGD), which is present in fibronectin, vitronectin, fibrinogen, and many others. Integrin-binding to laminins and collagens occurs at other recognition motifs, though cryptic RGD sites may become accessible after proteolytic cleavage of those proteins. Some integrins bind to counter receptors present on other cells, such as members of the ADAM (a disintegrin and metalloprotease) family or immunoglobulin-type receptors such as ICAM (intercellular adhesion molecule) and VCAM (vascular cell adhesion molecule) that are expressed on leukocytes and endothelial cells. Finally, bacterial polysaccharides and an increasing number of viruses are known to bind to integrins.

Various physical and functional interactions of integrins with other transmembrane receptors have been described. Some 11 integrins, in particular the laminin-binding integrins, associate with members of the tetraspanin family. It appears that such interactions do not modulate integrin-mediated adhesion but rather, that they affect the ability of cells to spread, migrate, and undergo morphological changes.[13] In addition, integrins can functionally interact with cell surface proteoglycans termed syndecans and they can transactivate receptor tyrosine kinases (see below for details).

Regulation of Integrin-Mediated Adhesion

For many biological processes, most notably hemostasis and immunity, it is important that integrin-mediated adhesion can be regulated. The number of integrin-ligand bonds can be regulated through changes in cellular shape, lateral diffusion of integrins in the membrane, and integrin clustering; aspects that can be controlled through cytoskeletal organization. Additionally, the intrinsic affinity of individual integrins for their ligands can be regulated from within the cell, a process referred to as "inside-out signaling".[5] Early studies on platelets had shown that thrombin binding to its receptor leads to a very rapid increase in fibrinogen binding by αIIbβ3. Since then, it has become clear that the α and β cytoplasmic tails are required to maintain integrins in the inactive state, that they associate with each other, and that this association is disrupted upon treatment with agonists known to cause integrin activation.[14]

One key player in regulation of integrin affinity appears to be the cytoskeletal adaptor talin. Binding of the talin head region to the integrin β cytoplasmic tail causes dissociation of the cytoplasmic tails and induces a conformational change in the extracellular region that increases integrin affinity.[15-17] How dissociation of the cytoplasmic tails leads to conformational changes all the way to the ligand-binding head domain is a subject of intense study. Based on electron microscopy, mutational analysis, structural analysis, and functional studies two models have been proposed. In the "switchblade model" inactive integrins are in a "bent" conformation with the head region facing the membrane. Dissociation of the cytoplasmic and transmembrane regions (e.g., upon talin binding) leads to dislocation of an EGF-like repeat in the β stalk, which causes the head region to extend upward in a switchblade-like motion.[7,18] In the "deadbolt model" the bent conformation is maintained in activated integrins but piston-like movements of the transmembrane regions (e.g., in response to talin binding) cause sliding of the stalks, which disrupts an interaction between the head region and the β stalk just above the membrane (the "deadbolt").[19,20] In both models, the described events during integrin activation ultimately cause conformational changes in the ligand-binding pocket of the head region that increase affinity for ligand.

The relative importance of conformational changes versus clustering in priming cells for adhesion versus post-ligand binding adhesion strengthening remains controversial.[21,22] In this respect, mutations in the β3 transmembrane region that cause constitutive αIIbβ3 activation have been reported to enhance the tendency of the β3 subunit to form homo-oligomers in one study but not in another, leaving open the question if clustering may in fact be a critical component of integrin activation.[23,24]

Integrin-Containing Adhesions

Integrin-engagement triggers the formation of membrane extensions that are required for cell spreading on ECM surfaces, for migration of cells into sheets of other cells, or for engulfment of particles or pathogens by phagocytic cells. Activation of Rho GTPases, the enzymes controlling the actin cytoskeleton, is involved in this process.[25] Ultimately, ligands, integrins, cytoskeletal proteins, and signaling molecules assemble in high local concentrations as aggregates on each side of the plasma membrane, forming "cell-matrix adhesions" in the case of integrins binding to ECM proteins[26] or forming complex intercellular junctions such as the "immunological synapse" in the case of leukocyte β2 integrins binding to counter receptors on other cells.[27] In contrast to the initial ligand-driven adhesion complexes, the formation of these more elaborate structures requires integrin-cytoskeletal linkage and is driven by cytoskeletal remodeling. Indeed, Rho family GTPases control the formation of these integrin-containing adhesions.[28]

The vast majority of integrins connect with filamentous (F)-actin in cell-matrix adhesions via cytoskeletal adaptor proteins such as talin, vinculin, paxillin, and others[26,29] (see below for details). In 2-dimensional culture systems these come in various flavors: cells can form "focal complexes" (small adhesions found in membrane protrusions of spreading and migrating cells), "focal adhesions" (larger adhesions connected by F-actin stress fibers that are derived from focal complexes in response to tension), and "fibrillar adhesions" (elongated adhesions associated with fibronectin matrix assembly)[30] (Fig. 2). In 3-dimensional cultures and in vivo, components of these structures coorganize in a mixed type of cell-matrix adhesions.[31] In osteoclasts, macrophages, some epithelial cells, and Rous sarcoma-transformed cells exceptional cell-matrix adhesions are found, termed "podosomes". These adhesions have been described mainly in vitro and consist of highly dynamic, dot-shaped F-actin-rich contacts whose function is not completely understood.[32] Finally, a unique type of cell-matrix adhesions, termed "hemidesmosomes", is formed by basal keratinocytes in the skin. Here, the unusually long cytoplasmic tail of the β4 subunit connects α6β4 not to the actin cytoskeleton but, instead, to the keratin cytoskeletal system. This association is not mediated by the proteins found in other cell-matrix adhesions (e.g., talin) but by a specialized protein complex including plectin, BP180, and BP230.[33]

The Integrin-Cytoskeleton Connection

Integrin function largely depends on the connection of integrins to the cytoskeleton.[34] The integrin cytoplasmic tails connect to the F-actin filaments through an exquisitely regulated multiprotein complex (Fig. 3). Some important players in this complex are discussed below.

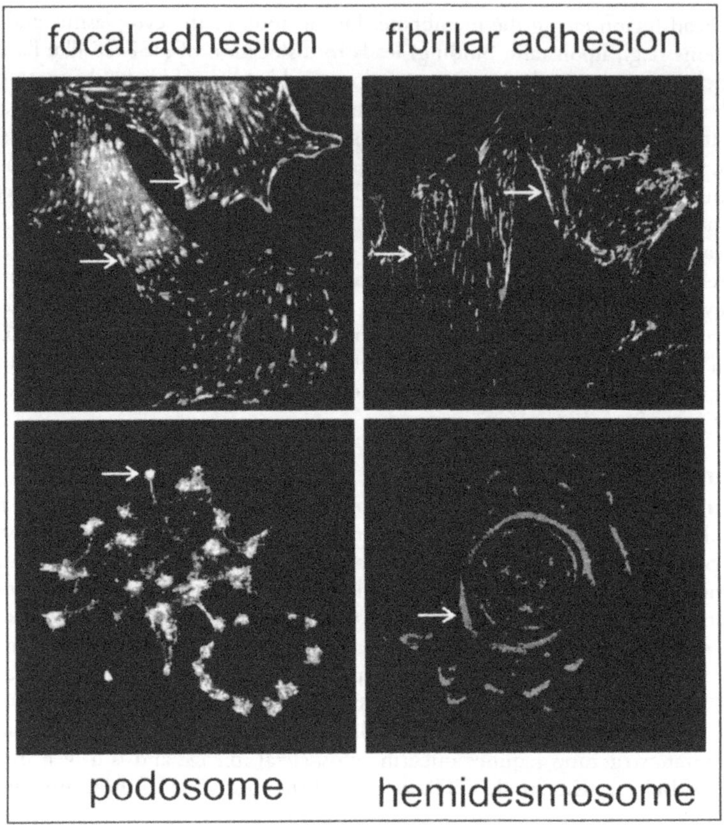

Figure 2. Cell-matrix adhesions. Immunofluorescence images showing four different types of cell-matrix adhesions. Arrows indicate clustering of paxillin in focal adhesions (upper left); tensin in fibrillar adhesions (upper right); activated Src in podosomes (lower left); and plectin in hemidesmosomes (lower right).

Talin

Talin is an antiparallel homodimeric actin cross-linking protein that can bind to integrin tails with the FERM domain in its globular head region and with a second, low affinity binding site in the tail.[35] The FERM domain also contains binding sites for F-actin and focal adhesion kinase (FAK). The talin tail region contains a conserved actin-binding site and multiple binding sites for vinculin. Through these interactions talin plays an important role in the assembly of cell-matrix adhesions. Talin knockout mice are embryonic lethal and undifferentiated talin null ES cells fail to assemble cell-matrix adhesions.[36,37] Cell-matrix-adhesion formation is restored following ES cell differentiation, which may be explained by the existence of a *talin-2* gene or by the usage of alternative integrin/actin binding proteins such as α-actinin or filamin. Biophysical studies have established that talin forms the initial link between microclusters of αvβ3 and the actin cytoskeleton.[38] The talin head FERM domain binds and activates PIPKIγ and thereby increases the local production of PIP2.[39,40] PIP2 in turn may activate several proteins present in adhesions, including talin resulting in reinforcement of the cytoskeletal linkage. Talin can be cleaved by the calcium-dependent protease calpain, which is important for the turn-over of cell-matrix adhesions.[41] Thus, besides its role in inside-out signaling discussed above, talin is also at the center of a multimolecular complex that connects integrins to the actin cytoskeleton.

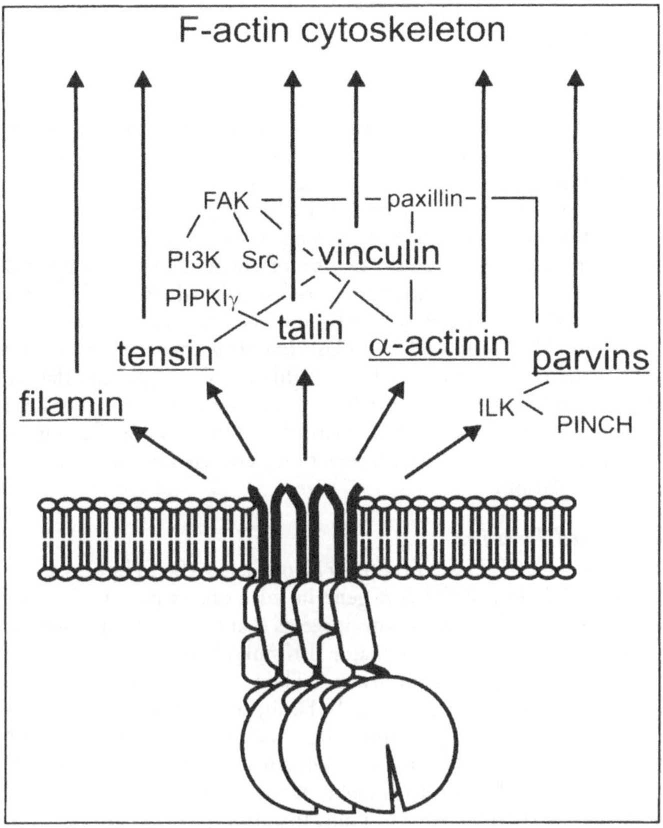

Figure 3. Integrin-F-actin connections. Schematic drawing of the various connections from integrin cytoplasmic tails to the actin cytoskeleton. Proteins involved in F-actin binding are underlined; those playing a regulatory role are indicated in small fonts.

α-actinin

Like talin, α-actinin is an antiparallel homodimeric actin cross-linking protein that can bind directly to integrin cytoplasmic tails or link integrins to the actin cytoskeleton indirectly through vinculin.[42] α-actinin is not dedicated to integrins but it also interacts with ICAMs, cadherins, syndecans, and other receptors. Four α-actinin isoforms exist in humans of which two are restricted to striated muscle. The interactions of α-actinin can be regulated by phospholipids and calcium and it may regulate cell-matrix adhesion disassembly in motile cells: it recruits MEKK1, a regulator of calpain, which in turn cleaves several cell-matrix adhesion components thereby regulating adhesion disassembly.[42,43]

Filamin

An alternative direct interaction from integrin cytoplasmic tails to F-actin can be mediated by filamins. In contrast to the antiparallel dimers talin and α-actinin, filamins are parallel homodimeric proteins that bind actin with their head domain and organize the actin cytoskeleton in a loose network.[44,45] With their rod domain they bind to transmembrane receptors including integrins and to Rho GTPases and several other proteins. The relative efficiencies of talin versus filamin-binding affect the ability of integrins to promote cell migration: mutations in the β1 cytoplasmic tail that increase binding to filamin-A inhibit membrane protrusion and cell migration whereas mutations in the β7 cytoplasmic tail that decrease binding to filamin-A have the opposite effect.[46]

Vinculin

Vinculin consists of a globular head domain that can bind talin and α-actinin and a tail region that can bind F-actin and paxillin.[35,47] Like for talin, vinculin's interactions can be regulated by PIP2 binding. Vinculin folds into an autoinhibited conformation through interactions between the head and the tail region. PIP2 binding transiently weakens this interaction as does F-actin binding to the vinculin tail, causing exposure of binding sites in the head region for talin and α-actinin. The open conformation of vinculin may be stabilized by binding to multiple proteins, which are concentrated in cell-matrix adhesions.[48,49] Vinculin knockout mice are embryonic lethal but this may be related to its function in cell-cell junctions as well as cell-matrix adhesions.[50] Vinculin null fibroblasts have few, small cell-matrix adhesions with an unusual high turnover rate and these cells are highly motile, suggesting that vinculin plays an important role in assembly of cell-matrix adhesions, although this does not hold through for all cell types and it is noteworthy that deletion of vinculin has no obvious effects in *Drosophila*.[35,50,51] The ability of vinculin to control the interaction between FAK and paxillin appears to be important in this regulation.[52]

ILK, PINCH, and Parvins

Integrin-linked kinase (ILK) was originally identified as a Ser/Thr kinase that binds integrin cytoplasmic tails.[53] Deletion of the *ILK* gene in mice causes peri-implantation lethality but based on genetic studies in flies and worms it seems that the kinase function of ILK is dispensable for normal development.[54-56] The major ILK null phenotype is an adhesion/polarization defect due to abnormal F-actin organization.[56] The C-terminal part of ILK that contains the putative kinase domain can bind paxillin and a family of F-actin-binding proteins identified in different laboratories under different names, together termed "parvins".[34] The N-terminal part of ILK contains 4 ankyrin repeats and a PH domain and it can bind PINCH1 and PINCH2, two adaptor proteins, which, in turn mediate interactions with several proteins involved in organization of the actin cytoskeleton.[57] Thus, this multiprotein complex can bind and regulate the actin cytoskeleton through multiple pathways.

Paxillin

Paxillin and its three related family members Hic-5, leupaxin, and PaxB are adaptor proteins that recruit multiple signaling intermediates to cell-matrix adhesions.[58] Paxillin contains four double zinc-finger motifs termed LIM domains in its C-terminal half, two of which are required for its localization in cell-matrix adhesions. It is not known which protein recruits paxillin in cell-matrix adhesions although direct binding to the α4 cytoplasmic tail has been demonstrated.[59] The N-terminal half of paxillin contains 5 LDxLLxxL (LD) motifs, which can bind vinculin, FAK, ILK, paxillin kinase linker (PKL), and α-parvin.[60,61] Paxillin can be phosphorylated on serine/threonine as well as on tyrosine (e.g., by FAK and Src), the latter generating SH2-binding sites that recruit the Src inhibitor, Csk and the SH2-SH3 adaptor protein, Crk. In turn, Crk couples to several signaling pathways including those in control of actin cytoskeletal organization.[62,63] Thus, paxillin brings together various regulatory pathways that control the dynamics of cell-matrix adhesions and the organization of the actin cytoskeleton.[64,65]

Tensin

Tensin is an actin filament cross-linking protein that localizes in cell-matrix adhesions and can bind the NPXY motif of integrin β tails through its PTB domain. Tensin comprises a multi-gene family, including the *tensin1*, *tensin2*, *tensin3*, and *cten* genes.[66] While most tensin family members are widely expressed, cten expression is restricted to epithelial cells in a few tissues, most notably the prostate. Tensin1 is important for proper kidney function and muscle regeneration while tensin3 is important for normal development of bone, lung, and intestine.[67,68] Tensin family members can be phosphorylated on tyrosine in response to growth factors and they can interact with other phosphotyrosine-containing proteins through their C-terminal SH2 domains (there is also a C-terminal PTB domain but apparently this does not bind tyrosine phosphorylated proteins).[66] Thus, tensin may represent an important signal integrator in cell-matrix adhesions. Two parts of the molecule can direct tensin localization to cell-matrix adhesions: the N-terminus through the actin-binding region and the C-terminus through the PTB domain. Although the cellular function of the different tensin members remains largely unclear, it has been implicated in fibronectin fibrilogenesis and tensin1 and 2 can promote cell migration, which requires cell-matrix adhesion targeting and phosphotyrosine-binding.[69,70]

FAK

Focal adhesion kinase is a tyrosine kinase whose activity is tightly regulated by cell adhesion but also by RTK signaling and oncogenes.[71] FAK localizes in cell-matrix adhesions through binding of its C-terminal "focal-adhesion targeting" region to paxillin and other, less-well characterized interactions. Clustering of integrins triggers the phosphorylation of Tyr397. Though FAK-mediated effects have been found to be independent of its kinase domain, this Tyr397 recognition site is essential for most, if not all functions of FAK. It forms a regulated binding site for Src and other SH2-containing proteins such as the p85 subunit of PI 3-kinase.[72] FAK-associated Src phosphorylates multiple other cell-matrix adhesion proteins as well as other tyrosine residues in FAK. FAK also binds the SH3 domain of the adaptor protein p130Cas, which, like paxillin, controls actin cytoskeletal organization through Crk.[73] FAK can also bind growth factor receptors and nonreceptor tyrosine kinases through its N-terminal FERM domain to regulate cell migration.[74,75] Thus, the FAK-Src complex may be a central regulator of adhesion dynamics in response to changes in the ECM and growth factor signaling. Indeed, fibroblasts deficient in FAK or Src family kinases show defects in cell-matrix adhesion turnover and cell motility.[65,76,77]

Biological Processes Involving Integrins

Matrix Assembly

Integrins participate in the assembly of various types of ECM. In the case of basement membranes, β1 integrins cooperate with the dystroglycan receptor to promote synthesis and polymerization of laminin chains into a multivalent network in which subsequently, collagen and other components are incorporated[78] (Fig. 4). Fibronectin-rich matrices are essential structures for cell migration during embryogenesis and wound healing.[79] The functional interaction of integrins with syndecans is important for fibronectin matrix assembly,[80] with α5β1 being the typical integrin involved although other integrins can compensate to some extent for its absence.[81,82] The high efficiency with which α5β1 promotes fibronectin matrix assembly may be related to the fact that binding of fibronectin through α5β1 efficiently stimulates the activity of RhoA,[83] a Rho GTPase that stimulates actomyosin-driven contractility, which is required for fibronectin fibrillogenesis.[84]

Cell Migration

Besides mediating stable adhesion, integrins play a crucial role in cellular motility.[85] Cell migration is essential for embryonic development, immune responses, and tissue repair, whereas deregulated migration characterizes metastatic cancers. Migrating cells establish a polarized morphology: the lamellipodium at the front contains an actin cytoskeletal meshwork associated with stationary focal complexes, whereas at the rear, F-actin stress fibers connect large, inward sliding focal contacts.[86] Stabilization of the lamellipodium requires integrin-mediated adhesion to the ECM or to other cells. By anchoring the actin cytoskeleton to the ECM in cell-matrix adhesions, integrins are also required for the generation of propulsive forces. It is presently unclear how spatial organization of the signals that control polarity of a migrating cell

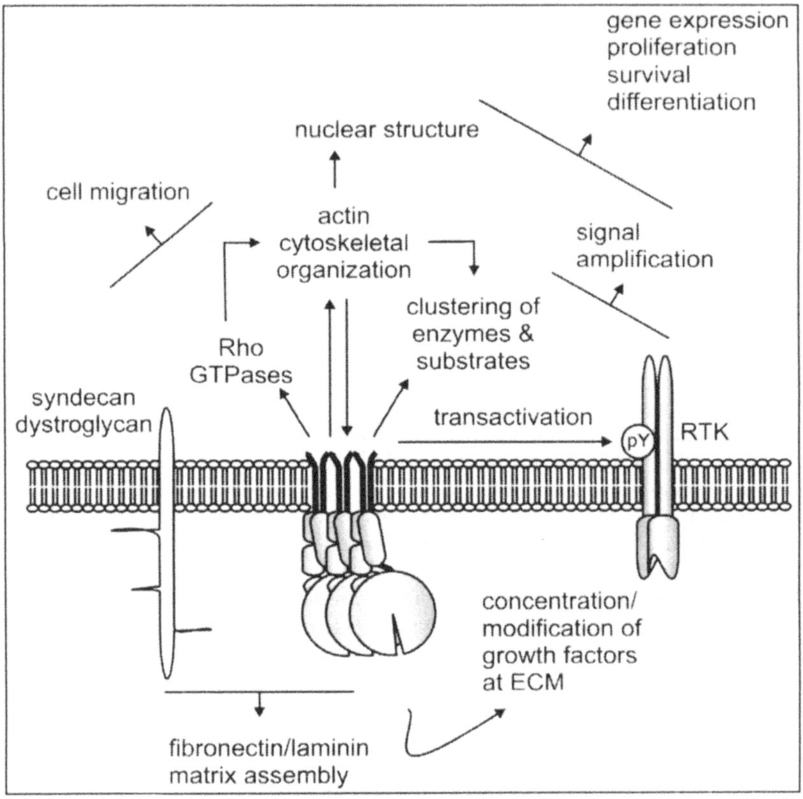

Figure 4. Functional implications of integrin-mediated adhesion. Schematic drawing of roles for integrins in ECM assembly, intracellular signaling pathways, and cell migration. See text for details.

is established, but concentration of distinct protein complexes at sites of integrin-mediated adhesion at the front and at the rear may be involved. Moreover, different integrins promote distinct modes of motility and cells may alter their motile behavior by expressing different integrins.[87]

Signaling

Integrin engagement can regulate intracellular signal transduction cascades that ultimately control differentiation, proliferation, and survival. For instance, for most cell types activation of the Raf—MEK—ERK signaling pathway is weak and transient when adhesion is perturbed but strong and sustained in the presence of integrin-mediated adhesion.[88,89] Through clustering of activated forms of signaling proteins such as FAK, Src, p130Cas, ILK, and ERK[26] cell-matrix adhesions may be the sites where amplification of signaling cascades takes place. It appears that oncogenes can uncouple this process of signal amplification from cell-anchorage: FAK activity is deregulated in transformed cells and the tyrosine kinase c-Abl phosphorylates paxillin in response to cell adhesion while its oncogenic variant, Bcr-Abl binds paxillin constitutively in a multi-phosphoprotein complex.[71,90,91] Even mRNA and ribosomes can be found associated with cell-matrix adhesions suggesting that they may locally connect signal transduction to protein synthesis.[92] Clustering and transactivation of PDGFR, EGFR, VEGFR and several other RTKs, is a more direct manner through which integrins can modulate RTK signaling.[93-95] Moreover, growth factors can be concentrated and modified by the ECM and integrin-mediated cell adhesion then allows their subsequent presentation to growth factor receptors.[96] Finally, through cytoskeletal connections with the nucleus the distribution and the stability of cell-matrix adhesions may even be connected physically to nuclear shape, chromatin structure, and gene expression.[97,98]

Through its effect on multiple signal transduction cascades, integrin-mediated cell adhesion regulates the G1 phase of the cell cycle. Integrins cooperate with RTKs to stimulate the cyclin E-cdk2 activity that drives S-phase entry.[99] Multiple different pathways have been described to connect integrins to G1 cell cycle progression but whichever pathway cells use, the organization of the actin cytoskeleton by integrins, is likely to be very important.[100] Most adherent cell types also depend on integrin-mediated adhesion for survival (apoptosis in response to loss of adhesion has been termed "anoikis"). Integrin-mediated cell adhesion in 2-dimensional culture systems stimulates PI3K-mediated PKB/AKT activity which mediates survival signals[101,102] and $\alpha6\beta4$ ligation supports NFκB-mediated survival signals in 3-dimensional cultures of mammary epithelial cells.[103] Integrins that are not ligand-bound can also trigger apoptosis of fully adherent cells by recruitment and activation of caspase-8[104,105] suggesting that a given integrin expression profile renders a cell dependent on a specific ECM environment for its survival. Finally, integrins also regulate the expression of genes related to differentiation. The synthesis of milk proteins by mammary epithelial cells,[106,107] the formation of contracting myotubes and expression of meromyosin by embryonic myoblasts,[108] the production of inflammatory cytokines by monocytes,[109] and the expression of involucrin in terminally differentiating keratinocytes[110] are all controlled in vitro by integrin-mediated adhesion.

Concluding Remarks

Here, some general features of the structure and function of integrins has been discussed. In the next three chapters we will learn in more detail how integrins are critically involved in extracellular matrix assembly, cell migration, and how integrins modulate many intracellular signaling cascades. Subsequently, the rest of this book will discuss what we can learn from genetic studies and from human diseases about the roles of integrins and associated proteins in animal development and tissue function.

Acknowledgements

I thank Stephan Huveneers and Sandy Litjens for providing images of podosomes and hemidesmosomes, and Arnoud Sonnenberg for critical reading of the manuscript. Funded by the Dutch Cancer Society (grant NKI 2003-2858).

References

1. Hynes RO, Zhao Q. The evolution of cell adhesion. J Cell Biol 2000; 150:F89-F96.
2. King N, Hittinger CT, Carroll SB. Evolution of key cell signaling and adhesion protein families predates animal origins. Science 2003; 301:361-363.
3. Harwood A, Coates JC. A prehistory of cell adhesion. Curr Opin Cell Biol 2004; 16:470-476.
4. Bokel C, Brown N. Integrins in development. Moving on, responding to, and sticking to the extracellular matrix. Dev Cell 2002; 3:311.
5. Hynes RO. Integrins: Bidirectional, allosteric signaling machines. Cell 2002; 110:673-687.
6. Xiong JP, Stehle T, Diefenbach B et al. Crystal structure of the extracellular segment of integrin alpha Vbeta3. Science 2001; 294:339-345.
7. Shimaoka M, Takagi J, Springer TA. Conformational regulation of integrin structure and function. Annu Rev Biophys Biomol Struct 2002; 31:485-516.
8. Liddington RC, Ginsberg MH. Integrin activation takes shape. J Cell Biol 2002; 158:833-839.
9. Hughes PE, Diaz-Gonzalez F, Leong L et al. Breaking the integrin hinge. A defined structural constraint regulates integrin signaling. J Biol Chem 1996; 271:6571-6574.
10. Calderwood DA, Fujioka Y, de Pereda JM et al. Integrin beta cytoplasmic domain interactions with phosphotyrosine-binding domains: A structural prototype for diversity in integrin signaling. Proc Natl Acad Sci USA 2003; 100:2272-2277.
11. de Melker AA, Sonnenberg A. Integrins: Alternative splicing as a mechanism to regulate ligand binding and integrin signaling events. Bioessays 1999; 21:499-509.

12. Mayer U. Integrins: Redundant or important players in skeletal muscle? J Biol Chem 2003; 278:14587-14590.
13. Hemler ME. Tetraspanin proteins mediate cellular penetration, invasion, and fusion events and define a novel type of membrane microdomain. Annu Rev Cell Dev Biol 2003; 19:397-422.
14. Calderwood DA. Integrin activation. J Cell Sci 2004; 117:657-666.
15. Vinogradova O, Velyvis A, Velyviene A et al. A structural mechanism of integrin alpha(IIb)beta(3) "inside-out" activation as regulated by its cytoplasmic face. Cell 2002; 110:587-597.
16. Garcia-Alvarez B, de Pereda JM, Calderwood DA et al. Structural determinants of integrin recognition by talin. Mol Cell 2003; 11:49-58.
17. Tadokoro S, Shattil SJ, Eto K et al. Talin binding to integrin beta tails: A final common step in integrin activation. Science 2003; 302:103-106.
18. Takagi J, Petre BM, Walz T et al. Global conformational rearrangements in integrin extracellular domains in outside-in and inside-out signaling. Cell 2002; 110:599-11.
19. Xiong JP, Stehle T, Goodman SL et al. New insights into the structural basis of integrin activation. Blood 2003; 102:1155-1159.
20. Adair BD, Xiong JP, Maddock C et al. Three-dimensional EM structure of the ectodomain of integrin {alpha}V{beta}3 in a complex with fibronectin. J Cell Biol 2005; 168:1109-1118.
21. Bazzoni G, Hemler ME. Are changes in integrin affinity and conformation overemphasized? Trends Biochem Sci 1998; 23:30-34.
22. Carman CV, Springer TA. Integrin avidity regulation: Are changes in affinity and conformation underemphasized? Curr Opin Cell Biol 2003; 15:547-556.
23. Li R, Mitra N, Gratkowski H et al. Activation of integrin alphaIIbbeta3 by modulation of transmembrane helix associations. Science 2003; 300:795-798.
24. Luo BH, Carman CV, Takagi J et al. Disrupting integrin transmembrane domain heterodimerization increases ligand binding affinity, not valency or clustering. Proc Natl Acad Sci USA 2005; 102:3679-3684.
25. DeMali KA, Burridge K. Coupling membrane protrusion and cell adhesion. J Cell Sci 2003; 116:2389-2397.
26. Geiger B, Bershadsky A, Pankov R et al. Transmembrane crosstalk between the extracellular matrix—cytoskeleton crosstalk. Nat Rev Mol Cell Biol 2001; 2:793-805.
27. Sims TN, Dustin ML. The immunological synapse: Integrins take the stage. Immunol Rev 2002; 186:100-117.
28. Nobes CD, Hall A. Rho, rac, and cdc42 GTPases regulate the assembly of multimolecular focal complexes associated with actin stress fibers, lamellipodia, and filopodia. Cell 1995; 81:53-62.
29. Burridge K, Chrzanowska-Wodnicka M. Focal adhesions, contractility, and signaling. Annu Rev Cell Dev Biol 1996; 12:463-518.
30. Zamir E, Geiger B. Molecular complexity and dynamics of cell-matrix adhesions. J Cell Sci 2001; 114:3583-3590.
31. Cukierman E, Pankov R, Stevens DR et al. Taking ceil-matrix adhesions to the third dimension. Science 2001; 294:1708-1712.
32. Linder S, Aepfelbacher M. Podosomes: Adhesion hot-spots of invasive cells. Trends Cell Biol 2003; 13:376-385.
33. Nievers MG, Schaapveld RQ, Sonnenberg A. Biology and function of hemidesmosomes. Matrix Biol 1999; 18:5-17.
34. Brakebusch C, Fassler R. The integrin-actin connection, an eternal love affair. EMBO J 2003; 22:2324-2333.
35. Critchley DR. Cytoskeletal proteins talin and vinculin in integrin-mediated adhesion. Biochem Soc Trans 2004; 32:831-836.
36. Priddle H, Hemmings L, Monkley S et al. Disruption of the talin gene compromises focal adhesion assembly in undifferentiated but not differentiated embryonic stem cells. J Cell Biol 1998; 142:1121-1133.
37. Monkley SJ, Zhou XH, Kinston SJ et al. Disruption of the talin gene arrests mouse development at the gastrulation stage. Dev Dyn 2000; 219:560-574.
38. Jiang G, Giannone G, Critchley DR et al. Two-piconewton slip bond between fibronectin and the cytoskeleton depends on talin. Nature 2003; 424:334-337.
39. Di Paolo G, Pellegrini L, Letinic K et al. Recruitment and regulation of phosphatidylinositol phosphate kinase type 1 gamma by the FERM domain of talin. Nature 2002; 420:85-89.
40. Ling K, Doughman RL, Firestone AJ et al. Type I gamma phosphatidylinositol phosphate kinase targets and regulates focal adhesions. Nature 2002; 420:89-93.
41. Franco SJ, Rodgers MA, Perrin BJ et al. Calpain-mediated proteolysis of talin regulates adhesion dynamics. Nat Cell Biol 2004; 6:977-983.

42. Otey CA, Carpen O. Alpha-actinin revisited: A fresh look at an old player. Cell Motil Cytoskeleton 2004; 58:104-111.
43. Cuevas BD, Abell AN, Witowsky JA et al. MEKK1 regulates calpain-dependent proteolysis of focal adhesion proteins for rear-end detachment of migrating fibroblasts. EMBO J 2003; 22:3346-3355.
44. Stossel TP, Condeelis J, Cooley L et al. Filamins as integrators of cell mechanics and signalling. Nat Rev Mol Cell Biol 2001; 2:138-145.
45. van der Flier A, Sonnenberg A. Structural and functional aspects of filamins. Biochim Biophys Acta 2001; 1538:99-117.
46. Calderwood DA, Huttenlocher A, Kiosses WB et al. Increased filamin binding to beta-integrin cytoplasmic domains inhibits cell migration. Nat Cell Biol 2001; 3:1060-1068.
47. Bailly M. Connecting cell adhesion to the actin polymerization machinery: Vinculin as the missing link? Trends Cell Biol 2003; 13:163-165.
48. Bakolitsa C, Cohen DM, Bankston LA et al. Structural basis for vinculin activation at sites of cell adhesion. Nature 2004; 430:583-586.
49. Izard T, Evans G, Borgon RA et al. Vinculin activation by talin through helical bundle conversion. Nature 2004; 427:171-175.
50. Xu W, Baribault H, Adamson ED. Vinculin knockout results in heart and brain defects during embryonic development. Development 1998; 125:327-337.
51. Volberg T, Geiger B, Kam Z et al. Focal adhesion formation by F9 embryonal carcinoma cells after vinculin gene disruption. J Cell Sci 1995; 108(Pt 6):2253-2260.
52. Subauste MC, Pertz O, Adamson ED et al. Vinculin modulation of paxillin-FAK interactions regulates ERK to control survival and motility. J Cell Biol 2004; 165:371-381.
53. Hannigan GE, Leung-Hagesteijn C, Fitz-Gibbon L et al. Regulation of cell adhesion and anchorage-dependent growth by a new beta 1-integrin-linked protein kinase. Nature 1996; 379:91-96.
54. Zervas CG, Gregory SL, Brown NH. Drosophila integrin-linked kinase is required at sites of integrin adhesion to link the cytoskeleton to the plasma membrane. J Cell Biol 2001; 152:1007-1018.
55. Mackinnon AC, Qadota H, Norman KR et al. C. elegans PAT-4/ILK functions as an adaptor protein within integrin adhesion complexes. Curr Biol 2002; 12:787-797.
56. Sakai T, Li S, Docheva D et al. Integrin-linked kinase (ILK) is required for polarizing the epiblast, cell adhesion, and controlling actin accumulation. Genes Dev 2003; 17:926-940.
57. Wu C. The PINCH-ILK-parvin complexes: Assembly, functions and regulation. Biochim Biophys Acta 2004; 1692:55-62.
58. Turner CE. Paxillin and focal adhesion signalling. Nat Cell Biol 2000; 2:E231-E236.
59. Liu S, Thomas SM, Woodside DG et al. Binding of paxillin to alpha4 integrins modifies integrin-dependent biological responses. Nature 1999; 402:676-681.
60. Nikolopoulos SN, Turner CE. Molecular dissection of actopaxin-integrin-linked kinase-Paxillin interactions and their role in subcellular localization. J Biol Chem 2002; 277:1568-1575.
61. Brown MC, Turner CE. Paxillin: Adapting to change. Physiol Rev 2004; 84:1315-1339.
62. Kiyokawa E, Hashimoto Y, Kobayashi S et al. Activation of Rac1 by a Crk SH3-binding protein, DOCK180. Genes Dev 1998; 12:3331-3336.
63. Valles AM, Beuvin M, Boyer B. Activation of Rac1 by paxillin-Crk-DOCK180 signaling complex is antagonized by Rap1 in migrating NBT-II cells. J Biol Chem 2004; 279:44490-44496.
64. Richardson A, Malik RK, Hildebrand JD et al. Inhibition of cell spreading by expression of the C-terminal domain of focal adhesion kinase (FAK) is rescued by coexpression of Src or catalytically inactive FAK: A role for paxillin tyrosine phosphorylation. Mol Cell Biol 1997; 17:6906-6914.
65. Webb DJ, Donais K, Whitmore LA et al. FAK-Src signalling through paxillin, ERK and MLCK regulates adhesion disassembly. Nat Cell Biol 2004; 6:154-161.
66. Lo SH. Tensin. Int J Biochem Cell Biol 2004; 36:31-34.
67. Lo SH, Yu QC, Degenstein L et al. Progressive kidney degeneration in mice lacking tensin. J Cell Biol 1997; 136:1349-1361.
68. Chiang MK, Liao YC, Kuwabara Y et al. Inactivation of tensin3 in mice results in growth retardation and postnatal lethality. Dev Biol 2005; 279:368-377.
69. Pankov R, Cukierman E, Katz BZ et al. Integrin dynamics and matrix assembly: Tensin-dependent translocation of alpha(5)beta(1) integrins promotes early fibronectin fibrillogenesis. J Cell Biol 2000; 148:1075-1090.
70. Chen H, Lo SH. Regulation of tensin-promoted cell migration by its focal adhesion binding and Src homology domain 2. Biochem J 2003; 370:1039-1045.
71. Parsons JT. Focal adhesion kinase: The first ten years. J Cell Sci 2003; 116:1409-1416.
72. Cary LA, Guan JL. Focal adhesion kinase in integrin-mediated signaling. Front Biosci 1999; 4:D102-D113.

73. Vuori K, Hirai H, Aizawa S et al. Introduction of p130cas signaling complex formation upon integrin-mediated cell adhesion: A role for Src family kinases. Mol Cell Biol 1996; 16:2606-2613.
74. Sieg DJ, Hauck CR, Ilic D et al. FAK integrates growth-factor and integrin signals to promote cell migration. Nat Cell Biol 2000; 2:249-256.
75. Chen R, Kim O, Li M et al. Regulation of the PH-domain-containing tyrosine kinase Etk by focal adhesion kinase through the FERM domain. Nat Cell Biol 2001; 3:439-444.
76. Arthur WT, Petch LA, Burridge K. Integrin engagement suppresses RhoA activity via a c-Src-dependent mechanism. Curr Biol 2000; 10:719-722.
77. Ren XD, Kiosses WB, Sieg DJ et al. Focal adhesion kinase suppresses Rho activity to promote focal adhesion turnover. J Cell Sci 2000; 113(Pt 20):3673-3678.
78. Li S, Edgar D, Fassler R et al. The role of laminin in embryonic cell polarization and tissue organization. Dev Cell 2003; 4:613-624.
79. George EL, Georges-Labouesse EN, Patel-King RS et al. Defects in mesoderm, neural tube and vascular development in mouse embryos lacking fibronectin. Development 1993; 119:1079-1091.
80. Schwarzbauer JE, Sechler JL. Fibronectin fibrillogenesis: A paradigm for extracellular matrix assembly. Curr Opin Cell Biol 1999; 11:622-627.
81. Wu C, Bauer JS, Juliano RL et al. The alpha 5 beta 1 integrin fibronectin receptor, but not the alpha 5 cytoplasmic domain, functions in an early and essential step in fibronectin matrix assembly. J Biol Chem 1993; 268:21883-21888.
82. Wennerberg K, Lohikangas L, Gullberg D et al. Beta 1 integrin-dependent and -independent polymerization of fibronectin. J Cell Biol 1996; 132:227-238.
83. Danen EH, Sonneveld P, Brakebusch C et al. The fibronectin-binding integrins {alpha}5{beta}1 and {alpha}v{beta}3 differentially modulate RhoA-GTP loading, organization of cell matrix adhesions, and fibronectin fibrillogenesis. J Cell Biol 2002; 159:1071-1086.
84. Zhong C, Chrzanowska-Wodnicka M, Brown J et al. Rho-mediated contractility exposes a cryptic site in fibronectin and induces fibronectin matrix assembly. J Cell Biol 1998; 141:539-551.
85. Ridley AJ, Schwartz MA, Burridge K et al. Cell migration: Integrating signals from front to back. Science 2003; 302:1704-1709.
86. Ballestrem C, Hinz B, Imhof BA et al. Marching at the front and dragging behind: Differential alphaVbeta3-integrin turnover regulates focal adhesion behavior. J Cell Biol 2001; 155:1319-1332.
87. Danen EH, van Rheenen J, Franken W et al. Integrins control motile strategy through a Rho-cofilin pathway. J Cell Biol 2005; 169:515-526.
88. Chen Q, Lin TH, Der CJ et al. Integrin-mediated activation of MEK and mitogen-activated protein kinase is independent of Ras [corrected]. J Biol Chem 1996; 271:18122-18127.
89. Renshaw MW, Ren XD, Schwartz MA. Growth factor activation of MAP kinase requires cell adhesion. EMBO J 1997; 16:5592-5599.
90. Salgia R, Sattler M, Pisick E et al. p210BCR/ABL induces formation of complexes containing focal adhesion proteins and the protooncogene product p120c-Cbl. Exp Hematol 1996; 24:310-313.
91. Lewis JM, Schwartz MA. Integrins regulate the association and phosphorylation of paxillin by c-Abl. J Biol Chem 1998; 273:14225-14230.
92. Chicurel ME, Singer RH, Meyer CJ et al. Integrin binding and mechanical tension induce movement of mRNA and ribosomes to focal adhesions. Nature 1998; 392:730-733.
93. Miyamoto S, Teramoto H, Gutkind JS et al. Integrins can collaborate with growth factors for phosphorylation of receptor tyrosine kinases and MAP kinase activation: Roles of integrin aggregation and occupancy of receptors. J Cell Biol 1996; 135:1633-1642.
94. Moro L, Venturino M, Bozzo C et al. Integrins induce activation of EGF receptor: Role in MAP kinase induction and adhesion-dependent cell survival. EMBO J 1998; 17:6622-6632.
95. Wang JF, Zhang XF, Groopman JE. Stimulation of beta 1 integrin induces tyrosine phosphorylation of vascular endothelial growth factor receptor-3 and modulates cell migration. J Biol Chem 2001; 276:41950-41957.
96. Faham S, Linhardt RJ, Rees DC. Diversity does make a difference: Fibroblast growth factor-heparin interactions. Curr Opin Struct Biol 1998; 8:578-586.
97. Maniotis AJ, Chen CS, Ingber DE. Demonstration of mechanical connections between integrins, cytoskeletal filaments, and nucleoplasm that stabilize nuclear structure. Proc Natl Acad Sci USA 1997; 94:849-854.
98. Lelievre SA, Weaver VM, Nickerson JA et al. Tissue phenotype depends on reciprocal interactions between the extracellular matrix and the structural organization of the nucleus. Proc Natl Acad Sci USA 1998; 95:14711-14716.
99. Assoian RK, Schwartz MA. Coordinate signaling by integrins and receptor tyrosine kinases in the regulation of G1 phase cell-cycle progression. Curr Opin Genet Dev 2001; 11:48-53.
100. Danen EH, Yamada KM. Fibronectin, integrins, and growth control. J Cell Physiol 2001; 189:1-13.

101. Khwaja A, Rodriguez-Viciana P, Wennstrom S et al. Matrix adhesion and Ras transformation both activate a phosphoinositide 3-OH kinase and protein kinase B/Akt cellular survival pathway. EMBO J 1997; 16:2783-2793.
102. King WG, Mattaliano MD, Chan TO et al. Phosphatidylinositol 3-kinase is required for integrin-stimulated AKT and Raf-1/mitogen-activated protein kinase pathway activation. Mol Cell Biol 1997; 17:4406-4418.
103. Weaver VM, Lelievre S, Lakins JN et al. beta4 integrin-dependent formation of polarized three-dimensional architecture confers resistance to apoptosis in normal and malignant mammary epithelium. Cancer Cell 2002; 2:205-216.
104. Varner JA, Emerson DA, Juliano RL. Integrin alpha 5 beta 1 expression negatively regulates cell growth: Reversal by attachment to fibronectin. Mol Biol Cell 1995; 6:725-740.
105. Stupack DG, Puente XS, Boutsaboualoy S et al. Apoptosis of adherent cells by recruitment of caspase-8 to unligated integrins. J Cell Biol 2001; 155:459-470.
106. Li ML, Aggeler J, Farson DA et al. Influence of a reconstituted basement membrane and its components on casein gene expression and secretion in mouse mammary epithelial cells. Proc Natl Acad Sci USA 1987; 84:136-140.
107. Edwards GM, Wilford FH, Liu X et al. Regulation of mammary differentiation by extracellular matrix involves protein-tyrosine phosphatases. J Biol Chem 1998; 273:9495-9500.
108. Menko AS, Boettiger D. Occupation of the extracellular matrix receptor, integrin, is a control point for myogenic differentiation. Cell 1987; 51:51-57.
109. Haskill S, Johnson C, Eierman D et al. Adherence induces selective mRNA expression of monocyte mediators and proto-oncogenes. J Immunol 1988; 140:1690-1694.
110. Adams JC, Watt FM. Fibronectin inhibits the terminal differentiation of human keratinocytes. Nature 1989; 340:307-309.

CHAPTER 2

Integrins in Extracellular Matrix Assembly

Roumen Pankov* and Katherine Clark

Integrin receptors are used by cells not only to bind, but also to actively organize extracellular matrices. They contribute in the in vivo polymerization of the major matrix constituents – fibronectin, laminin and collagen. Integrin participation in this process varies from indispensable, active stretching of the molecule and induction of fibronectin polymerization to more "passive" directing and facilitating of the formation of laminin polymers. The degree of integrin involvement, cellular input, and the sequence of each step in the assembly of different matrices depends primarily on the ability of matrix molecules to self-polymerize.

Introduction

The substance that fills the extracellular space and ensures the continuation of different tissues is known as the extracellular matrix (ECM). The old perception of this substance as an inert cellular "glue" is now long-gone. Important discoveries have demonstrated that the intercellular space is well organized, complex network of macromolecules. The integrity and composition of the ECM are vital for maintaining proper cellular functioning in the mature organisms, during development, and tissue repair.[1-3] Studies have documented four major classes of molecules as building constituents of the ECM in vertebrates – the collagens, proteoglycans, elastin, and glycoproteins.[4] With the exception of elastin, each class includes a large family of proteins with individual members that often demonstrate developmental-specific and tissue-specific distribution. These variations allow cells to craft unique environments by synthesizing and depositing specific members of each ECM family.

Secreted matrix molecules very often have to be modified by the cells in order to be properly incorporated into the ECM. These modifications can take place through deposition of proteases that proteolitically change the ECM proteins like collagens, or through direct interaction with the cells using specific ECM receptors, or both. The major class of these receptors is the integrin family of transmembrane heterodimeric glycoproteins. Integrin binding to the extracellular matrix proteins allows cells not only to acquire important information for the environment (see Chapter 4), to attach (see Chapter 1), and gain the necessary traction for migration (see Chapter 3), but also permits active cell-directed assembly and remodeling of the ECM. This chapter summarizes the current knowledge for the role of integrins in extracellular matrix organization with an emphasis on the strategies used by cells to fulfill this task.

Among the extracellular matrix ligands for integrins are the collagens and several members of the glycoprotein group – fibronectin, laminin, entactin/nidogen, tenascin, thrombospondin, von Willebrand factor and vitronectin.[5] The focus of this chapter will be on fibronectin, laminin and collagen since only these molecules are able to form polymeric structures, which act as a building framework of the different extracellular matrices.

The members of these three families of ECM macromolecules possess different abilities to polymerize. While both, laminin and collagen spontaneously form polymers in solution,

*Corresponding Author: Roumen Pankov—Craniofacial Developmental Biology and Regeneration Branch, National Institute of Dental and Craniofacial Research, National Institutes of Health, Bethesda, Maryland 20892-4370, U.S.A. Email: rpankov@biofac.uni-sofia.bg

Integrins and Development, edited by Erik H.J. Danen. ©2006 Landes Bioscience.

fibronectin dimers do not self-polymerize under physiologically relevant conditions and fibronectin fibrils are organized only on the surface of competent cells. Thus the integrin involvement in the formation of fibronectin, laminin and collagen matrices is a balance between the intrinsic ability of the ECM molecules to polymerize and the cellular input applied through the integrin receptors.

Integrins and Fibronectin Matrix Assembly

Fibronectin matrix deposition is upregulated at various times during development and in the adult organism to support the functions of cell adhesion, migration and differentiation. During development, neural crest cells migrate through the embryonic tissues in response to heterogeneity in the extracellular matrix. Molecules such as fibronectin and laminin act as permissive substrate, stimulating attachment and spreading of neural crest cell, whereas chondroitin sulphate proteoglycans are nonpermissive for migration. Taking various routes of migration through the embryo, neural crest cells reach their ultimate phenotype-specific sites and participate in the formation of a number of organ systems. In addition, fibronectin expression and deposition are markedly increased in tissues following injury, where it functions as a chemoattractant and adhesive molecule for migrating cells. During wound healing fibronectin promotes adhesion of keratinocytes and assist in their guidance across the wound base. In certain disease states, including atherosclerosis and pulmonary fibrosis, loss of fibronectin matrix regulation gives rise to either excess or inappropriate fibronectin deposition. Thus the fibronectin matrix appears to be a powerful and important regulator of normal and pathological cell behavior.

Fibronectin is a large glycoprotein found in blood at high concentrations (0.3 g/liter) as a soluble "plasma fibronectin". Wide varieties of cells secrete and organize considerably less soluble "cellular" form, which contains higher proportions of alternatively spliced variants than "plasma" fibronectin.[6] Soluble fibronectin is a compact dimer in a nonfunctional, closed conformation.[7-9] This form does not undergo fibril assembly, thus preventing the large quantities of circulating fibronectin from the disastrous consequences of accidental aggregation in the bloodstream. Fibronectin polymerization takes place only on the surface of competent cells like fibroblasts. This self-association process, termed fibronectin fibrillogenesis, or fibronectin matrix assembly,[10,11] is facilitated by multiple binding sites that have been identified along the molecule.[10,12] Some of these self-interaction sites are exposed and available for binding while others are cryptic and become accessible only after conformational changes. Exposure of the cryptic sites is obligatory for fibronectin fibrillogenesis to occur. Consequently, treatments that affect the conformation of the molecule increase fibronectin incorporation into the matrix.[13] Unfolding of the fibronectin molecule and uncovering of the hidden self-association binding sites is cell-regulated process dependant on integrin receptors and cellular contractility.

A considerable overlap exists in the binding of integrins to various ECM ligands. Therefore one integrin heterodimer can bind several adhesive proteins and one adhesive protein can be a ligand for several integrin receptors. Fibronectin is recognized by at least 12 different integrins,[5] which have diverse binding sites along the molecule,[12] but the majority of them are not capable of initiating formation of fibrils. Along with the "classical" fibronectin receptor $\alpha5\beta1$, integrins like $\alpha4\beta1$,[14] $\alpha V\beta3$,[15,16] and activated $\alpha IIb\beta3$[17] have been reported to support fibronectin matrix assembly, although not as efficiently as $\alpha5\beta1$. Numerous studies have demonstrated that $\alpha5\beta1$ integrin is the major receptor, responsible for fibronectin polymerization. For example transfection of $\alpha5$ integrin into Chinese hamster ovary cells leads to a large increase in fibronectin assembly.[18] Conversely, assembly of fibronectin by fibroblasts can be inhibited by antibodies against $\alpha5$ or $\beta1$ subunits[19] or by fibronectin fragments containing the $\alpha5\beta1$-binding domain.[20]

Fibronectin fibrillogenesis, mediated by $\alpha5\beta1$ integrin, is a multistep process (see 11) which begins with binding of fibronectin to the surface of a cell. Recognition is mediated by the primary adhesive (RGD) binding site,[21,22] recognized by the $\beta1$ integrin subunit and the synergy (PHSRN) sequence[23,24] that serves as a binding site for the α5 subunit. Both sites are located within the mid region of the fibronectin molecule known as the central cell-binding domain. However, binding of α5β1 is not restricted only to these sites. It can also occur at locations within the N-terminal portion of fibronectin.[25] Therefore, several integrin heterodimers can bind simultaneously to the same fibronectin molecule.

Binding of fibronectin can occur across the whole cell surface and the location of fibronectin polymerization can vary, depending on the cell type and density of the cell culture. For sparse fibroblasts, the preferential place for binding soluble fibronectin to the integrin receptors appears to be at focal adhesions[26] (Fig. 1A). These structures, act as cell adhesion sites that anchor bundles of actin microfilaments, thus providing support for the organization of the cytoskeleton (see Chapter 1). Interestingly, initial fibronectin secretion also occurs at these sites during cell spreading[26] and accumulation of nonfibrillar fibronectin aggregates can be found around the same structures on the surface of fully spread cells (R. Pankov, unpublished data). Such observations suggest that focal adhesions may act as places for gathering and preparation of the cell surface-bound fibronectin molecules prior to polymerization.

Binding of fibronectin to α5β1 integrin has several important consequences. The first outcome is the conformational change that may lead to exposure of at least some of the cryptic self-association sites within the fibronectin molecule.[11,14] Nevertheless, these conformational changes are not sufficient to induce fibrillogenesis since polymerization cannot occur without cellular contractility. Experimental approaches that prevent development of cellular tension,[27-29] or disrupt actin cytoskeleton,[17] successfully inhibit matrix assembly. A possible mechanism for exposing fibronectin cryptic sites, implies transformation of cellular tension into active translocation of α5β1 integrins out of focal adhesion and along bundles of actin filaments.[26,30] This movement provides the means for integrins to apply tensile forces to stretch fibronectin molecules and induce exposure of the cryptic self-association sites (Fig. 1B). Experimental stretching of fibronectin, adsorbed onto a silicone rubber substrate has been shown to increase binding of a 70 kDa fibronectin fragment, containing the assembly domain,[31] demonstrating that mechanical extension of the molecule can indeed promote self-assembly. Moreover, fibronectin fibrils on the surface of cultured cells are highly stretched. Using GFP-tagged fibronectin, Erickson and coworkers have shown that detachment of one end of a fibril from the cell surface resulted in a four fold shortening of the fibril length.[32] Stretching of the fibrils can be powered by integrin translocation since fibronectin fibrils have been observed to extend from focal adhesions[30] together with α5β1[26] and along the actin filament bundles toward the cell center.

While the integrin translocation, concomitant with fibronectin fibril extension is well documented, the mechanism that transforms cellular tension into integrin movement is still unclear. An attractive possibility involves changes in the mode of integrin interaction with the actin cytoskeleton, due to clustering. Clustering is another important consequence of fibronectin binding to α5β1 integrin. Both ligation and clustering are needed for organization of active protein complexes around integrin cytoplasmic tails.[33,34] As mentioned above, a fibronectin molecule contains several α5β1 binding sites and can easily induce formation of homogeneous aggregates of this integrin. These clusters of fibronectin-bound α5β1 integrins have the ability to translocate from focal adhesions and to generate a different adhesive structure, known as fibrillar adhesions.[35] In contrast to focal adhesions, these ECM contacts run parallel to actin filament bundles, contain only one representative of the integrin family – α5β1 integrin and have much simpler protein composition (see Chapter 1 and Fig. 1B). Thus the molecular complexes, formed around cytoplasmic tails of homogeneous clusters of fibronectin-occupied α5β1 integrins in fibrillar adhesions are distinct from those in integrin heterogeneous focal adhesions. It has been suggested[36] that the cytoskeletal proteins that link integrins to the cytoskeleton may vary depending on the integrins involved. Whereas focal adhesions act as substrate-anchoring sites for actin bundles and sustain actin-generated tension,[37] the complexes forming fibrillar adhesions are proposed to transform intracellular contractility or tension into directed movement along actin filaments.[26] Attractive speculative mechanisms for

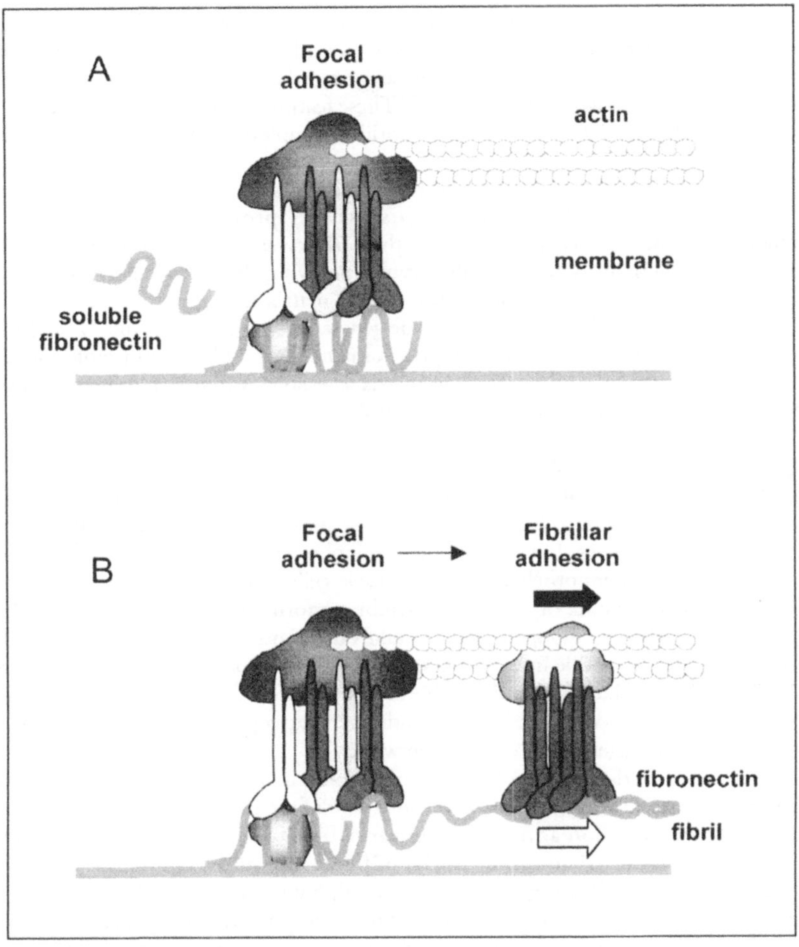

Figure 1. Model of early fibronectin fibrillogenesis driven by integrin translocation. A) Different integrin heterodimers (gray and light blue) together with integrin α5β1 (red) are clustered in focal adhesions. On the outside of the cell they interact with different ligands like vitronectin (dark gray) and collect and concentrate soluble fibronectin (green). Actin filaments (white) are anchored into the protein complexes that build up focal adhesions (dark blue) on the inner side of the plasma membrane. B) Binding of fibronectin to α5β1 integrins induces formation of homogeneous clusters of this integrin and organization of a new fibrillar adhesion protein complex (yellow). This complex translocates along actin filaments (black arrow), pulling clustered α5β1 integrins out of focal adhesions into new fibrillar adhesions. Translocating integrins in turn pull and stretch bound fibronectin molecules (white arrow), which unfolds and exposes cryptic self-association sites necessary for the formation of the fibronectin fibrils. To view color version of figure, please go to http://www.eurekah.com.

generating the motile force include myosin contractility and actin treadmilling. Long-range forces for translocation are probably generated by actin-myosin interactions, while short-range forces might be generated by actin treadmilling. This is supported by the capacity of contractility inhibitors such as H-7, ML-7, and BDM to block the movement of α5β1 integrins.[26,38] Newly incorporated actin appears first at focal adhesions and much later along stress-fiber bundles.[39] Capping of the growing (barbed) end could lead to treadmilling-driven shortening of the capped filaments, which together with release of actomyosin-based isometric tension, might result in active pulling of attached fibrillar adhesion complexes. The main cytoskeletal component of fibrillar adhesions – tensin has an actin barbed-end-capping domain together with two actin-binding domains. It has been suggested that its dimer could embrace the ends and block the growth of two actin filaments.[40] These features make tensin an attractive possible regulator, or mediator of the formation of fibrillar complexes that have the ability to move along actin filaments.[41]

Maturation of fibronectin fibrils, formed on the cell surface involves gradual conversion into a detergent-insoluble matrix.[13] This process probably involves the intrinsic disulphide-isomerase activity, found within the carboxy-end of the fibronectin molecule.[42] Interestingly, this activity is also partially cryptic implying that integrin-transmitted tension may also participate in this later stage of fibronectin matrix assembly.

In summary, fibronectin fibrils appear to be generated by a process of localized, directional, integrin translocation, which exposes cryptic domains on the fibronectin molecule, allowing polymerization into fibers and resulting in matrix assembly. An important prerequisite in this process is a well-organized, functional actin cytoskeleton.

Integrins in Laminin Matrix Assembly

Another extracellular matrix that is formed during embryogenesis is the basement membrane. It separates epithelium from the stroma and provides mechanical stability and barrier between different cell types. Diverse basement membranes are intimately associated with the basal part of epithelial, endothelial, fat, and muscle cells as well as with the peripheral nerve axons.[43,44] The constituents of basement membrane form sheet-like structures and include laminin, type IV collagen, nidogen, perlican and various other glycoproteins.[45] Both laminin and collagen can self-assemble into matrices, although in vivo polymerization occurs associated with the cell surface. Laminin polymerization is required for formation of a structurally recognisable basement membrane. Nidogen and other matrix proteins integrate into the laminin scaffold and form linkages to the collagen network.

Laminin is a cloverleaf shaped heterotrimer, which is formed through a long coiled-coil interaction of all three subunits forming the long arm of the protein. The remaining lengths of the subunits form three short arms. Several integrins bind to laminin in different regions of the protein, including many $\beta 1$ integrins ($\alpha 1\beta 1$, $\alpha 2\beta 1$, $\alpha 3\beta 1$, $\alpha 6\beta 1$, $\alpha 7\beta 1$) and $\alpha 6\beta 4$. In addition, laminin binds to dystroglycan, heparan sulphate and lipid sulphatide.[46,47] Unlike fibronectin, laminin is able to polymerize spontaneously into polygonal lattices through calcium-dependent interactions between its short arms.[48] This process occurs in vitro and has a critical concentration of 70-140 nM.[49] Polymerization below these critical levels has been found for laminin attached to synthetic phospholipid membranes.[50] Such observations have suggested that laminin-receptor interactions may act through generation of high local concentrations, thereby facilitating polymerization on the exterior of the cell. Indeed, in vivo laminin preferentially polymerizes when associated with the cell surface through integrins, dystroglycan and possibly other receptors for the laminin long arm. This polymerization has been termed "receptor facilitated self-assembly"[51] to indicate the cooperativity between the forming laminin polymer and receptor interactions. It has been demonstrated that $\beta 1$ integrins are required for basement membrane assembly around myotubes, ES and epithelial cells, and in the skin in $\beta 1$ and $\alpha 3\beta 1$ deficient mice.[52-55] Lack of these integrins makes cells unable to assemble proper basement membrane although they are surrounded by laminins, indicating that simple binding to the cells is not enough for laminin polymerization. A two phase process have been proposed to describe the mechanism by which integrins may influence laminin polymerization and the development of the basement membrane.[51,56] The first step includes binding of laminin to its receptors, leading to exceeding of the critical concentration on the cell surface (Fig. 2). This receptor facilitated laminin self-assembly is a mass-action driven process and does not require additional energy. The second step comprises laminin-cortical network reorganization. It involves the reorganization of laminin receptors and underlying cytoskeletal elements into

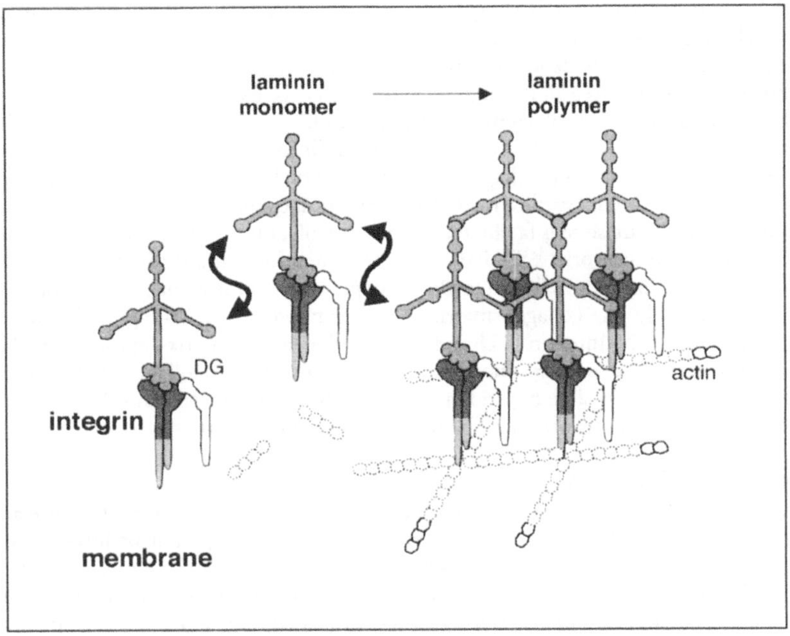

Figure 2. Schematic representation of receptor-facilitated laminin polymerization and laminin-cortical network formation on the muscle sarcolemal membrane. Laminin (green) becomes anchored to and concentrated on the outer surface of muscle cells. This is mediated by binding of the long laminin arm to integrin (red) and dystroglycan (yellow) receptors. Laminin polymerizes through short arm interactions (black curved arrows) which in turn leads to reorganization of the receptors. Ligand-occupied and reorganized receptors signal to the cell which results in reorganization of the underlying cytoskeleton (modified from refs. 51 and 56). To view color version of figure, please go to http://www.eurekah.com.

polygonal arrays in response to laminin polymer. This process appears to be active and requires actin filament remodelling and tyrosine phosphorylation, coordinated either by integrins or dystroglycan (Fig. 2).[51] Even though laminin can self-polymerize, it is clear that interaction with the cell surface, cell signalling, and the actin cytoskeleton are involved.

Further studies have revealed that β1 integrin is also required for α1 laminin chain expression,[57,58] and in the absence of the laminin α1 chain the laminin heterotrimer is unable to assemble and be secreted. β1 integrin knockout embryoid bodies are unable to assemble basement membranes, but are rescued when exogenous laminin is added and continue to develop normally.[59] Thus β1 integrins may have a dual action – being involved in the regulation of laminin α1 chain and participating in the receptor-facilitated laminin polymerization and these different mechanisms may operate in different cells.

Integrins in Collagen Matrix Assembly

Fibrillar collagens are synthesized as procollagen chains that form a heterotrimer made up from two or more different chains. The heterotrimer contains a region of triple helix and N- and C-propeptides. The propeptides are cleaved from the ends of the triple helices by specific proteases. Removal of the C-propeptide is essential for fibrillogenesis,[60] whereas the N-propeptide may remain, depending on the collagen, and become part of the matrix. Collagen is capable of spontaneous fibril formation, and after self-assembly the fibrils are stabilised and strengthened via crosslinking by lysyl oxidase. The fibril number and thickness of fibrillar collagen appears to be controlled by the interaction with various binding proteins, such as thrombospondin and tenascin.

Although the collagen matrix is capable of self-assembly, studies of its assembly in the presence of cells suggests that in vivo this process is more complicated. It has been found that collagen polymerization is dependent on its integrin receptors α2β1 and α11β1 as well as indirectly on the α5β1 integrin through fibronectin matrix assembly.[61,62] Fibronectin can bind directly to several different collagens and interdependence of their matrix assembly has been suggested. Mouse embryonic cells that do not express fibronectin and have no collagen binding integrins are unable to initiate collagen matrix assembly. Integrins α2β1 or α11β1 appear to nucleate collagen to the cell surface mediating formation of short soluble collagen fibres. Addition of fibronectin to these cells facilitated production of a fibronectin matrix and construction of a typical collagen network.[61] Similar results were obtained in a study by Li et al, in which collagen matrix formation was found to also be dependent on the actin cytoskeleton and RhoA activity.[62] Conversely, the collagen matrix is disorganised in the bone growth plate of mice lacking expression of β1 integrin in chondrocytes, whereas the matrix appears normal in those lacking fibronectin.[63] Currently, the role that integrins play in collagen matrix assembly is unclear. Although the possible dependence on the fibronectin matrix and therefore indirectly the integrins involved in its production, are potentially important.

Concluding Remarks

Fibronectin matrix and basement membrane are very different in their structure and composition; nevertheless they both are formed with active participation of integrin receptors. Accumulated data and the emerging models for the formation of these extracellular matrices show similar steps in the process of their assembly. Both involve ligand binding to integrins; ligand self-association; receptor clustering; and cytoskeletal reorganization. Despite these similarities there is also an important difference, rooted in the ability of the particular ligands to self-assemble. Fibronectin polymerization occurs as a result of the activity of a well-organized cytoskeleton, while laminin self-polymerization leads to cytoskeletal reorganization. This variation will trigger different integrin signaling cascades and is most probably related to the different tasks fulfilled by the fibroblast and epithelial cells.

References

1. Hay ED. Cell Biology of Extracellular Matrix. 2nd ed. New York: Plenum Press, 1991.
2. Trelstad RL. The Role of Extracellular Matrix in development. New York: Alan R Liss, 1984.
3. Brik DE, Silver FH, Trelstad RL. Matrix Assembly. In: Hay ED, ed. Cell Biology of Extracellular Matrix. 2nd ed. New York, London: Plenum Press, 1991:221-254.
4. Piez KA, Reddi HA. Extracellular Matrix Biochemistry. New York: Elsevier, 1984.
5. Plow EF, Haas TA, Zhang L et al. Ligand binding to integrins. J Biol Chem 2000; 275(29):21785-21788.
6. Yamada KM. Fibronectin and other cell interactive glycoproteins. In: Hay ED, ed. Cell Biology of Extracellular Matrix. 2nd ed. New York: Plenum Press, 1991.
7. Williams EC, Janmey PA, Ferry JD et al. Conformational states of fibronectin. Effects of pH, ionic strength, and collagen binding. J Biol Chem 1982; 257(24):14973-14978.
8. Erickson HP, Carrell NA. Fibronectin in extended and compact conformations. Electron microscopy and sedimentation analysis. J Biol Chem 1983; 258(23):14539-14544.
9. Rocco M, Carson M, Hantgan R et al. Dependence of the shape of the plasma fibronectin molecule on solvent composition. Ionic strength and glycerol content. J Biol Chem 1983; 258(23):14545-14549.
10. Geiger B, Bershadsky A, Pankov R et al. Transmembrane crosstalk between the extracellular matrix—cytoskeleton crosstalk. Nat Rev Mol Cell Biol 2001; 2(11):793-805.
11. Wierzbicka-Patynowski I, Schwarzbauer JE. The ins and outs of fibronectin matrix assembly. J Cell Sci 2003; 116(16):3269-3276.
12. Pankov R, Yamada KM. Fibronectin at a glance. J Cell Sci 2002; 115(Pt 20):3861-3863.
13. McKeown-Longo PJ, Mosher DF. Binding of plasma fibronectin to cell layers of human skin fibroblasts. J Cell Biol 1983; 97(2):466-472.
14. Schwarzbauer JE, Sechler JL. Fibronectin fibrillogenesis: A paradigm for extracellular matrix assembly. Curr Opin Cell Biol 1999; 11(5):622-627.

15. Wennerberg K, Lohikangas L, Gullberg D et al. Beta 1 integrin-dependent and -independent polymerization of fibronectin. J Cell Biol 1996; 132(1-2):227-238.
16. Wu C, Hughes PE, Ginsberg MH et al. Identification of a new biological function for the integrin alpha v beta 3: Initiation of fibronectin matrix assembly. Cell Adhes Commun 1996; 4(3):149-158.
17. Wu C, Keivens VM, O'Toole TE et al. Integrin activation and cytoskeletal interaction are essential for the assembly of a fibronectin matrix. Cell 1995; 83(5):715-724.
18. Giancotti FG, Ruoslahti E. Elevated levels of the alpha 5 beta 1 fibronectin receptor suppress the transformed phenotype of Chinese hamster ovary cells. Cell 1990; 60(5):849-859.
19. Akiyama SK, Yamada SS, Chen WT et al. Analysis of fibronectin receptor function with monoclonal antibodies: Roles in cell adhesion, migration, matrix assembly, and cytoskeletal organization. J Cell Biol 1989; 109(2):863-875.
20. McDonald JA, Quade BJ, Broekelmann TJ et al. Fibronectin's cell-adhesive domain and an amino-terminal matrix assembly domain participate in its assembly into fibroblast pericellular matrix. J Biol Chem 1987; 262(7):2957-2967.
21. Pierschbacher MD, Ruoslahti E. Cell attachment activity of fibronectin can be duplicated by small synthetic fragments of the molecule. Nature 1984; 309(5963):30-33.
22. Yamada KM, Kennedy DW. Dualistic nature of adhesive protein function: Fibronectin and its biologically active peptide fragments can autoinhibit fibronectin function. J Cell Biol 1984; 99(1 Pt 1):29-36.
23. Aota S, Nagai T, Yamada KM. Characterization of regions of fibronectin besides the arginine-glycine-aspartic acid sequence required for adhesive function of the cell-binding domain using site-directed mutagenesis. J Biol Chem 1991; 266(24):15938-15943.
24. Nagai T, Yamakawa N, Aota S et al. Monoclonal antibody characterization of two distant sites required for function of the central cell-binding domain of fibronectin in cell adhesion, cell migration, and matrix assembly. J Cell Biol 1991; 114(6):1295-1305.
25. Hocking DC, Sottile J, McKeown-Longo PJ. Activation of distinct alpha5beta1-mediated signaling pathways by fibronectin's cell adhesion and matrix assembly domains. J Cell Biol 1998; 141(1):241-253.
26. Pankov R, Cukierman E, Katz BZ et al. Integrin dynamics and matrix assembly: Tensin-dependent translocation of alpha(5)beta(1) integrins promotes early fibronectin fibrillogenesis. J Cell Biol 2000; 148(5):1075-1090.
27. Halliday NL, Tomasek JJ. Mechanical properties of the extracellular matrix influence fibronectin fibril assembly in vitro. Exp Cell Res 1995; 217(1):109-117.
28. Zhang Q, Checovich WJ, Peters DM et al. Modulation of cell surface fibronectin assembly sites by lysophosphatidic acid. J Cell Biol 1994; 127(5):1447-1459.
29. Zhang Q, Magnusson MK, Mosher DF. Lysophosphatidic acid and microtubule-destabilizing agents stimulate fibronectin matrix assembly through Rho-dependent actin stress fiber formation and cell contraction. Mol Biol Cell 1997; 8(8):1415-1425.
30. Ohashi T, Kiehart DP, Erickson HP. Dual labeling of the fibronectin matrix and actin cytoskeleton with green fluorescent protein variants. J Cell Sci 2002; 115(6):1221-1229.
31. Zhong C, Chrzanowska-Wodnicka M, Brown J et al. Rho-mediated contractility exposes a cryptic site in fibronectin and induces fibronectin matrix assembly. J Cell Biol 1998; 141(2):539-551.
32. Ohashi T, Kiehart DP, Erickson HP. Dynamics and elasticity of the fibronectin matrix in living cell culture visualized by fibronectin-green fluorescent protein. Proc Natl Acad Sci USA 1999; 96(5):2153-2158.
33. Miyamoto S, Akiyama SK, Yamada KM. Synergistic roles for receptor occupancy and aggregation in integrin transmembrane function. Science 1995; 267(5199):883-885.
34. Yamada KM, Miyamoto S. Integrin transmembrane signaling and cytoskeletal control. Curr Opin Cell Biol 1995; 7(5):681-689.
35. Zamir E, Katz BZ, Aota S et al. Molecular diversity of cell-matrix adhesions. J Cell Sci 1999; 112(Pt 11):1655-1669.
36. Critchley DR, Holt MR, Barry ST et al. Integrin-mediated cell adhesion: The cytoskeletal connection. Biochem Soc Symp 1999; 65:79-99.
37. Burridge K, Chrzanowska-Wodnicka M. Focal adhesions, contractility, and signaling. Annu Rev Cell Dev Biol 1996; 12:463-518.
38. Katz BZ, Zamir E, Bershadsky A et al. Physical state of the extracellular matrix regulates the structure and molecular composition of cell-matrix adhesions. Mol Biol Cell 2000; 11(3):1047-1060.
39. Small JV, Rottner K, Kaverina I et al. Assembling an actin cytoskeleton for cell attachment and movement. Biochim Biophys Acta 1998; 1404(3):271-281.
40. Lo SH, Janmey PA, Hartwig JH et al. Interactions of tensin with actin and identification of its three distinct actin-binding domains. J Cell Biol 1994; 125(5):1067-1075.

41. Zamir E, Katz M, Posen Y et al. Dynamics and segregation of cell-matrix adhesions in cultured fibroblasts. Nat Cell Biol 2000; 2(4):191-196.
42. Langenbach KJ, Sottile J. Identification of protein-disulfide isomerase activity in fibronectin. J Biol Chem 1999; 274(11):7032-7038.
43. Paulsson M. Basement membrane proteins: Structure, assembly, and cellular interactions. Crit Rev Biochem Mol Biol 1992; 27(1-2):93-127.
44. Schittny JC, Yurchenco PD. Basement membranes: Molecular organization and function in development and disease. Curr Opin Cell Biol 1989; 1(5):983-988.
45. Kalluri R. Basement membranes: Structure, assembly and role in tumour angiogenesis. Nat Rev Cancer 2003; 3(6):422-433.
46. Roberts DD, Rao CN, Magnani JL et al. Laminin binds specifically to sulfated glycolipids. Proc Natl Acad Sci USA 1985; 82(5):1306-1310.
47. Talts JF, Andac Z, Gohring W et al. Binding of the G domains of laminin alpha1 and alpha2 chains and perlecan to heparin, sulfatides, alpha-dystroglycan and several extracellular matrix proteins. Embo J 1999; 18(4):863-870.
48. Yurchenco PD, Cheng YS. Self-assembly and calcium-binding sites in laminin. A three-arm interaction model. J Biol Chem 1993; 268(23):17286-17299.
49. Yurchenco PD, Tsilibary EC, Charonis AS et al. Laminin polymerization in vitro. Evidence for a two-step assembly with domain specificity. J Biol Chem 1985; 260(12):7636-7644.
50. Kalb E, Engel J. Binding and calcium-induced aggregation of laminin onto lipid bilayers. J Biol Chem 1991; 266(28):19047-19052.
51. Colognato H, Winkelmann DA, Yurchenco PD. Laminin polymerization induces a receptor-cytoskeleton network. J Cell Biol 1999; 145(3):619-631.
52. DiPersio CM, Hodivala-Dilke KM, Jaenisch R et al. alpha3beta1 Integrin is required for normal development of the epidermal basement membrane. J Cell Biol 1997; 137(3):729-742.
53. Fassler R, Meyer M. Consequences of lack of beta 1 integrin gene expression in mice. Genes Dev 1995; 9(15):1896-1908.
54. Lohikangas L, Gullberg D, Johansson S. Assembly of laminin polymers is dependent on [beta]1-integrins. Experimental Cell Research 2001; 265(1):135-144.
55. Raghavan S, Bauer C, Mundschau G et al. Conditional ablation of beta1 integrin in skin. Severe defects in epidermal proliferation, basement membrane formation, and hair follicle invagination. J Cell Biol 2000; 150(5):1149-1160.
56. Colognato H, Yurchenco PD. Form and function: The laminin family of heterotrimers. Dev Dyn 2000; 218(2):213-234.
57. Aumailley M, Pesch M, Tunggal L et al. Altered synthesis of laminin 1 and absence of basement membrane component deposition in (beta)1 integrin-deficient embryoid bodies. J Cell Sci 2000; 113(Pt 2):259-268.
58. Sasaki T, Forsberg E, Bloch W et al. Deficiency of beta 1 integrins in teratoma interferes with basement membrane assembly and laminin-1 expression. Exp Cell Res 1998; 238(1):70-81.
59. Li S, Harrison D, Carbonetto S et al. Matrix assembly, regulation, and survival functions of laminin and its receptors in embryonic stem cell differentiation. J Cell Biol 2002; 157(7):1279-1290.
60. Kadler KE, Hojima Y, Prockop DJ. Assembly of collagen fibrils de novo by cleavage of the type I pC-collagen with procollagen C-proteinase. Assay of critical concentration demonstrates that collagen self-assembly is a classical example of an entropy-driven process. J Biol Chem 1987; 262(32):15696-15701.
61. Velling T, Risteli J, Wennerberg K et al. Polymerization of type I and III collagens is dependent on fibronectin and enhanced by integrins alpha 11beta 1 and alpha 2beta 1. J Biol Chem 2002; 277(40):37377-37381.
62. Li S, Edgar D, Fassler R et al. The Role of laminin in embryonic cell polarization and tissue organization. Developmental Cell 2003; 4(5):613-624.
63. Aszodi A, Hunziker EB, Brakebusch C et al. β1 integrins regulate chondrocyte rotation, G1 progression, and cytokinesis. Genes Dev 2003; 17:2465-2479.

Chapter 3

The Role of Integrins in Cell Migration

Bernhard Wehrle-Haller*

Introduction

Migration and Motility are essential components of the behavioral repertoire of a cell. During embryogenesis cells move in sheets or loosely attached populations to create complex tissues. In the adult, cell motility is crucial to maintain immunity, or repair damaged tissues. Uncontrolled cell adhesion and increased motility can result in pathological situations such as tumor cell dissemination and the formation of metastasis.

Cell migration occurs in response to chemokine or growth factor signals that are converted into cell shape changes, essentially mediated by structural changes of the actin cytoskeleton. Cell displacement is observed when the structural changes and forces created within the actin cytoskeleton are mechanically linked to the surrounding extracellular scaffold. This mechanical link needs to be highly regulated in order to allow the formation of new attachment sites at the cell front and the controlled dissociation of adhesion sites at the cell rear.

The understanding of the mechanisms that govern cell migration is therefore critically coupled to the question of how cellular receptors interact with their extracellular environment. Hence, members of the family of integrins have moved into the focus of attention. Integrins are heterodimeric membrane spanning receptors that play crucial roles in cell adhesion and migration.[1] This chapter will discuss general as well as specific aspects of how integrins are involved in the orchestration of the migration of individual cells.

Evolutionary Aspects of Integrin Dependent Cell Migration

When analyzing the appearance of integrin receptors during evolution, it is notable that unicellular highly mobile organisms, such as *Dictyostelium*, do not possess integrins. Apparently, the different types of substrates encountered in the soil, the natural habitat of these organisms, require weaker and less specific means of adhesions as provided by integrins. However, studies on the movement and chemotactic migration of amoeba demonstrated the fundamental role of actin polymerization and anterior to posterior polarization.[2-4] Moreover, since classical integrin adaptor molecules such as talin can be found in *Dictyostelium*,[5,6] we have to propose that integrins evolved in a cellular environment in which the entire cytoskeletal machinery that drives cell migration was already in place. Based on the appearance of integrins in metazoan cells, they might have played a role in the development of early multicellular organisms such as sponges and jellyfish.[7-10] Hence both the firm adhesion to extracellular matrix as well as the morphogenetic movements that occur during embryogenesis were supported by this new class of cell surface receptors.[11] Therefore, integrins have evolved to perform critical functions in cell adhesion as well as cell motility in complex multicellular organisms.

*Bernhard Wehrle-Haller—Department of Cellular Physiology and Metabolism, Centre Medical Universitaire, 1 Rue Michel-Servet, 1211 Geneva 4, Switzerland. Email: Bernhard.Wehrle-Haller@medecine.unige.ch

Integrins and Development, edited by Erik H.J. Danen. ©2006 Landes Bioscience.

The Role of Integrins in Cell Migration in Flies and Worms

Genetically well characterized biological systems such as *C. elegans* and *Drosophila* confirm the essential roles played by integrins in cell to matrix adhesion, such as the formation of the attachment sites for muscle cells or the adhesion between the dorsal and ventral wing epithelium (for reviews see refs. 12,13). Similar to the structural aspects of cell adhesion, the mechanisms that regulate cell migration have also been analyzed in these animals. Classical models of cell migration include the border cell migration, midgut migration and the development of the tracheal system in *Drosophila*, and neuronal and gonadal distal tip cell migration in *C. elegans*. The nonbiased genetic approach established roles for growth factor signaling during the development of the tracheal system,[14] cell-cell adhesion molecules in border cell migration[15] and integrins in midgut migration.[16] In addition, genetic analysis has established critical roles for the cytoplasmic integrin adaptor proteins talin and UNC-112 in cell adhesion and migration.[17-19] The importance of these additional molecular components reveal that the adhesive and signaling functions mediated by integrins during cell migration are embedded in a complex network of protein-protein interactions, intracellular signaling and mechanisms controlling polarity and intracellular protein transport.[20-22] In order to better understand the roles of integrins in migration, mutations that specifically affect cell migration but not cell adhesion are of particular interest. In *C. elegans*, the tyrosine to phenylalanine mutation of the two phylogenetically conserved tyrosine residues in the cytoplasmic tail of the unique β-integrin subunit (βpat-3), blocks the migration of gonadal distal tip cells but does not affect muscle function.[23] Thus, it is tempting to speculate that the role of integrins in cell migration involves the specific activation of intracellular signaling pathways as well as the dynamic interactions with molecules of the extracellular matrix. In fact, the challenge consists in understanding the mechanisms that allow a cell to reversibly engage integrins into cell-matrix adhesions sites and exploit them for adhesion as well as migration.

Definition of Cell Migration

Traditionally, fibroblasts have been used to study cell adhesion and migration.[24,25] Therefore the prototype of cellular motion has been established with these cells migrating on two-dimensional surfaces. As the study on other moving cells, such as leukocytes or epithelial cells has advanced, some of the hallmarks of cell adhesions need to be modified. Moreover, the role of the respective architecture of the 3-dimensional environment has to be taken into account in order to compare the movement of keratinocytes on top of a flat basement matrix with the slow tunneling of contractile fibroblasts within a three-dimensional collagen network.[26-29] Hence it is important to define the cellular processes that occur during cell migration in such a way that they are applicable for many different cell types within different tissue environments.

Cell migration is initiated by the development of cellular extensions such as filopodia and lamellipodia that define the front of migrating cells.[30] Within these cellular protrusions, small punctuate integrin-containing contacts form at sites of physical interaction with extracellular matrix components. These sites are referred to as "focal complexes". Focal complexes anchor the molecular scaffold formed by the polymerized actin cytoskeleton to the extracellular matrix. Once the actin cytoskeleton is linked to the extracellular matrix, new actin molecules that are added to the actin cytoskeletal scaffold at the tips or leading edges of filopodia and lamellipodia respectively push the plasma membrane forward.[31] In turn, the newly added extensions are stabilized by a new set of focal complexes formed in front of the previous ones.

The second phase of cell migration is initiated when focal complexes mature into "focal contacts". This maturation is dependent on intracellular or extracellular mechanical tension and localized actin polymerization and results in the compaction of the actin cytoskeleton and integrin receptors, creating a robust link between intracellular actin fibers (stress fibers) and the extracellular matrix.[32-34] In rapidly migrating leukocytes or fish keratinocytes, the formation of stress fibers and focal contacts is not observed, suggesting that the development of stress fibers and focal contacts is associated with cell anchorage rather than mobility.[35-38] During migration of fibroblasts, it has been observed that integrins and trailing parts of the cell remain on the substrate while the main cell body is moving forward.[39] Interestingly, the substrate affinity of the trailing cell rear, and hence the maximal speed of migration, depends on the concentration of extracellular ligands. This observation led to the important conclusion that maximal migration speeds of fibroblasts are obtained on low concentrations of ligands.[40,41] Hence the destabilization of stress fibers and their associated focal contacts may represent a target for intracellular signaling pathways that increase the speed of cell migration (see below).

Cell migration however cannot occur unless integrin dependent adhesion sites are dispersed at the rear of migrating cells. Before dispersal, focal contacts often demonstrate an apparent sliding in the direction of migration. This sliding is associated with a continuous reduction of focal contact size.[26,27,34,42] We recently proposed that the apparent sliding of focal contacts is the result of a polarized remodeling, created by the dispersal of integrin receptors at the distal and the recruitment of new integrin receptors at the proximal edge of sliding focal contacts respectively.[26,27]

In summary, cells of different origins migrate in an integrin dependent manner, all according to the same principles, involving (i) the formation of extensions at the cell front, (ii) integrin dependent focal complex formation, (iii) maturation into and plasticity of focal contacts and (iv) the controlled sliding and dispersal of focal contacts at the cell rear. In this chapter, we will use the terms focal complexes and focal contacts as defined above. The term "focal adhesions" will be used to refer to not further specified integrin dependent cell to matrix adhesion sites. We will now analyze the different roles of integrins applying this concept of cell migration.

Differential Roles of Specific Integrins in Cell Migration

24 different types of integrin heterodimers have been identified in vertebrates.[1] Due to their evolutionary and functional relationship, they can be grouped into sub-families performing different mechanical or signaling tasks, such as the recognition of a subset of extracellular ligands, being involved in the assembly of extracellular matrix or playing a regulatory role in cell migration.[8] The genetic inactivation of individual integrin chains and the resulting pathologies has established critical and distinct roles for each type of integrin, identifying roles in cell migration for most of them.

The β1-Integrin Family

The β1-integrin family is characterized by a total of 12 different heterodimers that all share a common β1-integrin subunit. While the β1-integrin chain provides the critical link to the actin cytoskeleton in each of these heterodimers, the associated α-chains determine ligand specificity, cellular localization and function in adhesion and/or migration.

α1β1, α2β1, α10β1 and *α11β1* integrins are receptors for collagen and support cell migration through collagen rich extracellular matrix. It has been demonstrated that the expression of α1 and α2 chains are increased upon VEGF secretion in response to injury and are required for lymphangiogenesis in vivo.[43] A similar increase in α1 expression and associated collagen remodeling is induced by growth factors such as TGFβ and PDGF.[44] In addition, an increase in α1 expression, in response to chemokines such as MCP-1 and IL-8, is induced in infiltrating lymphocytes.[45] In this context, it has been demonstrated that the infiltration and remodeling of collagen matrices is facilitated by the recruitment of the metalloproteinase MMP1 by the I-domain of the α2 integrin subunit.[46] α10β1 expression is limited to chondrocytes and provides binding to cartilage specific type II collagen.[47] Its genetic inactivation results in the dysfunction of growth plate chondrocytes.[48] α11β1 integrin is expressed in a subset of mesenchymal tissues and modulates PDGF dependent chemotaxis on collagen matrixes.[49,50]

α3β1, α6β1 and *α7β1* integrins bind to laminin, an important constituent of basement membranes. Although these integrins bind to various members of the laminin family of extracellular ligands, their differential expression in brain, epithelial and muscle tissues determines different roles in adhesion and migration.

The genetic inactivation of *α3β1* revealed a phenotype of weak skin blistering, kidney tubule defects, reduced branching during lung morphogenesis and a lamination defect in the neurocortex.[51] Of interest is the aberrant neuronal migration that occurs in the absence of α3-integrin and which results in altered actin cytoskeleton dynamics at the leading edge of migrating neurons.[52] During the migration of neurons from the ventricular zone through the cortex towards the pia mater, α3β1 integrin expression is tightly regulated. In *scrambler* mice the timely expression of α3-integrin is altered, resulting in prolonged neuronal migration and positioning defects. It has been demonstrated that the phosphorylation on tyrosines of disabled-1, the protein mutated in *scrambler* mice, induces the expression of α3 integrins in neurons. In response to the expression of α3β1 integrin at the cell surface, neurons migrating along radial glial cells, detach from these cells to adhere to the extracellular matrix protein laminin. The timely disconnection from the glial cell surface results in the correct positioning of the neuron within the neurocortex.[53]

In contrast to migration defects observed upon the loss of α3β1 integrin, its over-expression has been associated with increased formation of metastases of small cell lung carcinomas and colon carcinomas.[54-56]

α6β1 integrin is of importance for the migration and adhesion of epithelial cells on laminin containing basement membranes. The deletion of the α6-integrin chain results in a severe skin-blistering defect.[57] However, this defect of epithelial adhesion is attributed to the absence of the α6β4 integrin heterodimer.[58,59] It demonstrates the unique role of the β4-integrin subunit to link this particular laminin binding integrin to the intermediate filament system and not to the actin cytoskeleton.[60,61] Because of the crucial role of α6β4, the importance of the α6β1 integrin in morphogenesis and maintenance of epithelial tissues has not been clearly demonstrated. However, indirect evidence for the importance of the actin cytoskeleton-linked β1 integrin has been provided by the phenotype of the Kindler syndrome, in which affected individuals exhibit fragile skin.[62] The gene mutated in this syndrome is a β1-integrin adaptor protein named kindlerin belonging to the family of mig-2 related adaptor proteins, sharing homologies with UNC112 from *C. elegans*.[63] Kindlerin expression is increased in breast carcinomas upon TGFβ treatment and localizes to focal adhesions in keratinocytes.[64]

In other systems such as the developing cerebellar and cerebral cortex, α6β1 integrin and more generally the β1 integrins are required to define the structural entity of the tissue. In the absence of α6-integrins, neuronal migration along the surface of radial glial cells is not affected, suggesting that glial dependent neuronal migration is mediated by cell-cell adhesion receptors. However, in the absence of α6-integrin, the situation is similar to that of α3-integrin deficient mice, in which migrating neuronal precursors do not stop their migration along radial glial cells and form neuronal outgrowths beyond the pia mater.[65,66] In contrast, β1-integrins are required to physically anchor the endfeet of the radial glial cells, allowing these cells to span the entire neurocortex and thereby forming the neuronal migration pathways and contributing to the stabilization of the brain architecture.[67]

Another situation in which the regulation of the cell surface expression of α6β1 integrin is important for cell migration has been described in extravasating neutrophils. PECAM-1 dependent binding between neutrophils and endothelial cells results in the up-regulation of α6β1 integrins on the surface of neutrophils, which is critical for their transmigration through the perivascular basement membrane.[68]

Furthermore, in migrating cranial neural crest cells, the cycling of the α6β1 integrin between the cell surface and an intracellular pool, is a mechanism that promotes neural crest cell motility.[69]

α7β1 integrin is found expressed in muscle cells where its deletion results in a particular form of congenital muscular dystrophy.[70-72] Despite the important function for the maintenance of muscles, the specific function of α7β1 or their specific splice variants in myoblast migration during embryogenesis, the fusion of myotubes or muscle regeneration is not well studied.[73-76]

α4β1 and *α9β1* are dual specific integrins that recognize fibronectin at sites distinct from the major RGD containing cell binding site and VCAM-1 exposed on endothelial cells.[77,78] Although several members of the integrin family are involved in cell migration, the two structurally related α4β1 and α9β1 integrins appear to specifically enhance cell migration. For example, the initiation of α4β1 integrin expression in the neural crest coincides with the onset of neural crest migration. When the function of this integrin is blocked, both the number of neural crest cells and the distance of migration are reduced.[79,80]

During inflammation, leukocyte migration across the endothelium is facilitated by the interaction of α4β1 with VCAM-1. The pathological importance of this mechanism for the progression of auto-immune diseases such as multiple sclerosis or Crohn's disease has been recently demonstrated by the successful treatment of these conditions by a humanized anti-α4β1 integrin antibody (for a review see ref. 81).

It has been determined that the cytoplasmic tail of the α4-integrin subunit is responsible for the increase in migration of leukocytes.[82] Integrin α4β1 induced cell migration depends on the interaction of the α4 cytoplasmic tail with the focal adhesion adaptor protein paxillin.[83] However, in polarized migrating cells, the fraction of α4β1 integrin that localizes to the leading edge is not associated with paxillin. In fact, in this cellular location, paxillin binding to α4 is prevented by phosphorylation of a critical serine residue in the α4 cytoplasmic tail. Upon dephosphorylation of α4 integrin, paxillin can bind to α4β1 integrins, which results in their colocalization in focal complexes at the cell front and in focal contacts at the cell rear.[84] The reversible binding of paxillin to α4β1 integrins is critically required for cell migration, since cell migration is completely inhibited in cells transfected with an α4-integrin/paxillin fusion protein.[84] These data suggest that the recruitment of paxillin to α4β1 integrins has to follow a strict chronological order, required for the assembly of focal complexes at the cell front.[85] Furthermore, it has been demonstrated that paxillin localized to focal contacts located at the cell rear is able to prevent the formation of new lamellipodia by the recruitment of an ARF-GAP, and thus contributes to the stabilization of the anterior to posterior cell polarity required for efficient migration.[86] Another critical intracellular target of the α4 integrin /paxillin complex that could be responsible for the stabilization of cell polarity during migration, represents the focal adhesion kinase/proline-rich tyrosine kinase-2. The α4-integrin/paxillin dependent activation of this kinase stimulates αLβ2 dependent trans-endothelial migration.[87]

α9β1 integrin also interacts with paxillin, but it lacks the regulatory phosphorylation site and enhances cell motility independent of paxillin binding.[88] It was recently proposed that α9 dependent stimulation of cell migration is mediated by the specific recruitment of the spermidine/spermine N1-acetyltransferase to the α9 subunit.[89] However, the issue concerning the role of α9β1 integrin in stimulating cell migration remains puzzling. The genetic deletion of α9 integrin results in a fatal bilateral chylothorax, probable linked to an abnormal development of the lymphatic system.[90] Accordingly, it has recently been demonstrated that the lymphangiogenesis inducing growth factors VEGF-C and VEGF-D bind to α9β1 integrins.[91]

α5β1 integrin is an RGD-peptide dependent receptor for fibronectin and has been studied in great detail. This integrin supports adhesion of migrating cells to fibronectin, but is also critically involved in the assembly of fibronectin containing extracellular fibrils.[92,93] In fibroblasts cultured on a rigid 2-dimensional matrix, this integrin localizes in focal contacts at the cell periphery as well as to fibrillar adhesions located underneath the main cell body. Interestingly, an increase in the elasticity of the extracellular matrix reduces the phenotypic distinction between focal contacts and fibrillar adhesion.[94] It has been recently demonstrated that the α5 integrin subunit localized in fibrillar adhesions undergoes a structural change that induces the high affinity binding of the α5β1 integrin to extracellular fibronectin. The firm mechanical coupling across the plasma membrane enables fibroblasts to synthesize fibronectin fibrils by the acto-myosin dependent pulling of α5β1 integrins along stress fibers.[95]

By using a GFP-tagged form of α5-integrin, it became possible to follow the clustering as well as the recycling of α5β1 integrins in migrating fibroblasts. Importantly, it was observed that $\alpha5\beta1$ integrins were internalized at the cell rear and transported to the front.[96] Recently the dynamics of $\alpha5\beta1$ integrin in the plasma membrane of living cells was analyzed by image correlation microscopy.[97] It was demonstrated that $\alpha5\beta1$ integrins exists in submicroscopic clusters moving together with α-actinin. In addition, it was determined that $\alpha5\beta1$ integrins are loosely clustered within focal complexes, but densely packed and highly structured in focal contacts, confirming observations previously made for $\alpha v\beta3$ integrins.[34]

$\alpha8\beta1$ plays a crucial role during early steps of kidney morphogenesis, affecting the growth and branching of the ureteric bud epithelium.[98] Interestingly, $\alpha8\beta1$ integrin is expressed in the metanephric mesenchyme surrounding the ureteric bud. Its major ligand nephronectin is expressed in the ureteric bud epithelium from where it is deposited in the surrounding basement membrane.[99] Thus the role of $\alpha8\beta1$ integrin in the migration or movement of the ureteric bud epithelium is indirect and may be linked to $\alpha8\beta1$ integrin induced mesenchyme to epithelial signaling. Interestingly, it has recently been demonstrated that $\alpha8\beta1$ is able to bind in an RGD dependent manner to a latent form of TGFβ.[100] Nephronectin might therefore be able to compete for the binding of TGFβ making it available for morphogenetic events in the tissue. Although $\alpha8\beta1$ may not be directly involved in the stimulation of cell migration in a cell autonomous manner, its expression on vascular smooth muscle cells and myofibroblasts indicates a role in injury related scarring of the myocardium.[101]

$\alpha v\beta1$ integrin has specificity for RGD containing extracellular ligands. Because the αv subunit can bind several different β-integrin sububits, the αv family of integrins will be discussed below.

The $\beta2$ Integrin Family

The expression of integrins belonging to this family is the hallmark of cells of the immune system. The $\beta2$-integrin subunit forms various heterodimers with I-like containing α-integrin subunits (*$\alpha L\beta2$, $\alpha M\beta2$, $\alpha X\beta2$, $\alpha D\beta2$*), enabling their interactions with counter receptors of the IgG-domain family on the surface of activated endothelial cells or multiple ligands of the extracellular matrix.[102] These diverse interactions allow firm adhesion of chemokine-stimulated leukocytes to endothelial cells, followed by diapedesis and chemotaxis through damaged or infected tissues. Compared to other migrating cells, leukocytes demonstrate the most impressive motility. In contrast to fibroblasts, leukocytes lack stress fibers and focal contacts.

The best studied leukocyte integrin, the *$\alpha L\beta2$* integrin is switched to its high affinity form and recruited to the front of migrating leukocytes by the specific interaction with RAPL.[103,104] In order to be able to bind $\alpha L\beta2$ integrins, RAPL is stimulated by Rap1, a member of the small GTPase family of intracellular regulators that plays an important role in the activation of integrins.[105,106] It has been demonstrated that RAPL binds to the juxtamembrane region of the αL subunit, in such a way that it may provoke the unclasping of the transmembrane and cytoplasmic domains of the αL and $\beta2$ integrin subunits.[107-109] Thus RAPL has a dual role consisting in the activation of $\alpha L\beta2$ and in its recruitment to the leading edge of migrating leukocytes, preparing $\alpha L\beta2$ integrin for binding to ligands expressed on the endothelial cell surface.

Another unique feature of $\beta2$ integrins is their continuous recycling from the trailing uropod (cell rear of migrating leukocytes), to the cell front. This process requires the activity of myosin light chain kinase and intracellular Ca^{2+} transients. If either of these two intracellular signaling pathways are blocked, rear detachment of the uropod is prevented, leading to the arrest of leukocyte migration.[110-113] The ability of $\beta2$-integrins to undergo rapid recycling has been linked to a juxtamembrane localized tyrosine dependent YXXØ internalization motif.[114]

Interestingly, this motif is also present in the $\beta7$-integrin subunit, which like $\beta2$-integrin is expressed exclusively in cells of the hemaotopoietic system.[114] The $\beta7$-integrin subunit pairs with the $\alpha4$ and αE integrin chains forming $\alpha4\beta7$ and $\alpha E\beta7$ integrins. $\alpha4\beta7$ is critical for the morphogenesis of the gut-associated lymphatic system,[115] while $\alpha E\beta7$ regulates immuno-surveillance of the skin.[116] The $\beta7$-integrins are also involved in $CD8^+$ T-cell mediated graft rejection and consist of a promising therapeutic target to enhance the success of tissue transplantation.[117,118]

The *αMβ2* integrin is expressed in monocytes, granulocytes, macrophages and natural killer cells and has been implicated in diverse responses of these cells, including phagocytosis, cell-mediated killing and chemotaxis. These complex responses depend on the capacity of αMβ2 integrin to mediate leukocyte adhesion and migration, playing a central role in inflammation.[102] In contrast to αLβ2 integrin, αMβ2 and also *αXβ2* integrins recognize a large set of structurally unrelated ligands, including fibrinogen, complement fragment iC3b, ICAM-1, Facor X, JAM's and denatured proteins.[119,120] Recently it has been suggested that the reason for this large spectrum of ligands is the unusual high affinity of the αMβ2 and αXβ2 for carboxyl groups. Moreover, since both αMβ2 and αXβ2 integrins are colocalized with the receptor for the urokinase-type plasminogen activator to the leading edge of migrating cells,[121,122] it has been proposed that proteolytic degradation of extracellular matrix molecules such as fibrinogen creates new extracellular ligands specifically recognized by these two integrins.[120] Using these integrins as sensors for degraded extracellular matrix components would provide leukocytes with an additional tool to detect tissue damage or microbial invasion.

αDβ2 integrin is expressed on eosinophiles, monocyte/macrophages and neutrophils and is an alternative receptor for VCAM-1. Compared to the other β2-integrin family member its role in leukocyte migration is less studied. In an experimental model of spinal cord injury, it has been reported that the injection of monoclonal antibodies directed against αDβ2 reduces the number of lesion-induced accumulation of macrophages or neutrophils by approximately 50%.[123]

Besides αLβ2 recycling and RAPL dependent targeting to the cell front, it is not clear whether other β2 integrin subunit specific features are responsible for the high migratory capacity of leukocytes. When comparing the cytoplasmic domain of β2 integrins with that of other β-integrin chains, it is notable that the two NPXY motifs found in all other β-subunits exist as NPXF motifs in β2. Interestingly, when both NPXY to NPXF mutations are introduced into the cytoplasmic domain of the β3-integrin gene, the so mutated platelet integrin αIIbβ3 (see below) is no longer able to mediate blood clot retraction, leading to prolonged bleeding.[124] The same β-integrin mutation, when expressed in the context of the αvβ3 integrin, blocks shear stress induced reinforcement of cell adhesions[125] and inhibits βPat-3 dependent migration of gonadal distal tip cells in C. elegans.[23] Thus the capacity for fast cell migration and reliable transmigration of leukocytes at sites of chemokine release might be directly linked to the inability of β2 integrins to sense the shear forces constantly generated by the flowing blood.

The αv-Integrin Family

The αv-integrin subunit pairs with a series of different β-subunits *(αvβ1, αvβ3, αvβ5, αvβ6, αvβ8)* forming heterodimers with specificities for RGD peptide containing ligands. Among these different heterodimers, *αvβ3* integrin is the most studied in association with cell migration. Notably, the expression of αvβ3 integrin is up-regulated in endothelial cells that undergo angiogenesis, and is over-expressed in mammary carcinomas and melanomas.[126-128] In melanomas, the expression of αvβ3 is associated with the formation of metastasis and the protection against cell death in 3-D collagen matrixes.[128,129] Moreover, an increase in the formation of metastasis has been induced by the expression of a constitutively activated form of the αvβ3 integrin in breast carcinoma cells.[130] Due to its increased expression on endothelial cells during tumor induced angiogenesis, αvβ3 integrin has been the center of attention and the focus of the development of ligand mimetic small ligands, that are able to inhibit angiogenesis.[131,132] Interestingly one of these small soluble ligands, cyclic RGD peptide, is able to induce caspase-3 dependent apoptosis.[133] As a paradox however, the knockout of β3-integrin and the double knockout together with β5 integrin increases tumor-induced angiogenesis.[134] Hence αvβ3 integrin plays an essential role in the regulating of cell migration, in ways that have not been understood yet.

Based on the pathological importance of the αv-integrin family, it was surprising to see only minor defects in β3-integrin knockout animals. Similarly, the knockout of the *β5*-integrin demonstrated a slight increase in osteoclast maturation and activity, while other β5-integrin specific functions such as wound healing and adenovirus susceptibility were normal.[135,136] This may suggest that the β3 or β5-integrin can compensate for each other. However, several findings demonstrate that they nevertheless play different roles during cell migration. While αvβ3 expressed in CS-1 hamster melanoma cells was able to induce adhesion, spreading and migration, αvβ5 provided only adhesion. Interestingly, cell migration was induced when αvβ5 expressing cells were stimulated with cytokines. This difference in behavior was attributed to the extracellular domain of these different β-integrins.[137] While both integrins bind to vitronectin, only αvβ5 integrin is able to rapidly internalize and degrade soluble vitronectin.[138] It was recently demonstrated that the constitutive internalization of αvβ5 is of physiological importance. β5-integrin knockout mice develop an age related blindness that is due to a lack of phagocytic activity of the retinal pigment epithelium, normally involved in the uptake of defective photoreceptor rods that are shed by retinal ganglion neurons.[139]

The inactivation of the gene coding for the *β6*-integrin subunit results in the chronic inflammation of skin and lungs. Accordingly, keratinocyte migration on fibronectin and vitronectin is reduced.[140] The lung phenotype can be rescued by the constitutive expression of a β6-integrin transgene in bronchial and alveolar epithelial cells.[141] Since β6-integrin expression is increased during the repair of the airway epithelium, the epithelial expression of αvβ6 integrin could be crucial for the proliferation, adhesion and or migration of epithelial cells.

The expression of *β8*-integrin is elevated in the CNS compared to other organs. In the CNS, it is localized to dendritic spines and post-synaptic densities, however, a critical role in neuronal function has so far not been demonstrated.[142]

Although the size of the αv-integrin family is considerable, the inactivation of the αv-integrin gene does not result in embryonic lethality. However, newborn mice die shortly after birth due to intracerebral hemorrhages, probably due to a disruption of the mechanical association of the central nervous parenchyma with cerebral microvessels.[143]

The αIIbβ3 Integrin

This integrin is exclusively found expressed on the surface of megakaryocytes and platelets. It is essential for platelet aggregation and subsequent clot retraction. Therefore, this integrin provides firm adhesion and seems not to be involved in cell migration. However, during clot retraction, αIIbβ3 integrin dependent fibrin interactions is undergoing remodeling, reflecting in many ways the rearrangements of focal contacts at the rear of migrating cells (see below). Therefore, many aspects of integrin physiology first established for this integrin, such as activation induced unfolding, cytoplasmic adaptor binding and intracellular signaling, form the basis on which the behavior and function of other integrins can be predicted.[144-150]

Modulation of Integrin Affinity, Valency and Dynamics during Cell Migration

In order to understand the role of integrins during cell migration and to identify critical targets for therapeutical interventions, the behavior of integrin receptors during the formation, remodeling and dispersal of focal adhesions has to be determined and quantitatively analyzed. An important concept of integrin dependent adhesion encompasses the notion that integrin dependent adhesion is controlled by modulations of the affinity and the avidity of integrin receptors. Because the use of the term "avidity" has not been consistent,[108] it is important to define the different integrin specific parameters that are involved in the modulation of size and mechanical strength of focal adhesion sites. Three different integrin dependent mechanisms can be distinguished that involve (i) integrin *affinity* modulations, (ii) changes in the *valency* of integrin dependent adhesions and (iii) modifications of the *dynamic* exchange rates of integrins within focal adhesion sites.

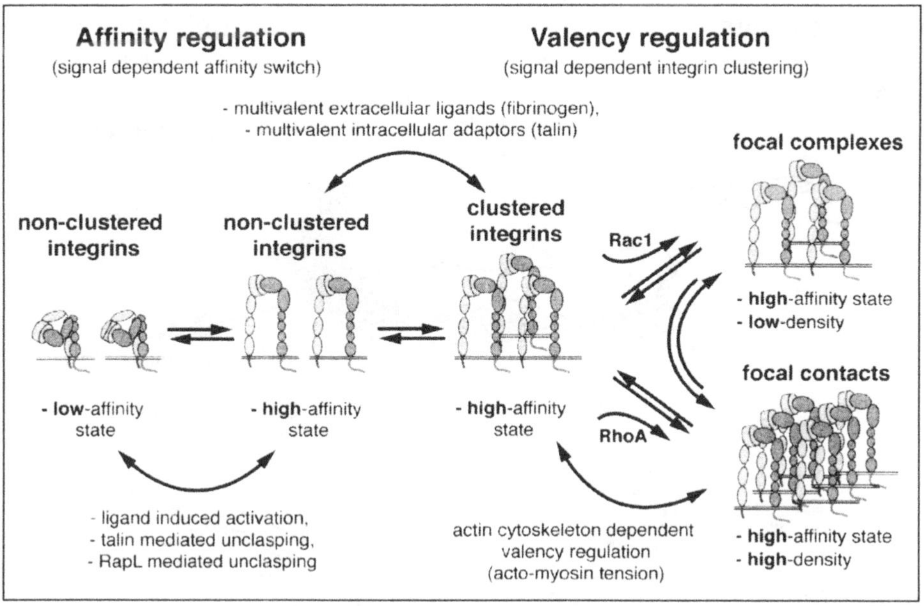

Figure 1. Signal dependent regulation of integrin affinity and valency. Schematic representation of the equilibrium between low and high-affinity integrins and its regulation by the binding of extracellular ligands or intracellular adaptors (left hand side). High affinity integrins will further undergo valency regulations (nano-clustering) that are controlled by the binding to multi-valent extracellular ligands or intracellular adaptors (center). Subsequent acto-myosin dependent valency regulations will determine integrin density and size of the respective focal adhesions (right hand side).

Integrin affinity modulations reflect the allosteric switch from the inactive integrin heterodimer to the fully unfolded extracellular ligand-binding receptor. Depending on the mode of integrin activation (inside-out versus outside-in), several differentially folded intermediates may exist in a dynamic equilibrium within the plasma membrane.[145]

Integrin valency modulations reflect the ability of integrin receptors to undergo lateral clustering within the plasma membrane in order to increase the number of individual integrin interactions and augment the overall cellular avidity towards extracellular ligands. Thus, the valency of integrin dependent binding is the result of "integrin packing density" multiplied by the "size" of focal adhesions and correlates with the overall strength and mechanical resistance of the link between the cell and the extracellular matrix.

Integrin dynamics reflect modulations of the temporal stability of integrins within focal adhesions. It is characterized by the time, expressed as a half-live, that integrin heterodimers persist within a given focal adhesion. The degree of integrin dynamics determines the remodeling capacity of focal adhesions.

In order for a cell to migrate and to form new focal complexes at the cell front and to remodel and disperse focal contacts at the cell rear, integrin affinity, valency and dynamics have to be regulated appropriately (see Figs. 1, 2). (i) At the cell front, integrins have to be in a high affinity state in order to bind to extracellular ligands encountered by filopodia and lamellipodia (high affinity). Once ligand is bound or recognized, rapid lateral clustering of integrins (increase in valency) is required to stabilize individual integrin contacts. In addition, a slow rate of integrin exchange (low dynamics) in newly formed focal complexes is essential to stabilize the actin cytoskeletal scaffold of filopodia and lamellipodia. (ii) Focal contacts that are localized at the level of the main cell body rapidly modulate their size and strength while maintaining a

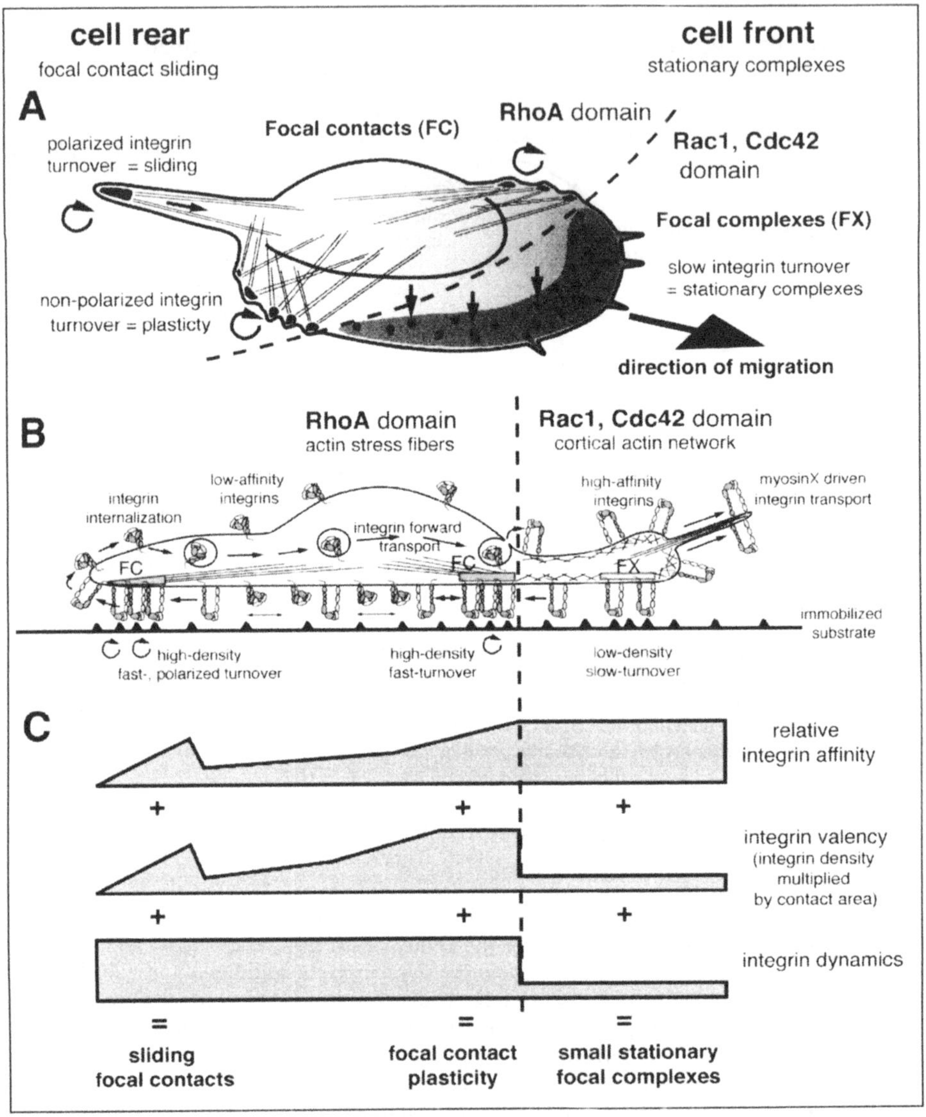

Figure 2. Modulation of integrin affinity, valency and dynamics between the front and the rear of migrating cells. A) Schematic representation of integrin behavior in focal complexes versus focal contacts in migrating cells. Analysis of integrin dynamics in focal complexes revealed slow integrin exchange in Rac1 controlled anterior regions of the cell (vertical arrows). In focal contacts localized to the posterior domain of the cell, which is controlled by RhoA activity, integrin exchange rates are fast (circling arrows), resulting in focal contact plasticity and sliding. B) Cartoon of a migrating cell expressing low-and high-affinity integrins within its cell membrane. The differential integrin densities, integrin dynamics (circling arrow) and organization of the actin cytoskeleton between the front and the cell rear are represented. In addition, integrin transport is depicted by anterograde vesicular traffic and myosin-X driven transport along filopodia. C) Graphic representation of anterior to posterior modulations of relative integrin affinity, integrin valency and integrin dynamics in the plasma membrane. The greater the thickness of the bar, the more important is the specific type of integrin behavior. The combination of the three parameters define and predict the behavior of focal complexes or focal contacts in migrating cells.

high degree of plasticity. Thus an elevated level of high-affinity integrins is required to rapidly respond to signals that modify the valency (size and density) of focal contacts. In addition, an increase in integrin dynamics provides focal contacts with the capacity to respond to mechanical changes in the cellular environment. (iii) At the cell rear, sliding and dispersal of focal contacts require an elevated level of integrin dynamics. The overall tendency of focal contacts to disperse at the rear of migrating cells, may suggest that the concentration of high-affinity integrins is reduced. In contrast, the polarized integrin assembly at the inner edge of sliding focal contacts requires a highly controlled regulation of integrin valency.

Integrin Affinity Modulations during Cell Migration

Affinity Modulations at the Cell Front

Cell migration through a complex tissue environment intuitively requires the need for receptors that can sense and explore the newly encountered extracellular matrix. In contrast, rear detachment of migrating cells suggests that integrins have lost their affinity for substrate. Surprisingly, despite the availability of antibody reagents that specifically recognize high affinity forms of integrins, the detection of increased concentration of high affinity integrins at the front of migrating cells has been difficult. Symptomatic for this, only one report has so far demonstrated the accumulation of high affinity $\alpha v\beta 3$ integrins at the front of migrating cell.[151] Interestingly, the activation of the small GTPase Rac1, which is involved in the formation of lamellipodia is required for this polarized accumulation of high-affinity integrins.[151] $\alpha 4\beta 1$ integrins are also found at the leading edge of migrating cells, however, it is not known whether this integrin is retained their in a high affinity state.[83,84] Similarly, it can only be hypothesized that the $\beta 1$-integrins that colocalize with talin at the leading edge of lamellipodia in fish keratinocytes are in their high affinity configuration.[152] In migrating leukocytes, RAPL that has been demonstrated to activate $\alpha L\beta 2$ integrins, is localized and recruited to the front of migrating cells.[104] Hence Rap1 dependent activation of RAPL in parallel with Rac1 signaling are involved in the polarized expression of high affinity integrins to the front of migrating cells.

Affinity Modulations at the Cell Rear

In dynamic focal contacts that are localized in the main cell body or towards the rear of the cell, the continuous incorporation and exchange of integrins suggests the presence of a local source of high affinity integrins. So far, it is not known whether integrins are laterally recruited from the surrounding plasma membrane or whether the newly incorporated high-affinity integrins are delivered to focal adhesions by vesicular transport.[153] However, based on several independent experimental findings, it is possible that focal adhesions are able to create their own local source of high affinity integrins. This idea is essentially based on the crucial role of talin in the induction of the high affinity state of integrins.[154,155] Integrins are activated by the binding of the head domain of talin to the β-integrin cytoplasmic tail.[156] Importantly, this process is regulated by intracellular signaling, involving calpain-mediated proteolysis of talin.[157] At the cell rear, increased rates of cell detachment are induced by MEKK signaling, resulting in calpain dependent proteolysis of focal adhesion proteins.[158] Consistently, the expression of a calpain resistant form of talin renders focal contacts more static, suggesting that calpain mediated talin proteolysis is responsible for integrin dynamics rather than integrin activation.[159] An alternative pathway of talin-dependent integrin activation involves phosphatidylinositol 4,5 bi-phosphate ($PI(4,5)P_2$) binding to the talin-head domain. It has been demonstrated that $PI(4,5)P_2$ binds to the head domain of talin which results in a change of the conformation of talin, resulting in integrin cytoplasmic tail binding and integrin activation.[160,161] Talin is also able to recruit the $PI(4,5)P_2$ synthesizing enzyme PIPKI γ to focal adhesions.[162,163] The recruitment of this kinase is critically required for focal adhesion formation. Thus, the local synthesis of $PI(4,5)P_2$ in focal contacts may be instrumental in talin dependent integrin activation as well as the $PI(4,5)P_2$ dependent formation of actin stress fibers.[164]

Valency Regulation of Integrins during Cell Migration

Valency Modulations in the Nano- and Micrometer Range

The modulation of the lateral association of integrin molecules during cell migration is crucial for the assembly of new focal adhesions at the cell front and their dispersal at the cell rear. Integrin clustering or an increase in valency, has been observed at two different scales. First, integrins form nanoclusters, consisting of a few integrin molecules that can only be demonstrated with specific methods relaying on resonance energy transfer such as FRET and BRET, or statistical methods such as image correlation microsocopy.[97,144] It has been demonstrated that nano-clustering of integrins can be induced by multivalent ligands such as the bi-valent fibrinogen.[144] Integrin nano-clusters may be critical for integrin dependent signaling and mechanical coupling of extracellular matrix to the actin cytoskeleton.[165] It has been proposed that the critical limits, required for a mechanical coupling of extracellular ligands to the actin cytoskeleton is the clustering of three integrins. This threshold has been determined by a recombinant tri-meric fibronectin molecule.[166]

A second class of valency regulation occurs at the micrometer level. In fact, the microscopic observation of focal complexes reveals adhesion surfaces that do not exceed 1 µm in diameter. In contrast, streak like focal contacts in stationary or migrating cells can easily extend to 5-10 µm in length. Interestingly, while the size of focal complexes is rather uniform, the size of focal contacts is dependent on the degree of intracellular tension.[34,167]

Valency Regulation by Modulation of Integrin Density

In addition to a difference in their size, focal complexes and focal contacts can be distinguished by a low versus a high integrin packing density, respectively.[34,97] In focal contacts, an acto-myosin dependent contraction of the actin-scaffold increases integrin density.[27,34] This observation also confirmed the notion that the formation of focal complexes does not require myosin based actin cytoskeleton contraction and can form in a tension free environment in advancing filopodia or lamellipodia.[33,34] Based on the different integrin packing densities between focal complexes and focal contacts, we have proposed that a limitation of the coating density of extracellular ligands would affect focal contact stability.[27] Accordingly, the critical threshold of monomeric ligand density required for focal contact formation was initially determined to be between 120 and 440 nm.[168] Recently, using an elegant method of ligand coating, this threshold was determined to be between 58 and 73 nm.[169] Thus, cells can interact with, and migrate on substrates that allow the spacing between two integrins to be 73 nm or larger, however these cells will only be able to form focal contacts when integrin spacing is smaller or equivalent to 58 nm.[169]

Therefore, the acto-myosin driven increase in integrin density requires a reduction in the spacing of extracellular ligands. Accordingly, an increase in the tension of the extracellular matrix, such as in a stretched collagen gel, would enlarge the spacing of integrin ligands. This would directly reduce integrin valency in focal contacts, potentially inducing intracellular signaling pathways. It has been demonstrated that a 15% increase in the surface of a flexible substrate can induce FAK and paxillin recruitment to focal contacts.[170] In turn, FAK activity is involved in the mechano-sensing of migrating fibroblasts.[171] If fibroblasts are exposed to cyclic mechanical stress, a change in gene expression is induced.[172] Interestingly, in this situation, Rho-kinase activity is critically required to transduce the externally mediated structural modulations of the actin cytoskeleton and focal contacts into altered gene expression.[173] This suggests that the link between individual integrins and the actin cytoskeleton is critically involved in valency mediated signaling or perception of the elasticity of the extracellular environment. Accordingly, cells respond to a gradient of extracellular elasticity by directed migration towards the stiffer substrate, a behavior that has been named "durotaxis".[174,175]

Hence, we can conclude that integrin valency modulations imposed by a nano or micro-patterned extracellular matrix are critically involved in cell spreading,[169,176] the capacity to respond to extracellular mechanical perturbations[165] and the sensing of the microelasticity of the extracellular environment.[171,174,175]

Dynamic Integrin Regulations during Cell Migration

Slow Integrin Dynamics at the Cell Front and Fast Dynamics at the Cell Rear

The first experiments that used dynamic methods to analyze the diffusion of integrin receptors in the plasma membrane were based on fluorescently labeled anti-integrin antibodies.[177] These noninterfering antibodies were added to living cells and the diffusion and clustering of the antibody/integrin complexes was measured by fluorescence recovery after photobleaching (FRAP).[178,179] In the combination with GFP labeled proteins, FRAP turned out to be an extremely powerful technique in order to analyze the association and dissociation dynamics of a complex macromolecular structure such as focal contacts. So far, only a few cytoplasmic adaptor proteins have been analyzed by this technique, nevertheless demonstrating the potential of this method to analyze the dynamic modulations of focal adhesions (e.g., ref. 180). In respect to integrins, we have used this technique to analyze integrin dynamics in the different types of focal adhesions that are formed in migrating cells.[27,34] It became apparent that the organization of the actin cytoskeleton plays a crucial role in the integrin exchange dynamics of focal adhesions. In low-density integrin adhesion sites, representing focal complexes, αvβ3 integrin exchange rates are slow.[34] In contrast, in focal contacts connected to actin stress fibers, αvβ3 integrin exchange is rapid, leading to a complete turnover of integrins within 10 minutes.[34] These observations have been made in focal contacts of highly mobile mouse melanoma cells. In contrast, the dynamics of αvβ3 integrin in focal contacts formed in stationary and highly contractile rat embryo fibroblasts is much slower.[181] This dynamics is even slower, when these cells begin to express the α-smooth muscle actin isoform,[181] assuming the contractile phenotype of myofibroblasts. In endothelial cells, another FRAP analysis of β3-GFP-integrin revealed that αvβ3 integrins trade places within a given focal contact. Interestingly, the intra-contact exchange rate is faster compared to the de-novo recruitment of integrins to focal contacts.[182] This suggests that within dynamic focal contacts, individual integrins exchange their place continuously, reducing the need for integrin recruitment.

Based on these data, it is apparent that the different half-lives of integrins are only partially linked to the overall stability or persistence of a given focal adhesion. For example at the front of fast migrating melanoma cells, the turnover of focal complexes is faster than the respective exchange rates of the αvβ3 integrin. In this situation, the rapid turnover of focal complexes can be explained by the depolymerization of the actin scaffold towards the inner edge of the lamellipodium. In the main cell body however, focal contacts persist much longer than the measured half-live of the integrin exchange. This demonstrates that the modulation of integrin dynamics is a tool to change the "quality" of focal adhesions, making them firm or plastic. Although integrin dynamics may not directly influence the turnover of focal contacts, it is likely that the degree of integrin dynamics determines the capacity and rate of assembly and remodeling of focal complexes and or focal contacts.

Integrin Dynamics in Sliding Focal Contacts

We have previously proposed that the sliding of focal contacts is a macroscopic representation of the recruitment of integrins to the inner edge and their dispersal at the outer edge of sliding focal contacts. Therefore, the maximal speed of focal contact sliding is determined by the rate of integrin recruitment to the inner edge of a sliding contact. Unfortunately, the regulatory mechanisms that control the recruitment of integrin to the inner edge of sliding focal contacts are not well understood. However, it is likely that this issue is directly linked to the mechanisms that control integrin clustering.

Are Integrin Affinity and Valency Modulations Linked?

At the cell front, changes in integrin affinity and valency (integrin clustering), often occur simultaneously, hence it would be possible that integrin activation is directly linked to or actively induces integrin clustering. In fact, based on in vitro studies such a scenario was proposed for the αIIbβ3 integrin. Li and coworkers suggested that the activation induced unclasping of the integrin transmembrane domains[183] allowed these isolated transmembrane domains to undergo homo-oligomerization with other α and β subunits. This transmembrane association would result in the formation of an inter-connected network of high affinity integrins, resulting in integrin clustering and increase in integrin valency. Thus, this result would predict that integrin affinity modulations for example induced by the binding of cytoplasmic adaptor proteins such as RAPL or talin (see above) would control the increase or decrease of integrin clustering in focal adhesions. Hence the control of the allosteric switch that regulates integrin activation would be of primordial importance for the behavior of individual focal adhesions.

Based on these data, we would predict that integrin activation by mutagenesis within the cytoplasmic or the extracellular domain would result in increased integrin clustering in living cells and modification of cell behavior. Accordingly the destruction of a salt bridge between the cytoplasmic domains of integrins (hinge region), results in integrin activation in CHO cells and increases tumor metastasis in carcinoma cells.[130,148] A mutation in the extracellular ligand-binding region of β1-integrin increases integrin affinity and induces the formation of sarcomas, suggesting the importance for integrin affinity regulations in cell migration and differentiation.[184] Similarly, the introduction of a "glycan wedge" that blocks the extracellular domain of αvβ3 integrin in the open configuration results in increased cell adhesion.[185]

Therefore, it is important to know whether affinity modulations of integrins have a direct or indirect impact on valency changes of integrins. Recently, the role of the integrin transmembrane domains in lateral integrin clustering has been analyzed by disulfide bond scanning of the exofacial side of the αIIbβ3 transmembrane domains. This analysis did not reveal a specific transmembrane mediated association of activated integrins in living cells.[147] Furthermore, integrin activation by disrupting the respective association between the transmembrane domains of the α- and β-integrin did not increase integrin clustering.[186] These observations are in apparent conflict with the visualized interaction between the stalk domains of purified integrins by electron-microscopy.[183,187] In these integrin preparations, integrins form small clusters apparently linked by a region of the proteins that includes the transmembrane domain. However, since the detergent, used to purify integrins had to be removed prior to EM observation by rotary shadowing, it is not clear to which extent these integrin oligomers are due to the hydrophobic association of the transmembrane domains in a detergent free buffer. In fact, when purified αIIbβ3 integrins are re incorporated into vesicles or planar membranes, no apparent clustering of activated αIIbβ3 integrins can be observed when analyzed by electronmicroscopy of negatively stained vesicles.[188] Nevertheless, within these reconstituted planar membranes, integrin clustering can be readily induced by the addition of the bivalent extracellular ligand fibrinogen.[188] In conclusion, these data suggest that integrin activation is a prerequisite but by itself insufficient for integrin clustering. In order to induce integrin clustering, the interaction with multivalent extracellular ligands and/or intracellular adaptors is required.

Lateral Integrin Clustering and Adaptor Recruitment at the Front of Migrating Cells

Chronology of Integrin Clustering and Adaptor Recruitment

At the leading edge of migrating cells, activated integrins cluster to form focal complexes. Focal complexes consist of integrins and various adaptor proteins that provide the link between the integrin and the actin cytoskeleton. Recently the chronology of αvβ3 integrin clustering at the leading edge and respective adaptor protein recruitment was analyzed in migrating endothelial cells.[85] This study revealed that clusters of αvβ3 integrins formed behind actin-rich regions in the lamellipodium. These integrin clusters, termed early focal complexes were positive for anti-phosphotyrosine antibodies, talin and paxillin but exhibited only low levels of vinculin and FAK.[85] Older focal complexes additionally recruited FAK, vinculin, VASP, α-actinin and the Arp2/3 complex while zyxin was only localized to focal contacts.[85] This temporal pattern of adaptor recruitment confirms older reports demonstrating the localization of talin and β1-integrins to the edge of protruding lamellipodia.[152,189] This suggests that the Arp2/3 and vinculin dependent link of the actin cytoskeleton to focal complexes[190] could be independently controlled from the clustering of integrins. Since talin, in addition to the integrin binding domains, exhibits also actin-binding sites, it could be the critical integrin adaptor that anchors extracellular ligand immobilized, clustered integrins to the actin cytoskeleton of the protruding lamellipodium.[191]

The Role of PI(4,5)P2 Lipids

However, there is still one missing piece in the puzzle of integrin clustering into focal complexes. Despite the importance of talin as a β3-integrin adaptor protein, talin is not sufficient to induce the clustering of activated integrins by itself. In fact, the sequestration of $PI(4,5)P_2$ by the overexpression of the PH domain of PLCδ[161] inhibits talin recruitment to focal adhesions. Hence talin may require the $PI(4,5)P_2$ lipid as a cofactor to bind to activated integrins and to link them to $PI(4,5)P_2$ containing lipid domains. Moreover, the activation of vinculin and the cross-linking of F-actin to the plasma membrane by ERM proteins requires $PI(4,5)P_2$.[164] Furthermore, the phosphatidylinositol 4-phosphate 5-kinase α is a downstream effector of the small GTPase ARF6.[192] The parallel activation of Rac1 and ARF6 are required for the formation of lamellipodia.[193,194] Thus the combination of $PI(4,5)P_2$ synthesis and actin polymerization at the leading edge of migrating cells create the subcellular environment that allows talin activation, integrin clustering and membrane associated actin polymerization, resulting in the formation of focal complexes.

Regulation of Integrin Clustering at the Cell Rear

In contrast to the cell front, which is under the influence of Rac1 and ARF6 signaling, RhoA signaling controls the formation of actin stress fibers and the formation of focal contacts. As mentioned above the integrin exchange rate in focal contacts is faster than in focal complexes. Because integrin dynamics are not influenced by integrin activation, it is likely that the rate of integrin turnover is regulated at the level of the linkage between the cytoplasmic tail of integrins and the actin cytoskeleton. Several intracellular signaling pathways have been associated with an increase in focal adhesion dynamics and increased migration rate. Integrin dependent FAK activation, c-src signaling and the MAPK pathway have all been implicated in the regulation of focal adhesion turnover.[158,195-197] One of the crucial elements of this remodeling pathway is the activation of cytoplasmic protease μ-calpain.[158,198-200] The calpain induced proteolytic cleavage of talin into an integrin binding head and actin binding tail domain may result in the destabilization of focal contacts. Interestingly, the expression of calpain insensitive talin stabilizes the cycle of focal contact assembly and disassembly.[159] It will be interesting to determine, whether integrin dynamics are also slowed down in focal contacts containing protease resistant talin.

In conclusion, several intracellular signaling pathways converge at the cell front to firmly link activated integrins to the actin cytoskeleton, a process which requires adaptor proteins and $PI(4,5)P_2$. At the cell rear, the continuous remodeling of the link between integrins and the actin cytoskeleton, allows a precise control over cell migration and regulated rear detachment.

Intracellular Integrin Transport

So far, the discussion of the role of integrins in cell migration has been limited to changes in integrin affinity, clustering and dynamics. Generally these mechanisms are differentially regulated between the front and the rear of migrating cells. In order for cell migration to persist, the maintenance of the anterior to posterior polarity is crucial. One important concept of cell polarity is the directed transport of proteins and membranes to the front of the cell. It has been proposed that cell motility is regulated by the transport of membranes and adhesion receptors from the rear to the front of migrating cells.[201] As discussed above, some integrin receptors have been observed to get internalized at the cell rear and to be transported to the cell front.[96]

Detailed data have been gathered on the recycling of $\alpha v\beta 3$ integrins in growth factor stimulated cells. It has been demonstrated that the recycling pathway of integrins is accelerated in PDGF treated cells. Instead of the recruitment of integrins from Rab11 positive late endosomes in nonstimulated cells, PDGF treatment induces the recycling of $\alpha v\beta 3$ integrins from Rab4 positive early endosomes.[153] Furthermore, it has been demonstrated that the PDK1 kinase that associates with the cytoplasmic tail of $\beta 3$-integrins is required for this rapid integrin recycling, enabling cell spreading and migration.[202] In addition, growth factor induced integrin recycling to the cell surface, requires the phosphorylation and inhibition of glycogen synthase kinase 3 (GSK-3) by protein kinase B/Akt.[203] These data demonstrate the important link between growth factor induced formation of lamellipodia and the accelerated recycling of $\alpha v\beta 3$ integrin to the front of migrating cells.

During PDGF stimulated integrin recycling, integrins reappear at the dorsal surface and accumulate only later at the cell front.[153] It is therefore not clear whether integrins reach the front of migrating cells by vesicular transport, by lateral diffusion or by a different transport mechanism. The recent identification of the specific interaction of the cytoplasmic tail of $\beta 3$ integrins with the FERM domain of myosin-X has suggests the existence of a powerful, myosin based integrin traffic. The over-expression of myosin-X induces the formation of filopodia in which integrins are colocalized with myosin-X to their tips.[204] This suggests that $\alpha v\beta 3$ integrins have been transported to the tips of filopodia by this motor protein. Since the integrin interacting FERM domain of myosin-X has homologies to the integrin binding FERM domain of talin, it is likely, that integrins can only be transported when their cytoplasmic tails are in an unclasped, active conformation. The subsequent release of $\alpha v\beta 3$ integrins from myosin-X may be in response to a local competition with other integrin adaptor proteins such as talin, or may occur due to the mechanical interaction of the transported integrins with immobilized extracellular ligands. Myosin-X driven integrin delivery to the front of migrating cells may therefore occur along actin filaments bundles within filopodia. Interestingly, focal complexes in migrating endothelial cells form predominantly at the base of actin rich ridges,[85] which could represent bona-fide myosin-X dependent integrin delivery pathways. Based on these data, specific integrin recycling and targeted delivery to the front of migrating cells are processes that are critically involved in the establishment of new integrin dependent contacts with the extracellular matrix.

Conclusion and Perspectives

We have learned that integrins play critical roles to control cell migration through extracellular matrix. Interestingly, different types of motile cells have acquired specific strategies to harvest the capacities of different types of integrins to undergo reversible ligand binding, in order to develop highly specific and tightly controlled mechanisms to direct cell migration through a complex tissue environment. It will remain a challenge to dissect the multitude of signaling pathways and to identify the hierarchies of protein-protein or protein-lipid interactions that control integrin clustering, dispersal and transport, essential steps for cell migration to occur.

Dedication

In memory of Martin Pfaff.

Acknowledgements

I would like to thank Drs Beat Imhof, Frederic Saltel and Michel Aurrand-Lions for discussions and sharing ideas. A special thanks to Monique Wehrle-Haller for helping me with the manuscript and the Swiss National Science Foundation for supporting my research (31-64000.00 and 3100A0-103805).

References

1. Hynes RO. Integrins: Bidirectional, allosteric signaling machines. Cell 2002; 110(6):673-687.
2. Park KC, Rivero F, Meili R et al. Rac regulation of chemotaxis and morphogenesis in Dictyostelium. EMBO J 2004; 23(-21):4177-4189.
3. Funamoto S, Meili R, Lee S et al. Spatial and temporal regulation of 3-phosphoinositides by PI 3-kinase and PTEN mediates chemotaxis. Cell 2002; 109(5):611-623.
4. Bretschneider T, Jonkman J, Kohler J et al. Dynamic organization of the actin system in the motile cells of Dictyostelium. J Muscle Res Cell Motil 2002; 23(7-8):639-649.
5. Kreitmeier M, Gerisch G, Heizer C et al. A talin homologue of Dictyostelium rapidly assembles at the leading edge of cells in response to chemoattractant. J Cell Biol 1995; 129(1):179-188.
6. Niewohner J, Weber I, Maniak M et al. Talin-null cells of Dictyostelium are strongly defective in adhesion to particle and substrate surfaces and slightly impaired in cytokinesis. J Cell Biol 1997; 138(2):349-361.
7. Reber-Muller S, Studer R, Muller P et al. Integrin and talin in the jellyfish Podocoryne carnea. Cell Biol Int 2001; 25(8):753-769.
8. Hughes AL. Evolution of the integrin alpha and beta protein families. J Mol Evol 2001; 52(1):63-72.
9. Brower DL, Brower SM, Hayward DC et al. Molecular evolution of integrins: Genes encoding integrin beta subunits from a coral and a sponge. Proc Natl Acad Sci USA 1997; 94(17):9182-9187.
10. Pancer Z, Kruse M, Muller I et al. On the origin of Metazoan adhesion receptors: Cloning of integrin alpha subunit from the sponge Geodia cydonium. Mol Biol Evol 1997; 14(4):391-398.
11. Whittaker CA, Hynes RO. Distribution and evolution of von Willebrand/integrin A domains: Widely dispersed domains with roles in cell adhesion and elsewhere. Mol Biol Cell 2002; 13(10):3369-3387.
12. Brown NH. Cell-cell adhesion via the ECM: Integrin genetics in fly and worm. Matrix Biol 2000; 19(3):191-201.
13. Bokel C, Brown NH. Integrins in development: Moving on, responding to, and sticking to the extracellular matrix. Dev Cell 2002; 3(3):311-321.
14. Ribeiro C, Petit V, Affolter M. Signaling systems, guided cell migration, and organogenesis: Insights from genetic studies in Drosophila. Dev Biol 2003; 260(1):1-8.
15. Niewiadomska P, Godt D, Tepass U. DE-Cadherin is required for intercellular motility during Drosophila oogenesis. J Cell Biol 1999; 144(3):533-547.
16. Devenport D, Brown NH. Morphogenesis in the absence of integrins: Mutation of both Drosophila beta subunits prevents midgut migration. Development 2004; 131(21):5405-5415.
17. Rogalski TM, Mullen GP, Gilbert MM et al. The UNC-112 gene in Caenorhabditis elegans encodes a novel component of cell-matrix adhesion structures required for integrin localization in the muscle cell membrane. J Cell Biol 2000; 150(1):253-264.
18. Brown NH, Gregory SL, Rickoll WL et al. Talin is essential for integrin function in Drosophila. Dev Cell 2002; 3(4):569-579.
19. Cram EJ, Clark SG, Schwarzbauer JE. Talin loss-of-function uncovers roles in cell contractility and migration in C. elegans. J Cell Sci 2003; 116(Pt 19):3871-3878.
20. Starz-Gaiano M, Montell DJ. Genes that drive invasion and migration in Drosophila. Curr Opin Genet Dev 2004; 14(1):86-91.
21. Somogyi K, Rorth P. Evidence for tension-based regulation of Drosophila MAL and SRF during invasive cell migration. Dev Cell 2004; 7(1):85-93.
22. Pinheiro EM, Montell DJ. Requirement for Par-6 and Bazooka in Drosophila border cell migration. Development 2004; 131(21):5243-5251.
23. Lee M, Cram EJ, Shen B et al. Roles for beta(pat-3) integrins in development and function of Caenorhabditis elegans muscles and gonads. J Biol Chem 2001; 276(39):36404-36410.
24. Weston JA, Abercrombie M. Cell mobility in fused homo- and heteronomic tissue fragments. J Exp Zool 1967; 164(3):317-323.
25. Abercrombie M, Dunn GA, Heath JP. The shape and movement of fibroblasts in culture. Soc Gen Physiol Ser 1977; 32:57-70.
26. Wehrle-Haller B, Imhof BA. Actin, microtubules and focal adhesion dynamics during cell migration. Int J Biochem and Cell Biol 2003; 35:39-50.
27. Wehrle-Haller B, Imhof BA. The inner lives of focal adhesions. Trends Cell Biol 2002; 12(8):382-389.
28. Friedl P, Borgmann S, Brocker EB. Amoeboid leukocyte crawling through extracellular matrix: Lessons from the Dictyostelium paradigm of cell movement. J Leukoc Biol 2001; 70(4):491-509.
29. Cukierman E, Pankov R, Yamada KM. Cell interactions with three-dimensional matrices. Curr Opin Cell Biol 2002; 14(5):633-639.

30. Small JV, Stradal T, Vignal E et al. The lamellipodium: Where motility begins. Trends Cell Biol 2002; 12(3):112-120.
31. Borisy GG, Svitkina TM. Actin machinery: Pushing the envelope. Curr Opin Cell Biol 2000; 12(1):104-112.
32. Geiger B, Bershadsky A. Assembly and mechanosensory function of focal contacts. Curr Opin Cell Biol 2001; 13(5):584-592.
33. Riveline D, Zamir E, Balaban NQ et al. Focal contacts as mechanosensors: Externally applied local mechanical force induces growth of focal contacts by an mDia1-dependent and ROCK- independent mechanism. J Cell Biol 2001; 153(6):1175-1186.
34. Ballestrem C, Hinz B, Imhof BA et al. Marching at the front and dragging behind: Differential alphaVbeta3- Integrin turnover regulates focal adhesion behavior. J Cell Biol 2001; 155(7):1319-1332.
35. Oliver T, Dembo M, Jacobson K. Separation of propulsive and adhesive traction stresses in locomoting keratocytes. J Cell Biol 1999; 145(3):589-604.
36. Lee J, Leonard M, Oliver T et al. Traction forces generated by locomoting keratocytes. J Cell Biol 1994; 127(6 Pt 2):1957-1964.
37. Dembo M, Wang YL. Stresses at the cell-to-substrate interface during locomotion of fibroblasts. Biophys J 1999; 76(4):2307-2316.
38. Anderson KI, Cross R. Contact dynamics during keratocyte motility. Curr Biol 2000; 10(5):253-260.
39. Chen WT. Mechanism of retraction of the trailing edge during fibroblast movement. J Cell Biol 1981; 90(1):187-200.
40. Palecek SP, Huttenlocher A, Horwitz AF et al. Physical and biochemical regulation of integrin release during rear detachment of migrating cells. J Cell Sci 1998; 111(Pt 7):929-940.
41. Palecek SP, Loftus JC, Ginsberg MH et al. Integrin-ligand binding properties govern cell migration speed through cell-substratum adhesiveness. Nature 1997; 385(6616):537-540.
42. Smilenov LB, Mikhailov A, Pelham RJ et al. Focal adhesion motility revealed in stationary fibroblasts [In Process Citation]. Science 1999; 286(5442):1172-1174.
43. Hong YK, Lange-Asschenfeldt B, Velasco P et al. VEGF-A promotes tissue repair-associated lymphatic vessel formation via VEGFR-2 and the alpha1beta1 and alpha2beta1 integrins. FASEB J 2004; 18(10):1111-1113.
44. Kagami S, Kondo S, Loster K et al. Alpha1beta1 integrin-mediated collagen matrix remodeling by rat mesangial cells is differentially regulated by transforming growth factor-beta and platelet-derived growth factor-BB. J Am Soc Nephrol 1999; 10(4):779-789.
45. Ferrero E, Fabbri M, Poggi A et al. Tumor-driven matrix invasion by infiltrating lymphocytes: Involvement of the alpha1 integrin I-domain. Eur J Immunol 1998; 28(8):2530-2536.
46. Stricker TP, Dumin JA, Dickeson SK et al. Structural analysis of the alpha(2) integrin I domain/procollagenase-1 (matrix metalloproteinase-1) interaction. J Biol Chem 2001; 276(31):29375-29381.
47. Camper L, Holmvall K, Wangnerud C et al. Distribution of the collagen-binding integrin alpha10beta1 during mouse development. Cell Tissue Res 2001; 306(1):107-116.
48. Bengtsson T, Aszodi A, Nicolae C et al. Loss of alpha10beta1 integrin expression leads to moderate dysfunction of growth plate chondrocytes. J Cell Sci 2005; 118(Pt 5):929-936.
49. Popova SN, Rodriguez-Sanchez B, Liden A et al. The mesenchymal alpha11beta1 integrin attenuates PDGF-BB-stimulated chemotaxis of embryonic fibroblasts on collagens. Dev Biol 2004; 270(2):427-442.
50. Tiger CF, Fougerousse F, Grundstrom G et al. alpha11beta1 integrin is a receptor for interstitial collagens involved in cell migration and collagen reorganization on mesenchymal nonmuscle cells. Dev Biol 2001; 237(1):116-129.
51. Schmid RS, Anton ES. Role of integrins in the development of the cerebral cortex. Cereb Cortex 2003; 13(3):219-224.
52. Schmid RS, Shelton S, Stanco A et al. alpha3beta1 integrin modulates neuronal migration and placement during early stages of cerebral cortical development. Development 2004; 131(24):6023-6031.
53. Sanada K, Gupta A, Tsai LH. Disabled-1-regulated adhesion of migrating neurons to radial glial fiber contributes to neuronal positioning during early corticogenesis. Neuron 2004; 42(2):197-211.
54. Yoshimasu T, Sakurai T, Oura S et al. Increased expression of integrin alpha3beta1 in highly brain metastatic subclone of a human nonsmall cell lung cancer cell line. Cancer Sci 2004; 95(2):142-148.
55. Tsuji T, Kawada Y, Kai-Murozono M et al. Regulation of melanoma cell migration and invasion by laminin-5 and alpha3beta1 integrin (VLA-3). Clin Exp Metastasis 2002; 19(2):127-134.
56. Sordat I, Decraene C, Silvestre T et al. Complementary DNA arrays identify CD63 tetraspanin and alpha3 integrin chain as differentially expressed in low and high metastatic human colon carcinoma cells. Lab Invest 2002; 82(12):1715-1724.
57. Pulkkinen L, Kimonis VE, Xu Y et al. Homozygous alpha6 integrin mutation in junctional epidermolysis bullosa with congenital duodenal atresia. Hum Mol Genet 1997; 6(5):669-674.

58. Pulkkinen L, Rouan F, Bruckner-Tuderman L et al. Novel ITGB4 mutations in lethal and nonlethal variants of epidermolysis bullosa with pyloric atresia: Missense versus nonsense. Am J Hum Genet 1998; 63(5):1376-1387.
59. Pulkkinen L, Uitto J. Mutation analysis and molecular genetics of epidermolysis bullosa. Matrix Biol 1999; 18(1):29-42.
60. Koster J, van Wilpe S, Kuikman I et al. Role of binding of plectin to the integrin beta4 subunit in the assembly of hemidesmosomes. Mol Biol Cell 2004; 15(3):1211-1223.
61. Jonkman MF, Pas HH, Nijenhuis M et al. Deletion of a cytoplasmic domain of integrin beta4 causes epidermolysis bullosa simplex. J Invest Dermatol 2002; 119(6):1275-1281.
62. Ashton GH, McLean WH, South AP et al. Recurrent mutations in kindlin-1, a novel keratinocyte focal contact protein, in the autosomal recessive skin fragility and photosensitivity disorder, Kindler syndrome. J Invest Dermatol 2004; 122(1):78-83.
63. Siegel DH, Ashton GH, Penagos HG et al. Loss of kindlin-1, a human homolog of the Caenorhabditis elegans actin-extracellular-matrix linker protein UNC-112, causes Kindler syndrome. Am J Hum Genet 2003; 73(1):174-187.
64. Kloeker S, Major MB, Calderwood DA et al. The Kindler syndrome protein is regulated by transforming growth factor-beta and involved in integrin-mediated adhesion. J Biol Chem 2004; 279(8):6824-6833.
65. De Arcangelis A, Georges-Labouesse E. Integrin and ECM functions: Roles in vertebrate development. Trends Genet 2000; 16(9):389-395.
66. Georges-Labouesse E, Mark M, Messaddeq N et al. Essential role of alpha 6 integrins in cortical and retinal lamination. Curr Biol 1998; 8(17):983-986.
67. Graus-Porta D, Blaess S, Senften M et al. Beta1-class integrins regulate the development of laminae and folia in the cerebral and cerebellar cortex. Neuron 2001; 31(3):367-379.
68. Dangerfield JP, Wang S, Nourshargh S. Blockade of alpha6 integrin inhibits IL-1beta- but not TNF-alpha-induced neutrophil transmigration in vivo. J Leukoc Biol 2005; 77(2):159-165.
69. Strachan LR, Condic ML. Cranial neural crest recycle surface integrins in a substratum-dependent manner to promote rapid motility. J Cell Biol 2004; 167(3):545-554.
70. Hayashi YK, Chou FL, Engvall E et al. Mutations in the integrin alpha7 gene cause congenital myopathy. Nat Genet 1998; 19(1):94-97.
71. Mayer U, Saher G, Fassler R et al. Absence of integrin alpha 7 causes a novel form of muscular dystrophy. Nat Genet 1997; 17(3):318-323.
72. Pegoraro E, Cepollaro F, Prandini P et al. Integrin alpha 7 beta 1 in muscular dystrophy/myopathy of unknown etiology. Am J Pathol 2002; 160(6):2135-2143.
73. Kaariainen M, Kaariainen J, Jarvinen TL et al. Integrin and dystrophin associated adhesion protein complexes during regeneration of shearing-type muscle injury. Neuromuscul Disord 2000; 10(2):121-132.
74. Kaariainen M, Nissinen L, Kaufman S et al. Expression of alpha7beta1 integrin splicing variants during skeletal muscle regeneration. Am J Pathol 2002; 161(3):1023-1031.
75. Burkin DJ, Wallace GQ, Milner DJ et al. Transgenic expression of alpha7beta1 integrin maintains muscle integrity, increases regenerative capacity, promotes hypertrophy, and reduces cardiomyopathy in dystrophic mice. Am J Pathol 2005; 166(1):253-263.
76. Burkin DJ, Wallace GQ, Nicol KJ et al. Enhanced expression of the alpha 7 beta 1 integrin reduces muscular dystrophy and restores viability in dystrophic mice. J Cell Biol 2001; 152(6):1207-1218.
77. Arroyo AG, Yang JT, Rayburn H et al. Alpha4 integrins regulate the proliferation/differentiation balance of multilineage hematopoietic progenitors in vivo. Immunity 1999; 11(5):555-566.
78. Taooka Y, Chen J, Yednock T et al. The integrin alpha9beta1 mediates adhesion to activated endothelial cells and transendothelial neutrophil migration through interaction with vascular cell adhesion molecule-1. J Cell Biol 1999; 145(2):413-420.
79. Kil SH, Krull CE, Cann G et al. The alpha4 subunit of integrin is important for neural crest cell migration. Dev Biol 1998; 202(1):29-42.
80. Testaz S, Duband JL. Central role of the alpha4beta1 integrin in the coordination of avian truncal neural crest cell adhesion, migration, and survival. Dev Dyn 2001; 222(2):127-140.
81. von Andrian UH, Engelhardt B. Alpha4 integrins as therapeutic targets in autoimmune disease. N Engl J Med 2003; 348(1):68-72.
82. Kassner PD, Alon R, Springer TA et al. Specialized functional properties of the integrin alpha 4 cytoplasmic domain. Mol Biol Cell 1995; 6(6):661-674.
83. Liu S, Thomas SM, Woodside DG et al. Binding of paxillin to alpha4 integrins modifies integrin-dependent biological responses. Nature 1999; 402(6762):676-681.

84. Goldfinger LE, Han J, Kiosses WB et al. Spatial restriction of alpha4 integrin phosphorylation regulates lamellipodial stability and alpha4beta1-dependent cell migration. J Cell Biol 2003; 162(4):731-741.
85. Zaidel-Bar R, Ballestrem C, Kam Z et al. Early molecular events in the assembly of matrix adhesions at the leading edge of migrating cells. J Cell Sci 2003; 116(Pt 22):4605-4613.
86. Nishiya N, Kiosses WB, Han J et al. An alpha(4) integrin-paxillin-Arf-GAP complex restricts Rac activation to the leading edge of migrating cells. Nat Cell Biol 2005; 7(4):343-352.
87. Rose DM, Liu S, Woodside DG et al. Paxillin binding to the alpha 4 integrin subunit stimulates LFA-1 (integrin alpha L beta 2)-dependent T cell migration by augmenting the activation of focal adhesion kinase/proline-rich tyrosine kinase-2. J Immunol 2003; 170(12):5912-5918.
88. Young BA, Taooka Y, Liu S et al. The cytoplasmic domain of the integrin alpha9 subunit requires the adaptor protein paxillin to inhibit cell spreading but promotes cell migration in a paxillin-independent manner. Mol Biol Cell 2001; 12(10):3214-3225.
89. Chen C, Young BA, Coleman CS et al. Spermidine/spermine N1-acetyltransferase specifically binds to the integrin alpha9 subunit cytoplasmic domain and enhances cell migration. J Cell Biol 2004; 167(1):161-170.
90. Huang XZ, Wu JF, Ferrando R et al. Fatal bilateral chylothorax in mice lacking the integrin alpha9beta1. Mol Cell Biol 2000; 20(14):5208-5215.
91. Vlahakis NE, Young BA, Atakilit A et al. The lymphangiogenic vascular endothelial growth factors VEGF-C and -D are ligands for the integrin alpha9beta1. J Biol Chem 2005; 280(6):4544-4552.
92. Zamir E, Katz M, Posen Y et al. Dynamics and segregation of cell-matrix adhesions in cultured fibroblasts. Nat Cell Biol 2000; 2(4):191-196.
93. Pankov R, Cukierman E, Katz BZ et al. Integrin dynamics and matrix assembly: Tensin-dependent translocation of alpha(5)beta(1) integrins promotes early fibronectin fibrillogenesis. J Cell Biol 2000; 148(5):1075-1090.
94. Zajaczkowski MB, Cukierman E, Galbraith CG et al. Cell-matrix adhesions on poly(vinyl alcohol) hydrogels. Tissue Eng 2003; 9(3):525-533.
95. Clark K, Pankov R, Travis MA et al. A specific alpha5beta1-integrin conformation promotes directional integrin translocation and fibronectin matrix formation. J Cell Sci 2005; 118(Pt 2):291-300.
96. Laukaitis CM, Webb DJ, Donais K et al. Differential dynamics of alpha 5 integrin, paxillin, and alpha-actinin during formation and disassembly of adhesions in migrating cells. J Cell Biol 2001; 153(7):1427-1440.
97. Wiseman PW, Brown CM, Webb DJ et al. Spatial mapping of integrin interactions and dynamics during cell migration by image correlation microscopy. J Cell Sci 2004; 117(Pt 23):5521-5534.
98. Muller U, Wang D, Denda S et al. Integrin alpha8beta1 is critically important for epithelial-mesenchymal interactions during kidney morphogenesis. Cell 1997; 88(5):603-613.
99. Brandenberger R, Schmidt A, Linton J et al. Identification and characterization of a novel extracellular matrix protein nephronectin that is associated with integrin alpha8beta1 in the embryonic kidney. J Cell Biol 2001; 154(2):447-458.
100. Lu M, Munger JS, Steadele M et al. Integrin alpha8beta1 mediates adhesion to LAP-TGFbeta1. J Cell Sci 2002; 115(Pt 23):4641-4648.
101. Bouzeghrane F, Mercure C, Reudelhuber TL et al. Alpha8beta1 integrin is upregulated in myofibroblasts of fibrotic and scarring myocardium. J Mol Cell Cardiol 2004; 36(3):343-353.
102. Imhof BA, Aurrand-Lions M. Adhesion mechanisms regulating the migration of monocytes. Nat Rev Immunol 2004; 4(6):432-444.
103. Katagiri K, Ohnishi N, Kabashima K et al. Crucial functions of the Rap1 effector molecule RAPL in lymphocyte and dendritic cell trafficking. Nat Immunol 2004; 5:1045-1051.
104. Katagiri K, Maeda A, Shimonaka M et al. RAPL, a Rap1-binding molecule that mediates Rap1-induced adhesion through spatial regulation of LFA-1. Nat Immunol 2003; 4:741-748.
105. Bos JL, de Rooij J, Reedquist KA. Rap1 signalling: Adhering to new models. Nat Rev Mol Cell Biol 2001; 2(5):369-377.
106. de Bruyn KM, Rangarajan S, Reedquist KA et al. The small GTPase Rap1 is required for Mn(2+)- and antibody-induced LFA-1- and VLA-4-mediated cell adhesion. J Biol Chem 2002; 277(33):29468-29476.
107. Tohyama Y, Katagiri K, Pardi R et al. The critical cytoplasmic regions of the alphaL/beta2 integrin in Rap1-induced adhesion and migration. Mol Biol Cell 2003; 14(6):2570-2582.
108. Carman CV, Springer TA. Integrin avidity regulation: Are changes in affinity and conformation underemphasized? Curr Opin Cell Biol 2003; 15(5):547-556.
109. Kim M, Carman CV, Springer TA. Bidirectional transmembrane signaling by cytoplasmic domain separation in integrins. Science 2003; 301(5640):1720-1725.
110. Niggli V. Rho-kinase in human neutrophils: A role in signalling for myosin light chain phosphorylation and cell migration. FEBS Lett 1999; 445(1):69-72.

111. Eddy RJ, Pierini LM, Matsumura F et al. Ca2+-dependent myosin II activation is required for uropod retraction during neutrophil migration. J Cell Sci 2000; 113(Pt 7):1287-1298.
112. Lawson MA, Maxfield FR. Ca(2+)- and calcineurin-dependent recycling of an integrin to the front of migrating neutrophils. Nature 1995; 377(6544):75-79.
113. Pierini LM, Lawson MA, Eddy RJ et al. Oriented endocytic recycling of alpha5beta1 in motile neutrophils. Blood 2000; 95(8):2471-2480.
114. Fabbri M, Fumagalli L, Bossi G et al. A tyrosine-based sorting signal in the beta2 integrin cytoplasmic domain mediates its recycling to the plasma membrane and is required for ligand-supported migration. EMBO J 1999; 18(18):4915-4925.
115. Wagner N, Lohler J, Kunkel EJ et al. Critical role for beta7 integrins in formation of the gut-associated lymphoid tissue. Nature 1996; 382(6589):366-370.
116. Schon MP, Schon M, Warren HB et al. Cutaneous inflammatory disorder in integrin alphaE (CD103)-deficient mice. J Immunol 2000; 165(11):6583-6589.
117. Kellersmann R, Lazarovits A, Grant D et al. Monoclonal antibody against beta7 integrins, but not beta7 deficiency, attenuates intestinal allograft rejection in mice. Transplantation 2002; 74(9):1327-1334.
118. Sun X, Qiao H, Shi J et al. Beta7 integrins contribute to skin graft rejection. Transplantation 2002; 74(8):1202-1203.
119. Davis GE. The Mac-1 and p150,95 beta 2 integrins bind denatured proteins to mediate leukocyte cell-substrate adhesion. Exp Cell Res 1992; 200(2):242-252.
120. Vorup-Jensen T, Carman CV, Shimaoka M et al. Exposure of acidic residues as a danger signal for recognition of fibrinogen and other macromolecules by integrin alphaXbeta2. Proc Natl Acad Sci USA 2005; 102(5):1614-1619.
121. Kindzelskii AL, Laska ZO, Todd IIIrd RF et al. Urokinase-type plasminogen activator receptor reversibly dissociates from complement receptor type 3 (alpha M beta 2' CD11b/CD18) during neutrophil polarization. J Immunol 1996; 156(1):297-309.
122. Kindzelskii AL, Eszes MM, Todd IIIrd RF et al. Proximity oscillations of complement type 4 (alphaX beta2) and urokinase receptors on migrating neutrophils. Biophys J 1997; 73(4):1777-1784.
123. Mabon PJ, Weaver LC, Dekaban GA. Inhibition of monocyte/macrophage migration to a spinal cord injury site by an antibody to the integrin alphaD: A potential new anti-inflammatory treatment. Exp Neurol 2000; 166(1):52-64.
124. Law DA, DeGuzman FR, Heiser P et al. Integrin cytoplasmic tyrosine motif is required for outside-in alphaIIbbeta3 signalling and platelet function. Nature 1999; 401(6755):808-811.
125. Boettiger D, Huber F, Lynch L et al. Activation of alpha(v)beta3-vitronectin binding is a multistage process in which increases in bond strength are dependent on Y747 and Y759 in the cytoplasmic domain of beta3. Mol Biol Cell 2001; 12(5):1227-1237.
126. Brooks PC, Clark RA, Cheresh DA. Requirement of vascular integrin alpha v beta 3 for angiogenesis. Science 1994; 264(5158):569-571.
127. Eliceiri BP, Cheresh DA. Role of alpha v integrins during angiogenesis. Cancer J 2000; 6(Suppl 3):S245-249.
128. Filardo EJ, Brooks PC, Deming SL et al. Requirement of the NPXY motif in the integrin beta 3 subunit cytoplasmic tail for melanoma cell migration in vitro and in vivo. J Cell Biol 1995; 130(2):441-450.
129. Bao W, Stromblad S. Integrin alphav-mediated inactivation of p53 controls a MEK1-dependent melanoma cell survival pathway in three-dimensional collagen. J Cell Biol 2004; 167(4):745-756.
130. Felding-Habermann B, O'Toole TE, Smith JW et al. Integrin activation controls metastasis in human breast cancer. Proc Natl Acad Sci USA 2001; 98(4):1853-1858.
131. Brooks PC, Montgomery AM, Rosenfeld M et al. Integrin alpha v beta 3 antagonists promote tumor regression by inducing apoptosis of angiogenic blood vessels. Cell 1994; 79(7):1157-1164.
132. Gutheil JC, Campbell TN, Pierce PR et al. Targeted antiangiogenic therapy for cancer using Vitaxin: A humanized monoclonal antibody to the integrin alphavbeta3. Clin Cancer Res 2000; 6(8):3056-3061.
133. Buckley CD, Pilling D, Henriquez NV et al. RGD peptides induce apoptosis by direct caspase-3 activation. Nature 1999; 397(6719):534-539.
134. Reynolds LE, Wyder L, Lively JC et al. Enhanced pathological angiogenesis in mice lacking beta3 integrin or beta3 and beta5 integrins. Nat Med 2002; 8(1):27-34.
135. Lane NE, Yao W, Nakamura MC et al. Mice lacking the integrin beta5 subunit have accelerated osteoclast maturation and increased activity in the estrogen-deficient state. J Bone Miner Res 2005; 20(1):58-66.
136. Huang X, Griffiths M, Wu J et al. Normal development, wound healing, and adenovirus susceptibility in beta5-deficient mice. Mol Cell Biol 2000; 20(3):755-759.

137. Filardo EJ, Deming SL, Cheresh DA. Regulation of cell migration by the integrin beta subunit ectodomain. J Cell Sci 1996; 109(Pt 6):1615-1622.
138. Memmo LM, McKeown-Longo P. The alphavbeta5 integrin functions as an endocytic receptor for vitronectin. J Cell Sci 1998; 111(Pt 4):425-433.
139. Nandrot EF, Kim Y, Brodie SE et al. Loss of synchronized retinal phagocytosis and age-related blindness in mice lacking alphavbeta5 integrin. J Exp Med 2004; 200(12):1539-1545.
140. Huang X, Wu J, Spong S et al. The integrin alphavbeta6 is critical for keratinocyte migration on both its known ligand, fibronectin, and on vitronectin. J Cell Sci 1998; 111(Pt 15):2189-2195.
141. Huang X, Wu J, Zhu W et al. Expression of the human integrin beta6 subunit in alveolar type II cells and bronchiolar epithelial cells reverses lung inflammation in beta6 knockout mice. Am J Respir Cell Mol Biol 1998; 19(4):636-642.
142. Nishimura SL, Boylen KP, Einheber S et al. Synaptic and glial localization of the integrin alphavbeta8 in mouse and rat brain. Brain Res 1998; 791(1-2):271-282.
143. McCarty JH, Monahan-Earley RA, Brown LF et al. Defective associations between blood vessels and brain parenchyma lead to cerebral hemorrhage in mice lacking alphav integrins. Mol Cell Biol 2002; 22(21):7667-7677.
144. Buensuceso C, de Virgilio M, Shattil SJ. Detection of integrin alpha IIbbeta 3 clustering in living cells. J Biol Chem 2003; 278(17):15217-15224.
145. Xiao T, Takagi J, Coller BS et al. Structural basis for allostery in integrins and binding to fibrinogen-mimetic therapeutics. Nature 2004; 432:59-67.
146. Vinogradova O, Velyvis A, Velyviene A et al. A structural mechanism of integrin alpha(IIb)beta(3) "inside-out" activation as regulated by its cytoplasmic face. Cell 2002; 110(5):587-597.
147. Luo BH, Springer TA, Takagi J. A specific interface between integrin transmembrane helices and affinity for ligand. PLoS Biol 2004; 2:e153.
148. Hughes PE, Diaz-Gonzalez F, Leong L et al. Breaking the integrin hinge. A defined structural constraint regulates integrin signaling. J Biol Chem 1996; 271(12):6571-6574.
149. Bertoni A, Tadokoro S, Eto K et al. Relationships between Rap1b, affinity modulation of integrin alpha IIbbeta 3, and the actin cytoskeleton. J Biol Chem 2002; 277(28):25715-25721.
150. Buensuceso CS, Obergfell A, Soriani A et al. Regulation of outside-in signaling in platelets by integrin-associated protein kinase C beta. J Biol Chem 2005; 280(1):644-653.
151. Kiosses WB, Shattil SJ, Pampori N et al. Rac recruits high-affinity integrin alphavbeta3 to lamellipodia in endothelial cell migration. Nat Cell Biol 2001; 3(3):316-320.
152. Lee J, Jacobson K. The composition and dynamics of cell-substratum adhesions in locomoting fish keratocytes. J Cell Sci 1997; 110:2833-2844.
153. Roberts M, Barry S, Woods A et al. PDGF-regulated rab4-dependent recycling of alphavbeta3 integrin from early endosomes is necessary for cell adhesion and spreading. Curr Biol 2001; 11(18):1392-1402.
154. Tadokoro S, Shattil SJ, Eto K et al. Talin binding to integrin beta tails: A final common step in integrin activation. Science 2003; 302:103-106.
155. Calderwood DA. Talin controls integrin activation. Biochem Soc Trans 2004; 32:434-437.
156. Garcia-Alvarez B, de Pereda JM, Calderwood DA et al. Structural determinants of integrin recognition by talin. Cell 2003; 11(1):49-58.
157. Yan B, Calderwood DA, Yaspan B et al. Calpain cleavage promotes talin binding to the beta 3 integrin cytoplasmic domain. J Biol Chem 2001; 276(30):28164-28170.
158. Cuevas BD, Abell AN, Witowsky JA et al. MEKK1 regulates calpain-dependent proteolysis of focal adhesion proteins for rear-end detachment of migrating fibroblasts. EMBO J 2003; 22(13):3346-3355.
159. Franco SJ, Rodgers MA, Perrin BJ et al. Calpain-mediated proteolysis of talin regulates adhesion dynamics. Nat Cell Biol 2004; 6(10):977-983.
160. Isenberg G, Goldmann WH. Peptide-specific antibodies localize the major lipid binding sites of talin dimers to oppositely arranged N-terminal 47 kDa subdomains. FEBS Lett 1998; 426:165-170.
161. Martel V, Racaud-Sultan C, Dupe S et al. Conformation, localization, and integrin binding of talin depend on its interaction with phosphoinositides. J Biol Chem 2001; 276(24):21217-21227.
162. Di Paolo G, Pellegrini L, Letinic K et al. Recruitment and regulation of phosphatidylinositol phosphate kinase type 1 gamma by the FERM domain of talin. Nature 2002; 420(6911):85-89.
163. Ling K, Doughman RL, Firestone AJ et al. Type I gamma phosphatidylinositol phosphate kinase targets and regulates focal adhesions. Nature 2002; 420(6911):89-93.
164. Yin HL, Janmey PA. Phosphoinositide regulation of the actin cytoskeleton. Annu Rev Physiol 2003; 65:761-789.
165. Koo LY, Irvine DJ, Mayes AM et al. Coregulation of cell adhesion by nanoscale RGD organization and mechanical stimulus. J Cell Sci 2002; 115(Pt 7):1423-1433.

166. Coussen F, Choquet D, Sheetz MP et al. Trimers of the fibronectin cell adhesion domain localize to actin filament bundles and undergo rearward translocation. J Cell Sci 2002; 115(Pt 12):2581-2590.
167. Balaban NQ, Schwarz US, Riveline D et al. Force and focal adhesion assembly: A close relationship studied using elastic micropatterned substrates. Nat Cell Biol 2001; 3(5):466-472.
168. Massia SP, Hubbell JA. An RGD spacing of 440 nm is sufficient for integrin alpha V beta 3-mediated fibroblast spreading and 140 nm for focal contact and stress fiber formation. J Cell Biol 1991; 114(5):1089-1100.
169. Arnold M, Cavalcanti-Adam EA, Glass R et al. Activation of integrin function by nanopatterned adhesive interfaces. Chemphyschem 2004; 5(3):383-388.
170. Sawada Y, Sheetz MP. Force transduction by Triton cytoskeletons. J Cell Biol 2002; 156(4):609-615.
171. Wang HB, Dembo M, Hanks SK et al. Focal adhesion kinase is involved in mechanosensing during fibroblast migration. Proc Natl Acad Sci USA 2001; 98(20):11295-11300.
172. Trachslin J, Koch M, Chiquet M. Rapid and reversible regulation of collagen XII expression by changes in tensile stress. Exp Cell Res 1999; 247(2):320-328.
173. Chiquet M, Sarasa-Renedo A, Tunc-Civelek V. Induction of tenascin-C by cyclic tensile strain versus growth factors: Distinct contributions by Rho/ROCK and MAPK signaling pathways. Biochim Biophys Acta 2004; 1693(3):193-204.
174. Lo CM, Wang HB, Dembo M et al. Cell movement is guided by the rigidity of the substrate. Biophys J 2000; 79(1):144-152.
175. Pelham Jr RJ, Wang Y. Cell locomotion and focal adhesions are regulated by substrate flexibility. Proc Natl Acad Sci USA 1997; 94(25):13661-13665.
176. Lehnert D, Wehrle-Haller B, David C et al. Cell behavior on micropatterned substrates: Limits of extracellular matrix geometry for spreading and adhesion. J Cell Sci 2003; in press.
177. Duband JL, Nuckolls GH, Ishihara A et al. Fibronectin receptor exhibits high lateral mobility in embryonic locomoting cells but is immobile in focal contacts and fibrillar streaks in stationary cells. J Cell Biol 1988; 107(4):1385-1396.
178. Axelrod D, Koppel DE, Schlessinger J et al. Mobility measurement by analysis of fluorescence photobleaching recovery kinetics. Biophys J 1976; 16(9):1055-1069.
179. White J, Stelzer E. Photobleaching GFP reveals protein dynamics inside live cells. Trends Cell Biol 1999; 9(2):61-65.
180. Danen EHJ, van Rheenen J, Franken W et al. Integrins control motile strategy through a Rho-cofilin pathway. J Cell Biol 2005; 169(3):515-526.
181. Hinz B, Dugina V, Ballestrem C et al. Alpha-smooth muscle actin is crucial for focal adhesion maturation in myofibroblasts. Mol Biol Cell 2003; 14(6):2508-2519.
182. Tsuruta D, Gonzales M, Hopkinson SB et al. Microfilament-dependent movement of the beta3 integrin subunit within focal contacts of endothelial cells. FASEB J 2002; 16(8):866-868.
183. Li R, Mitra N, Gratkowski H et al. Activation of integrin alphaIIbbeta3 by modulation of transmembrane helix associations. Science 2003; 300(5620):795-798.
184. Evans RD, Perkins VC, Henry A et al. A tumor-associated beta 1 integrin mutation that abrogates epithelial differentiation control. J Cell Biol 2003; 160(4):589-596.
185. Luo BH, Springer TA, Takagi J. Stabilizing the open conformation of the integrin headpiece with a glycan wedge increases affinity for ligand. Proc Natl Acad Sci USA 2003; 100:2403-2408.
186. Luo BH, Carman CV, Takagi J et al. Disrupting integrin transmembrane domain heterodimerization increases ligand binding affinity, not valency or clustering. Proc Natl Acad Sci USA 2005; 102(10):3679-3684.
187. Litvinov RI, Nagaswami C, Vilaire G et al. Functional and structural correlations of individual alphaIIbbeta3 molecules. Blood 2004; 104(13):3979-3985.
188. Erb EM, Tangemann K, Bohrmann B et al. Integrin alphaIIb beta3 reconstituted into lipid bilayers is nonclustered in its activated state but clusters after fibrinogen binding. Biochemistry 1997; 36(24):7395-7402.
189. DePasquale JA, Izzard CS. Accumulation of talin in nodes at the edge of the lamellipodium and separate incorporation into adhesion plaques at focal contacts in fibroblasts. J Cell Biol 1991; 113(6):1351-1359.
190. DeMali KA, Barlow CA, Burridge K. Recruitment of the Arp2/3 complex to vinculin: Coupling membrane protrusion to matrix adhesion. J Cell Biol 2002; 159(5):881-891.
191. Nayal A, Webb DJ, Horwitz AF. Talin: An emerging focal point of adhesion dynamics. Curr Opin Cell Biol 2004; 16(1):94-98.
192. Honda A, Nogami M, Yokozeki T et al. Phosphatidylinositol 4-phosphate 5-kinase alpha is a downstream effector of the small G protein ARF6 in membrane ruffle formation. Cell 1999; 99(5):521-532.

193. Santy LC, Casanova JE. Activation of ARF6 by ARNO stimulates epithelial cell migration through downstream activation of both Rac1 and phospholipase D. J Cell Biol 2001; 154(3):599-610.
194. Radhakrishna H, Al-Awar O, Khachikian Z et al. ARF6 requirement for Rac ruffling suggests a role for membrane trafficking in cortical actin rearrangements. J Cell Sci 1999; 112(Pt 6):855-866.
195. Fincham VJ, Frame MC. The catalytic activity of Src is dispensable for translocation to focal adhesions but controls the turnover of these structures during cell motility. EMBO J 1998; 17(1):81-92.
196. Frame MC, Fincham VJ, Carragher NO et al. v-Src's hold over actin and cell adhesions. Nat Rev Mol Cell Biol 2002; 3(4):233-245.
197. Westhoff MA, Serrels B, Fincham VJ et al. SRC-mediated phosphorylation of focal adhesion kinase couples actin and adhesion dynamics to survival signaling. Mol Cell Biol 2004; 24(18):8113-8133.
198. Dourdin N, Bhatt AK, Dutt P et al. Reduced cell migration and disruption of the actin cytoskeleton in calpain-deficient embryonic fibroblasts. J Biol Chem 2001; 276(51):48382-48388.
199. Carragher NO, Frame MC. Calpain: A role in cell transformation and migration. Int J Biochem Cell Biol 2002; 34(12):1539-1543.
200. Huttenlocher A, Palecek SP, Lu Q et al. Regulation of cell migration by the calcium-dependent protease calpain. J Biol Chem 1997; 272(52):32719-32722.
201. Bretscher MS, Aguado-Velasco C. Membrane traffic during cell locomotion. Curr Opin Cell Biol 1998; 10(4):537-541.
202. Woods AJ, White DP, Caswell PT et al. PKD1/PKCmicro promotes alphavbeta3 integrin recycling and delivery to nascent focal adhesions. EMBO J 2004; 23(13):2531-2543.
203. Roberts MS, Woods AJ, Dale TC et al. Protein kinase B/Akt acts via glycogen synthase kinase 3 to regulate recycling of alpha v beta 3 and alpha 5 beta 1 integrins. Mol Cell Biol 2004; 24(4):1505-1515.
204. Zhang H, Berg JS, Li Z et al. Myosin-X provides a motor-based link between integrins and the cytoskeleton. Nat Cell Biol 2004; 6(6):523-531.

Chapter 4

The Essence of Integrin Signal Transduction:
Assembly of Dynamic Scaffolds and Cross-Talk with Other Receptors

Sara Cabodi and Paola Defilippi*

Introduction

Integrin signalling has a critical function in organizing cells in tissues during both embryonic development and tissue repair. Following their binding to the extracellular ligands, the intracellular signalling pathways triggered by integrins are directed to two major functions: organisation of the actin cytoskeleton and regulation of cell behaviour including survival, differentiation and growth. Basic research conducted in the past twelve years has lead to remarkable breakthroughs in this field. Integrins are catalytically inactive and translate positional cues into biochemical signals by direct and/or functional association with intracellular adaptors, cytosolic tyrosine kinases or growth factor and cytokine receptors. The purpose of this chapter is to highlight recent experimental and conceptual advances in integrin signalling with particular emphasis to the ability of integrins to regulate Fak/Src family kinases (SFKs) activation and to cross-talk with receptors for soluble growth factors and cytokines.

Regulation of Outside-In Signalling

Early Evidences for Integrin Signal Transduction

Integrins are enzymatically inactive receptors, which link with cellular components to elicit signal transduction. Integrin ability to transduce signals inside the cells upon ligand binding, the so called outside-in signalling, was inferred from a series of simple observations. Organisation of the actin or intermediate filament cytoskeleton during cell adhesion clearly requires intracellular signals that trigger polymerisation of cytoskeletal proteins. Immunofluorescence experiments with anti-phosphotyrosine antibodies showed that most of the tyrosine phosphorylated proteins are localized at cell-substratum and cell-cell contacts in both normal and transformed cells,[1] suggesting that the membrane-cytoskeleton interaction is crucially regulated by these events. In addition, the ability of cell matrix interactions to control cell proliferation[2] and differentiation,[3-6] strongly implied the existence of matrix-dependent signalling events. Direct evidence for a matrix-mediated intracellular signalling was provided in the late eighties by a series of experiments showing that gene expression is altered in response to cell adhesion. Transcription of early response c-jun and c-fos genes are induced in quiescent fibroblasts upon fibronectin interaction.[7] In monocytes, transcription of genes related to the inflammatory response is also induced by specific matrix proteins.[8]

A number of reports subsequently showed that more up-stream events, such as elevation of intracellular pH,[9] Ca^{2+} transients[10] and activation of tyrosine kinases[11] can be triggered by

*Corresponding Author: Paola Defilippi—Department of Genetics, Biology and Biochemistry, University of Torino, Via Santena 5 bis, 10126, Torino, Italy. Email: paola.defilippi@unito.it.

Integrins and Development, edited by Erik H.J. Danen. ©2006 Landes Bioscience.

integrins. The first evidence that integrin occupancy by ligands induces tyrosine phosphorylation came from studies of the fibrinogen receptor alphaIIbbeta3 in platelets.[11-13] Further work showed that interaction of normal or transformed cells with several matrix proteins or integrin receptor clustering by specific antibodies triggers tyrosine phosphorylation of intracellular proteins. A major breakthrough came from the cloning and characterization of the Focal adhesion Kinase (Fak), a tyrosine kinase originally identified as a v-Src substrate (see below). Fak is activated by integrin-mediated adhesion as well as growth factors and other agonists, providing evidence that integrins are indeed signalling receptors as well as mediators of cell adhesion. The list of integrin-mediated signalling events has now been expanded to comprise most of the known pathways. These include stimulation of the phosphatidyl inositol phosphate metabolism, Ras and MAPK activation, activation of protein kinase C and regulation of the Rho family of monomeric GTPases; for excellent reviews see references 14-16. These signals largely overlap with those activated by growth factor and cytokine receptors. However, detailed analysis of the extent and duration of each signals vary depending on whether the integrin, the RPTKs or both are engaged by the ligand, indicating that signals from integrin and growth factor and cytokine receptors are properly integrated. This view reflects the fact that in vivo, cells must integrate multiple stimuli from adhesions, growth factors, hormones and mechanical stresses to determine the appropriate response.

Interaction of Integrins with Cytoskeletal Molecules and Its Regulation by Phosphoinositide Signalling

Interaction of integrins with cytoskeletal molecules was originally inferred from immunolocalization studies. In cultured cells integrins specifically localize to focal adhesions, which are sites of close apposition with the growth substratum where actin filament ends are anchored to the plasma membrane.[17] At these sites, integrins codistribute with cytoskeletal proteins such as talin, vinculin and alpha actinin, which are part of the actin bundle-membrane linkage.[18,19] The cytoplasmic domain of the beta subunit contains all necessary information to recognize components of the focal adhesions, while the alpha subunits can modulate the interaction of the integrin heterodimer with the cytoskeleton. Thus, the role of the alpha subunit cytoplasmic domain is to mask a focal adhesion localization default signal present in the beta subunit so that the un-ligated receptor remains free in the membrane. Upon ligand binding the constraint imposed by the alpha subunit is released and the heterodimer can interact with focal adhesion components via the beta subunit domain.

Talin, alpha actinin and vinculin, prominent integrin-associated structural proteins, bind F-actin, thus recruiting actin filaments to integrin clusters. In addition these integrin-associated molecules are multi-functional scaffolding protein, that by means of protein-protein interactions bring kinases, phophatases and their substrates together, which, in turn, regulates the dynamics of integrin-cytoskeleton connection (Fig. 1).

Talin is a large major actin-binding protein,[20,21] which binds strongly to beta1A, beta1D, beta2, beta 3 and beta5. Recent data underline how this binding represents a final common step in integrin activation pathways. The major integrin-binding site lies within the talin FERM (Four-point-one, Ezrin, Radixin, Moesin) domain.[22,23] FERM domains are present in structural proteins as well as in signalling proteins such as the JAK family of tyrosine kinases and several tyrosine phosphatases and mediate protein-protein interactions either by docking the cytoplasmic tails of transmembrane proteins or by allowing intramolecular or homophilic intermolecular interactions. Talin FERM domain binding to the beta1 integrin occurs via a variant of the classical PTB domain (phosphotyrosine-binding domain)-NPxY interaction. Talin binding to integrins disrupts an intracellular salt bridge between the alpha and the beta subunit, leading to increased integrin affinity which strengthens the interaction with the ECM.[24,25] The critical role played by formation of the talin-integrin complex in integrin activation has been confirmed by the fact that mutations, in either talin or integrin beta tails, which disrupt complex formation, inhibit integrin activation. Furthermore, use of RNA interference

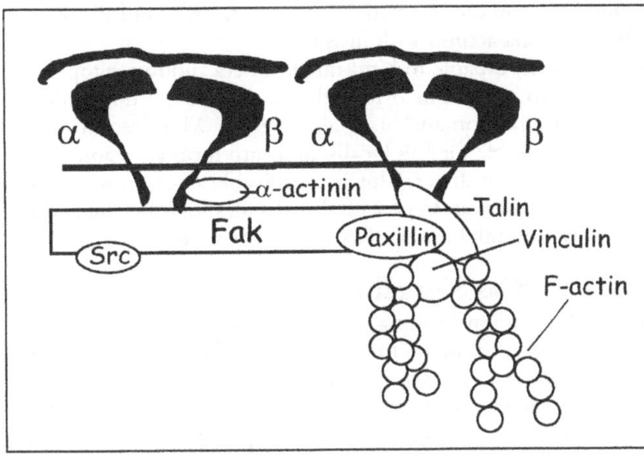

Figure 1. Interaction of integrins with cytoskeletal molecules. Overview of the different molecules by which integrins can link to the actin cytoskeleton. Talin, alpha-actinin, vinculin and Fak binds to integrin complexes and recruits actin filaments to integrin clusters. Additional molecules such as paxillin and Src take part to the complex.

to knockdown talin expression selectively reveals that talin is essential for beta1 and beta3 integrin activation in response to physiological agonists.[26] Talin binding to beta 1 integrin is also critical for building early mechanical linkages to the cytoskeleton. Deletion of talin1 blocked early formation of adhesion sites in response to force, even though Src family kinases, focal adhesion kinase, and spreading were activated normally. Recruitment of vinculin and paxillin to sites of force application also required talin1,[27,28] indicating that this molecule plays a major role in strengthening fibronectin-integrin-cytoskeleton connections. Mammals have two highly similar isoforms of talin, suggesting redundant functions in vivo. This redundancy has been confirmed by the observation that talin1-deficient embryonic cells, although showing reduced adhesion, spreading and ability to form focal adhesions,[29] organise vinculin-containing focal adhesions after differentiation.[30] Whether this is due to the expression of talin 2 will require the analysis of double knock-outs of talin1 and 2. Phosphoinositides such as PIP2 have been shown to increase binding of talin to beta1 integrin,[31] by inducing a conformational change that un-masks the tail-binding site within the talin FERM domain. PIP2 production may, in part, occur through integrin-dependent stimulation of the phosphoinositide 5-kinase isoform, through a Rho-dependent mechanism.[32] Recent data has revealed an other pathway. Two groups have found that talin regulates the local production of PIP2 in focal adhesions. Upon integrin clustering, talin is recruited to focal complexes and binds and activates one isoform of the PIP2-producing enzymes, the phosphatidylinositol phosphate kinase type 1 gamma (PIPK I gamma).[33,34] Therefore talin can stimulate increased local production of PIP2, which associates with talin and other proteins, facilitating the formation of focal adhesions. Interaction of talin with PIPK I gamma requires integrin and growth factor signalling as well as Fak-dependent phosphorylation of PIPK I gamma.[35] PIP2 also regulates the binding of talin with vinculin, an ubiquitous cytoskeletal protein. On binding PIP2, vinculin exposes criptic binding sites for talin and this facilitates its binding to actin.[36]

Alpha-actinin is a cell-adhesion and cytoskeletal protein that crosslinks actin filaments to bundles and networks and links these filaments directly to the integrin beta subunit cytoplasmic domain. In addition to integrins, alpha actinin interacts with vinculin, zyxin, Erk1/2, MEKK1, protein kinase N and the p85 subunit of phosphoinositide 3-kinase (PI-3K) (for review see ref. 37), behaving as an important scaffolding protein. These interactions are not constitutive but regulated, mostly by phosphoinositides binding. As already mentioned for talin, PIP2 binding increases the interaction with actin, PKN and PI-3K,[38-40] while PIP3 decreases the binding of alpha actinin with integrins.[41]

One additional protein capable to bind integrin beta subunit cytoplasmic domain is Fak, which has been shown to interact in vitro with the cytoplasmic domain of the beta1 integrin through its amino-terminal domain[42] (see also below). More recent data indicate that the amino-terminus is dispensable for Fak localisation into focal adhesions,[43] suggesting that the integrin-Fak interaction is not direct. This is conceivable since, in addition to being a kinase, Fak is an adaptor protein that can bind several molecules such as talin and paxillin,[44,45] which can constitute the physical link between Fak and integrin tails.

The SFK-Fak-p130Cas Signalling

As mentioned before, upon binding to the matrix, integrins undergo a conformational change and their cytoplasmic domains interact through linker molecules with the cytoskeleton and with signalling proteins to convey information to the nucleus. Tyrosine phosphorylation of proteins represents a preferential way to transduce signals throughout the cell and it is also a primary response to integrin stimulation. Several components that are tyrosine phosphorylated in response to integrin stimulation have been identified. Among them, the Src Family Kinases (SFKs), Fak and the adaptor molecule p130Cas play a prominent role in integrin signalling.

Focal Adhesion Kinase Structure

Fak is a 116kDa protein[46,47] belonging to a family of nonreceptor protein tyrosine kinases including Pyk2. It is expressed in most tissues and cell types and evolutionary conserved toward mammalian species as well as lower eukaryotic organisms, including Drosophila and zebrafish.[48] Fak consists of three major domains: an amino-terminal region, a central catalytic domain, and a carboxy-terminal portion.[49,50] The amino-terminal domain contains the tyrosine residue (Tyr) 397, which is the primary site of auto catalytic phosphorylation.[51,52] By sequence comparison, this region exhibits homology with FERM domains. Several binding partners of the Fak FERM domain have been identified such as the cytoplasmic tails of the beta1 integrin subunit, activated growth factor receptors, the cytoplasmic tyrosine kinase Etk/Bmx, and the FERM domain of ezrin.[42,53,54] In addition, intra molecular interactions involving the FERM domain and the catalytic domain have been proposed to regulate Fak kinase catalytic activity and its downstream signalling[55,56] (Fig. 2).

The carboxy-terminal region is characterized by two proline-rich domains that bind to SH3 containing proteins and a tyrosine residue at 925 that binds to SH2 domains when phosphorylated. In addition, this domain contains the Focal Adhesion Targeting (FAT) sequence, that is required for the localization of Fak to newly formed and existing adhesion complexes.[48] Many signalling proteins such as talin, paxillin, p130Cas, ASAP1 and Graf-2 binds to specific sites in the FAT domain underlaying its importance in regulating Fak signalling.[57] Graf is a GTPase-activating (GAP) protein, whose activity is directed to RhoA and Cdc42, while ASAP1 is a GAP for Arf1 and Arf6.[58,59] Through carboxy-terminal proline-rich regions Fak associates in vitro and in vivo to the p130Cas SH3 domain (for review see refs. 60, 61). The details of this interaction and its contribution to integrin signalling will be discussed below.

Structurally, as recently revealed by X-rays cristallography and NMR analysis, the FAT domain is characterized by the presence of four-helix bundles that resemble those found in other adhesion molecules such as p130Cas, vinculin and alpha catenin[48] and that can mediate the docking of Fak in adhesion complexes.

Integrin-Dependent Activation of Fak

Upon integrin engagement, the recruitment of Fak to focal adhesions and its interaction with other proteins induce its trans-auto phosphorylation on Tyr397.[62] This event leads to an increased activation of Fak,[63] which, in turn, regulates the phosphorylation of Tyr576 and 577 within a conserved region of the kinase domain and of proteins associated to focal adhesion sites.[49] The phosphorylation of Fak on Tyr576 and Tyr577 might confer an active

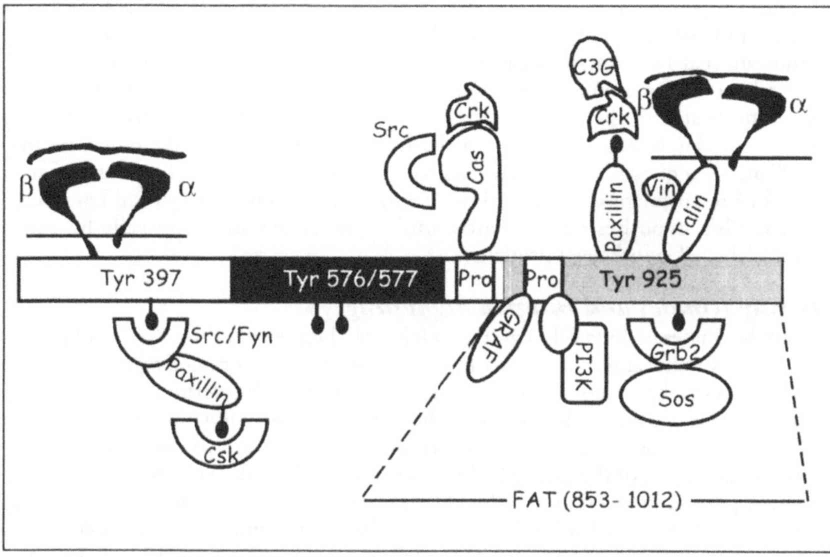

Figure 2. Structure of Fak molecule and its potential binding sites. Fak is composed by the central kinase domain (black box), flanked by large aminoterminal and carboxyterminal sequences. The aminoterminal domain contains the binding site for the integrin cytoplasmic domain and the tyrosine residue 397, which upon phosphorylation can associate to the SH2 domain of Src kinases. This region exhibits also homology to the FERM domain. The catalytic domain contains two tyrosines residues, Tyr 576 and Tyr 577, whose phosphorylation regulates the kinase activity of the molecule. The carboxyterminal part of the molecule contains the FAT (Focal Adhesion Targeting) region, two proline-rich regions and Tyr 925, which form binding sites for downstream transducing molecules.

conformation to this domain for stable interactions with protein substrates and thus for the initiation of downstream signalling. Phosphorylation of Tyr397 creates sites of high affinity interaction for other proteins containing SH2 or PTB domains.[64] The auto-phosphorylation of Tyr397 is necessary to create a binding site for the SH2 domain of Src family kinases, since mutation of Tyr397 to Phe abolishes Src binding.[51] The p85 regulatory subunit of PI-3K binds also to this site and its binding is considered as a mechanism of adhesion-induced activation of class Ia PI-3K.[65] The same site mediates also a direct interaction with the SH2 domain of phospholipase C gamma, leading to adhesion-dependent association of the two molecules and increased inositol phosphate production in mouse embryo fibroblasts.[66] Tyr397 is a binding site also for the adapter protein Grb7,[67] suggesting that Grb7 might be an important downstream mediator of integrin-Fak signal pathways in the regulation of cell migration, although the molecular mechanisms are still not well understood.

Src-mediated phosphorylation of Tyr925 following adhesion to fibronectin promotes the association of Fak with the SH2 domain of the Grb2 adaptor protein.[68] Upon fibronectin stimulation, Grb2, as in a classical receptor tyrosine kinase pathway, promotes signalling to MAPK linking Fak to Sos, a guanine nucleotide exchange factor, which converts inactive Ras-GDP to active Ras-GTP, representing one of the multiple pathways upstream of Ras (for review see ref. 69).

In addition to the associations established between tyrosine phosphorylated residues and SH2 domain-containing proteins, Fak binds integrins and several molecules, such as paxillin, talin, p130Cas, Graf-2, ASAP1 through structural motifs present in its sequence (Fig. 2). The binding of Fak with ASAP and/or GRAF has been suggested to be important to link adhesion signalling to regulation of small-GTP binding proteins that are crucial regulators of cytoskeleton organization. Therefore, Fak plays an important role as a scaffold to recruit cytoskeletal and signaling molecules involved in integrin-dependent events. The data discussed above indicate that Fak contains distinct and partially overlapping sequences able to interact with several molecules, some of which drive localization of Fak to the adhesion sites, while others are implicated in the downstream signalling. What is driving the choice between the different interacting molecules or the contribution of each of them in the downstream signalling is still an open question. An interesting hypothesis would be that the timing of activation of Fak and of recruitment of distinct downstream effectors might vary in different cell compartments. This hypothesis is currently under evaluation in living cells by the use of fluorescent-labelled cytoskeletal proteins.

Src Family Kinases and Integrin Signalling

Src family kinases include Blk, Fgr, Fyn, Hck, Lck, Lyn, Src, Yes and Yrk[70] and is characterized by several structural domains: a carboxy-terminal tail containing a regulatory tyrosine in position 527 in chicken, four SH domains and an unique amino-terminal domain. The SH1 domain represents the kinase domain with the Tyr auto phosphorylation site in position 416, the SH2 domain acts as a docking site for tyrosine phosphorylated proteins and contributes to the negative regulation of the kinase by binding to Tyr 527 in the tail; the SH3 domain regulates protein-protein interaction and the SH4 domain contains a myristylation site that is important for membrane localization.[71] SFKs can exist in two conformations: closed or inactive and open or active. When the Tyr527 in the tail is phosphorylated the kinase domain is covered,[72] when it is de-phosphorylated the molecule opens up and is free for auto-phosphorylation and for interactions with downstream effectors. Several mechanisms that lead to Src activation have been described: the activation might be the result of a de-phosphorylation event performed by PTPases such as Csk, PTPalpha and SHP1/2,[73,74] but it might also occur in response to high levels of its expression.[75] Ligand-induced activation of growth factor receptors can result in increased c-Src activation, probably by disrupting the intra molecular interactions that are responsible for the closed conformation.[71] In addition, Src can be activated by the direct binding of phosphorylated tyrosines or proline rich regions of Fak or p130Cas respectively to its SH2 and SH3 domains,[57] thus competing out the intermolecular interactions that hold the molecule in an inactive state.

It is now clear that cellular localization of c-Src depends on its status of activation: while inactive Src has mainly a perinuclear localization, its activation allows the indirect association of the SH3 domain with actin. Thus activated Src is translocated at the cell periphery at sites of cell adhesion, where it becomes associated to plasma membrane through its myristoylated domain[76] and close to transmembrane tyrosine kinase receptors and integrins. In epithelial cells, active c-Src phosphorylated on Tyr-416 in the c-Src activation loop, associates with integrins in a macromolecular complex in the early phases of cell adhesion.[77] In platelets c-Src has been shown to bind constitutively and selectively to beta3 integrins through an interaction involving the c-Src SH3 domain and the carboxy-terminal region of the beta3 cytoplasmic tail. These results invoke a model whereby Src is primed for activation by direct interaction with an integrin beta tail, and integrin clustering stabilizes activated Src by inducing intermolecular auto-phosphorylation.[78] Using bioluminescence resonance energy transfer that allows to determine interactions at nanometer proximity in living cells, the platelet integrin alphaIIbbeta3 has been shown to interact with Src kinase in a selective and dynamic manner during sequential phases of alphaIIbbeta3 signalling to the actin cytoskeleton.[79]

Upon adhesion, activated Src can phosphorylate Fak at residues Tyr576, 577 and 925, creating additional SH2-domain binding sites (Fig. 2). Src-dependent phosphorylation on the regulatory loop tyrosines in Fak kinase domain further promote Fak catalytic activity and thus autophosphorylation at Tyr397. In SFK-/- cells the extent of phosphorylation of Tyr397 is decreased,[80] further supporting the idea that Src not only binds to Fak and then phosphorylates it, but it might be also responsible for the phosphorylation of Fak Tyr397 activating site. These observations suggest a dual mechanism through which Src faces Fak: one as a

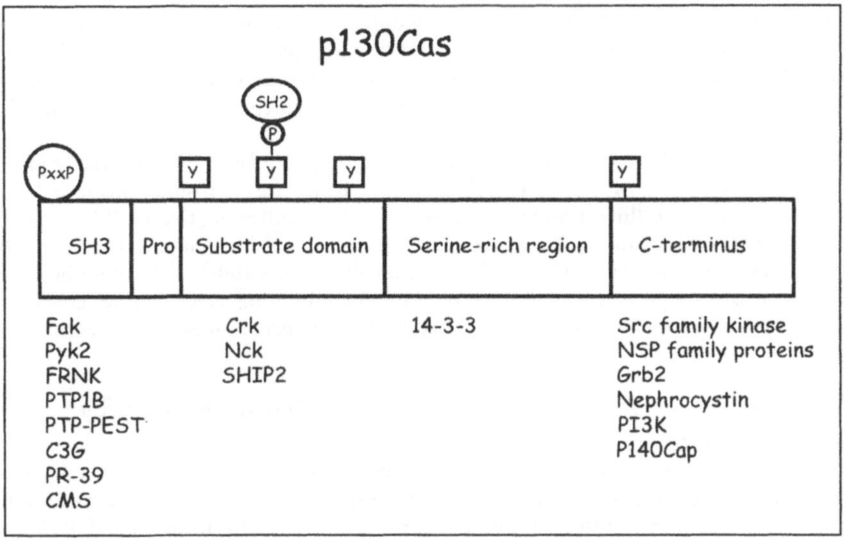

Figure 3. Schematic structure of the p130Cas adaptor. p130Cas is divided in several regions, which are depicted here. Each domain is known to bind to specific effectors, the most relevant being indicated on the bottom of the figure.

downstream effector and the other one as an upstream modulator. Upon integrin engagement Src is recruited to focal adhesions, likely through interactions with beta1 or beta3 integrin and it binds to Fak as well as p130Cas, creating a trimeric complex Src/Fak/p130Cas. Another important mediator of integrin-dependent Src signalling is the adaptor protein paxillin that is phosphorylated by Src at position Tyr31 and 118, which in turn recruits the adaptor protein Crk, contributing to downstream signalling.[81]

The p130Cas Adaptor Protein

p130Cas (Crk-associated substrate) is an adaptor protein originally identified as the major substrate of v-Src and v-Crk (for reviews see refs. 60, 61, 82). Structurally, p130Cas is characterized by several interacting domains including an SH3 domain, a substrate-binding domain (SD) characterised by 15 tyrosine residue motifs (Tyr-X-X-Pro, YXXP) whose phosphorylation generates SH2-binding sites, a serine-rich region and a carboxy-terminus containing a Src binding domain (Fig. 3). Due to its ability to associate with multiple signalling partners, p130Cas has been shown to be involved in several signalling pathways controlling adhesion, migration, apoptosis and transformation. Upon integrin clustering, p130Cas is localized to adhesion complexes and it is phosphorylated on tyrosine residues,[83,84] thus propagating the signalling originated from integrins to downstream effectors. In addition to Src and Fak, p130Cas can associate with the kinases Bmx and PI3K, the tyrosine phosphatases PTP1B, PTP-PEST, and SHIP2, the adaptor proteins Crk, Grb2, Nck and 14-3-3, the guanine nucleotide exchange factors C3G and AND-34, and the recently identified p140Cap[60,61,85] (Fig. 3).

Several evidences imply a role for p130Cas in mechanisms of cell transformation such as de-regulation of adhesion, enhanced cell motility and migration, proliferation and apoptosis. c-Src is the kinase responsible for integrin-dependent p130Cas phosphorylation, as shown by the use of c-Src-/- cells, while Fak has a very poor capacity to phosphorylate p130Cas.[86,87] However, Fak expression along with Src is important for achieving high levels of p130Cas tyrosine phosphorylation. Therefore the major function of Fak is to act as a docking protein to recruit Src which phosphorylates p130Cas. A secondary Fak-independent mechanism involves

Src directly bound to the p130Cas Src-binding domain.[87] p130Cas thereof recruits the adaptor protein Crk, that in turn binds to DOCK180 and affects Rac and c-Jun NH_2-terminal kinase (JNK) activation.[88,89] Crk also binds to C3G, a GTP-exchange factor (GEF), that leads to activation of Rap1 thereby contributing to stability and integrity of integrin-based adhesive structures.[90]

It has been recently reported that over-expression of p130Cas is able to directly influence the extent of activation of c-Src kinase,[91] as well as to increase tyrosine phosphorylation of multiple endogenous cellular proteins, such as cortactin,[92] indicating that p130Cas may functions as a positive upstream regulator of c-Src kinase activity. A likely mechanism is that binding of p130Cas to the SH3 domain of c-Src contributes to stabilize the kinase in its active conformation. Since p130Cas contains several potential binding sites for other effector molecules, such as tyrosine PTPases, the role of additional molecules in Src kinase activation can not be excluded.

SFK-Fak-p130Cas as a Signalling Scaffold in Migration, Invasion and Proliferation

The SFK-Fak-p130Cas signaling scaffold is involved in the organization of focal adhesions and actin cytoskeleton as well as in activation of the Ras-MAPK pathway, gene expression and apoptosis, ultimately regulating complex processes such as motility, invasion and proliferation (Fig. 4).

Fak and Src are clearly implicated in focal adhesion turnover associated with cell migration and invasion. Focal adhesions are dynamic structures that are disassembled in physiological conditions such as cellular movement and mitosis or pathological events like transformation and metastasis formation. Cell spreading and focal adhesion remodelling during cell motility or cell detachment are controlled by Fak and Src activities.[93,94] Consequently, cells prepared from Fak or Src/Yes/Fyn -/- mice exhibit impaired migration,[95,96] as well as over-expression of the Fak dominant negative FRNK molecule reduces cell motility.[97] Consistently, over-expression of Fak leads to increased ability of cells to migrate on fibronectin, an effect that can be abolished by mutation of Tyr397, indicating that tyrosine phosphorylation is required for this activity, likely through the recruitment of Src to focal adhesions. The use of live-cell imaging and expression of fluorescently labelled focal adhesion components to measure the rate constants for focal adhesion disassembly in live cells,[98] indicates that Fak and Src in combination with p130Cas and paxillin are all required for efficient disassembly of focal adhesions. Mutational analysis indicates that c-Src-induced cell spreading is independent from kinase activity and requires SH2 and SH3 domains. The SH2 domain is necessary for cell spreading, while SH3 domain is required for focal adhesion localization.[99] Once bound to Fak, however, Src can phosphorylate paxillin on Tyr31 and Tyr118. These two phosphorylated tyrosines are sites for Crk binding. Expression of a mutant paxillin protein lacking Tyr31 and Tyr118 cannot restore migration of paxillin -/- cells,[100] demonstrating an important role of the paxillin-Crk complex in Src-dependent cell motility.

The role of Src in focal adhesion turn over is highlighted by studies looking at the activation of the oncogenic v-Src. v-Src-transformed fibroblasts exhibit reduced adherence and altered cell morphologies. In these cells, v-Src localizes in residual focal contacts and in specific adhesive structures called podosomes. v-Src transformation correlates with disassembly of focal adhesions and with hyper-phosphorylation of integrins and of cytoskeletal proteins such as paxillin, tensin and Fak. Recently v-Src induced disruption of focal adhesions has been correlated to degradation of Fak kinase, immediately before the onset of cell rounding and detachment induced by v-Src expression.[101,102] These data suggest that aberrant activation of Src kinases can profoundly affect the organization of focal adhesions leading to loss of cell spreading. However, further analysis of the different kinase activities for transforming and nontransforming Src molecules and the changes in tyrosine phosphorylation of effector proteins located in focal adhesions may lead to better understanding of the role for these PTKs in the regulation of actin assembly during cell adhesion.

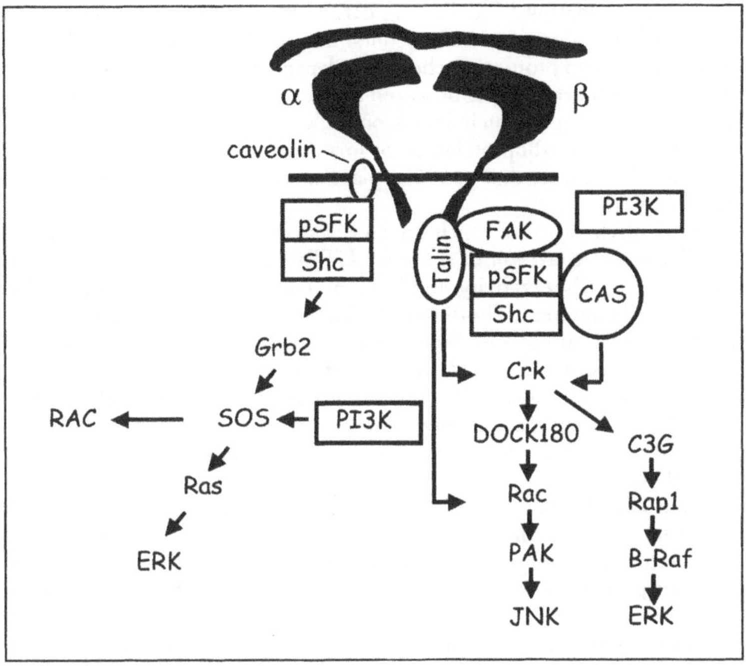

Figure 4. Signalling pathways emanating from integrins. The beta subunit binds to talin and Fak, through which is associated to the SFK and p130Cas. p130Cas can recruit Crk leading to Rac and rap activation. The alpha subunit in turn, binds to SFK and Shc, leading to activation of the Ras/Erk pathway. See text for details.

Recent data underline that the role of Fak in Src-dependent cell motility is distinct from that in invasion: v-Src transfected into Fak-/- cells increases cell motility, but not cell invasion, indicating that Fak is required for the ability of v-Src to induce cellular invasiveness.[103] Cell invasion requires transient Fak accumulation at lamellipodia, formation of a Fak–Src-p130Cas–Dock180 signaling complex, elevated Rac and JNK activation, and increased matrix metalloproteinase expression and activity. These data support the hypothesis that Fak recruitment to lamellipodia/invadopodia promotes cell invasion and Fak localization to focal contacts promotes integrin-stimulated cell motility, each in part through the activation of distinct signaling pathways.

In line with the fact that p130Cas is a primary c-Src substrate, cells derived from p130Cas -/- embryos showed significant defects in cell movement after mechanical wounding and in cell migration toward fibronectin.[104] p130Cas involvement in cell migration does account on its tyrosine phosphorylation by Src, on its association with Crk SH2 domain and Rac activation via DOCK180 (see above ref. 105). However, it is unclear which or how many of the fifteen p130Cas YXXP tyrosine motifs present in the substrate domain are critically involved. Recently, using site-directed mutagenesis combined with tryptic phosphopeptide mapping as well as mass spectrometry, specific tyrosines in YXXP motifs 6-15, with particular emphasis on Tyr253, have been shown to be the major sites of p130Cas substrate domain phosphorylation by Src, suggesting that cell migration might be regulated by phosphorylation of specific sites.[106,107] PTP-PEST is a protein tyrosine phosphatase that dephosphorylates p130Cas:[108] consistently its over-expression, which decreases the level of tyrosine phosphorylation of p130Cas and Fak, causes migration defects[109] (see also below). p130Cas has been also implicated in the process of Src-mediated oncogenic transformation. Expression of an active Src mutant in p130Cas-/- cells is unable to induce anchorage-independent cell growth and full morphological transformation,[104] indicating that p130Cas is a major player in these events. The carboxy terminal region of p130Cas containing the Src binding domain is sufficient to recover the ability of Src to promote anchorage-independent growth,[110] further supporting the idea that the association between p130Cas and Src increases Src kinase activation. Recent data, showing that p130Cas reexpression in knock-out cells enhances the capacity of Src-transformed cells to produce matrix metalloproteinases, to form podosomes, and to invade through basement membranes,[111] indicate that p130Cas plays an essential role in the development of an invasive cell phenotype.

An additional mechanism by which Src and Fak can control focal adhesion organization and cell motility is the regulation of the small GTPase Rho. Integrin engagement leads to a transient depression in RhoA activity, and it has been argued that this promotes lamellipodial extension during cell migration. The dip in RhoA activity requires Src and Fak, by affecting the status of p190RhoGAP. Simply put, Src-dependent phosphorylation of p190RhoGAP would cause a transient inactivation of RhoA, relieving the contractile forces at the sites of integrin engagement thus leading to focal adhesion disruption.[112] An other regulator of this event is paxillin, which binds to p120RasGap preventing it to associate and inhibit p190RhoGap. As a consequence of this, paxillin can contribute to the suppression of RhoA activity in the initial phase of cell-adhesion.[113]

In conclusion a model can be proposed in which c-Src localizes at focal adhesions by an SH3-mediated interaction. Once in focal adhesions, by an SH2-mediated interaction with Fak, c-Src can induce tyrosine phosphorylation of target molecules such as Fak itself, p130Cas and paxillin, thus promoting actin skeleton organization, focal adhesion dynamics, cell motility and cell invasion.

Integrins, in cooperation with soluble ligands and their receptors, RPTKs (see also below), regulate the signal transduction pathways that ultimately control the G1 phase cyclin-dependent kinases[114] (Figs. 4, 5). It is now clear that integrins regulate and enhance the sequential activation of Ras and Erk/JNK MAPK pathways leading to cyclin D1 expression and activation of cyclin D-dependent kinases. Integrin signalling activates Erk MAPK through the SFK, which in turn regulate Fak or Shc signalling. The SFK/Fak signalling is activated through the cytoplasmic portion of the beta1 integrin;[115,116] while the SFK/Shc pathway is activated by the cytoplasmic domain of specific alpha subunits.[117] The SFK/Fak signalling mediates activation of Rac,[116] likely through regulation of the subcellular distribution of lipid rafts, which in turn drive Rac targeting to the plasma membrane and its coupling to the downstream effector molecule PAK.[118,119] The major role of Rac in regulation of the Erk1/2 MAPK signalling has been the control of nuclear translocation of Erk.[116] Interestingly, Fak participates also in integrin-dependent activation of JNK, through a Fak/p130Cas/Rac pathway.[88,89] The JNK MAPK is also activated by the alpha6beta4 integrin through sequential induction of Shc/Ras/PI-3K and Rac pathways.[120]

Tyrosine PTPases Implicated in Integrin-Mediated Signalling

Cellular phosphotyrosine levels are controlled by the coordinate and competing actions of protein tyrosine kinases and phosphotyrosine phosphatases (PTPases).[121] Early work showed that reduction of cell adhesion by trypsin treatment induces PTPase activity.[1] Accordingly, PTPase activity, measured as ability to dephosphorylate Fak, is inhibited when stress fibers and focal adhesions are assembled,[122] confirming that integrin-dependent adhesion results in PTPase activity reduction. On the other hand, treatment of cells with PTPase inhibitors leads to increased tyrosine phosphorylation of Fak and focal adhesion assembly.[123,124]

Although little is known about the role of PTPases as positive and negative regulators of integrin signalling, recent data highlight a prominent role played by specific PTPases in linking integrin signalling to SFK-Fak-p130Cas signalling and cell migration.

PTPalpha is a widely distributed trans-membrane molecule that has been involved as a physiological upstream regulator of SFKs. Under specific circumstances PTPalpha is constitutively phosphorylated on Tyr789, which binds to the SFK SH2 domain, displacing the internal

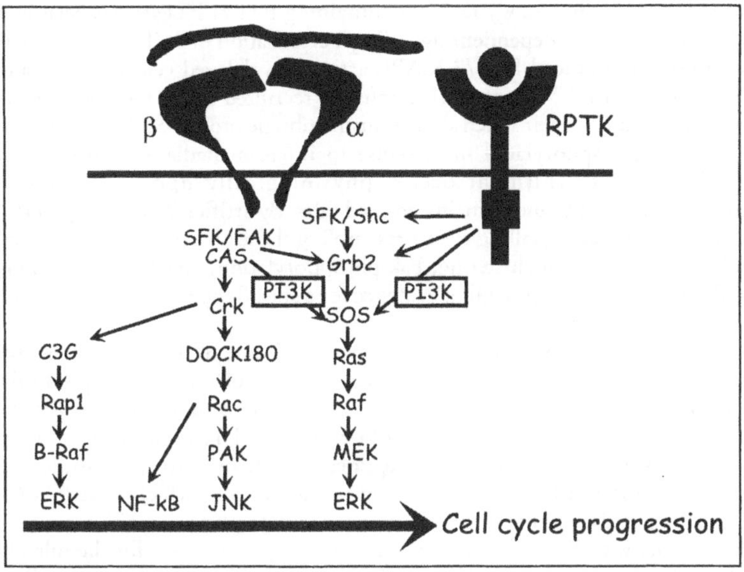

Figure 5. Cooperative pathways between integrins and growth factor receptors. Integrins and growth factor receptors (RPTK) coordinately control signalling cascades, leading to cell cycle progression.

constrain and leading to kinase activation,[125-127] for review.[128] Alternatively, PTPalpha has been shown to directly dephosphorylate Src regulatory Tyr527 on the carboxy-terminal tail both in vitro and in vivo.[74] As expected, in PTPalpha knock-out animals Src and Fyn activities are strongly affected,[127] confirming the physiological role of PTPalpha in regulating Src kinase family activity. PTPalpha is detected in focal adhesions.[129] Interestingly mutation of tyrosine 789 does not allow association of PTPalpha with focal adhesions. As a possible consequence of its role on Src activity, the PTPalpha -/- embryonic fibroblasts show reduced or abolished integrin-mediated formation of Src-Fak and Fyn-Fak complexes and decreased integrin-dependent phosphorylation of Fak, particularly on Tyr397. These cells also show reduced phosphorylation of p130Cas, and reduced Erk activation,[130] establishing that PTPalpha activity is required for early integrin-proximal events as an Src kinase activator. PTPalpha has also been found physically associated with the integrin alphavbeta3 during cell spreading on fibronectin and upon this binding acquire the ability to activate Fyn kinase.[131] PTPalpha is required for the force-dependent reinforcement of alphavbeta3 integrin adhesion. These data together with the decrease in focal adhesions seen in PTPalpha -/-, suggest a role for PTPalpha and Fyn in the late activation of Rho downstream to integrin clustering.

The LAR protein also resides at focal adhesions[132] where it can interact with laminin-nidogen.[133] LAR-laminin-nidogen binding may play a role in regulating cell signalling induced by laminin-nidogen, resulting in adhesion-dependent cell morphological changes and giving rise to long cellular processes.

Coming to the cytoplasmic protein tyrosine phosphatases, PTP-PEST associates with the focal adhesion proteins p130Cas[134] and paxillin[135] and has recently been implicated in cell migration. PTP-PEST is activated in an adhesion-dependent manner and localizes to the tips of membrane protrusions in spreading fibroblasts. The expression level of PTP-PEST modulates the activity of the small GTPase, Rac1, suppressing activation of Rac1 in response to both integrin-mediated adhesion or growth factor stimulation. PTP-PEST is activated by integrins and localized to regions where it can control motile events at the leading edge through inhibition of the small GTPase Rac1.[109]

Also the SHP-2 (Src homology region 2 containing PTP2) PTPase has been involved as a positive effector in integrin-dependent signalling. Perturbation of SHP-2 levels or activity leads to decreased integrin-induced Erk1/2 MAPK activation, delayed cell spreading and reduced migration on fibronectin.[136] To act, SHP-2 must be recruited to the membrane by binding to phosphorylated docking proteins such as the transmembrane protein SHP substrate 1 (SHPS-1), which is tyrosine phosphorylated in response to integrin-mediated adhesion by Fak and SFKs.[137,138] SHP-2 recruitment occurs physiologically upon adhesion mostly to cholesterol-enriched membrane domains, termed rafts. By artificially targeting SHP-2 to rafts, non attached cells display signalling events resembling those observed after fibronectin attachment, such as beta1 integrin clustering, Fak phosphorylation, and ERK activation, and also increases Rho-GTP levels,[139] providing evidence for a specific spatial organization of SHP-2 in integrin signalling network.

A relevant player in integrin signalling is also PTEN, one of the most frequently mutated tumor suppressors in human cancers. PTEN is a lipid phosphatase which specifically dephosphorylates phosphatidylinositol (PtdIns) (3,4,5)P3 on the 3' phosphate to generate PtdIns (4,5) P2, thereby antagonizing the action of PI3-K. In vitro PTEN can also remove phosphate residues from phosphotyrosine containing peptides or proteins. Two cytoplasmic phosphoprotein substrates are the Fak kinase and the adaptor protein Shc, which links PTEN action to integrin signalling. Although the in vivo data on protein de-phosphorylation are controversial, PTEN can interact with Fak and Shc. This property may be relevant for the role of PTEN in anoikis, the induction of apoptosis in cells after loss of contact with the extracellular matrix. Anoikis as anchorage dependence of survival, is defective in many transformed and malignant cells. Reconstitution of PTEN in cells that have PTEN mutations restores anoikis:[140,141] cells can acquire partial resistance to anoikis when Fak is over-expressed along with PTEN, indicating a role for Fak in this event. PTEN reconstitution or over-expression inhibits cell migration,[141,142] with a transient effect on cell adhesion and spreading, reduced number of focal contacts and alterations in actin cytoskeleton organisation. The effects of PTEN overexpression can be due to direct effects on the Fak-p130Cas signalling[143] as well as to variations of PtdIns (3,4,5)P3 levels,[41] both of which can affect cell motility.

Very recently the existence of a redox circuitry that regulates oxidative inhibition of PTPases upon cell adhesion has been revealed. Reactive oxygen species (ROS) induce reversible oxidation of PTPases, whose activity is dependent on the redox state of a low pKa active site cysteine. Thus, production of ROS facilitates a tyrosine phosphorylation-dependent cellular signalling response by transiently inactivating those PTPs that normally suppress the signal. A variety of mitogenic signals, including those released by receptor tyrosine kinase ligands involve the intracellular generation of ROS as a critical downstream event. Recent data indicate that intracellular ROS are generated after integrin engagement and that these oxidant intermediates are necessary for integrin signalling during fibroblast adhesion and spreading. Integrin-induced ROS are required to oxidize/inhibit a specific low molecular weight PTPase, thereby preventing the enzyme from dephosphorylating and inactivating Fak. Accordingly, Fak and Src phosphorylation and other downstream events, including focal adhesion formation, cell spreading and MAPK phosphorylation, are all significantly attenuated by inhibition of redox signaling,[144] suggesting a prominent role for ROS in integrin signalling.

Integrin Cross-Talk with RPTKs and Cytokine Receptors

In addition to the pathways described above, growth factor receptors are integrin partners in assembling a transduction machinery. Increasing evidence indicates that integrins cooperate with receptor protein tyrosine kinases (RPTKs) in transducing proliferative signals and in regulating survival and migration. Regulation of cell cycle is a prototypic event occurring by joint integrin/RPTKs signalling.[114] Strong evidence in fact, indicates that integrin- and RPTK-dependent signals need to be integrated at various levels, in order to induce cell proliferation. Normal diploid cells need to adhere to the matrix to progress through the mid-G1 phase of the cell cycle upon mitogen treatment.[145] In adherent cells, soluble ligands, such as EGF or PDGF, stimulate RPTK activity and a cascade of downstream events which cooperates with those induced by integrins. Joint integrin/RPTKs signalling causes expression of cyclin D, activation of cyclin D-dependent kinases and degradation of Cdk inhibitor p27.[146] Integrins regulate RPTKs functions including receptor transactivation, coordination and compartmentalization and downstream signalling. Reciprocal cross-talk between integrins and the RPTKs consists in at least three hierarchical mechanisms, the first consisting of integrin ability to trigger ligand-independent RPTKs activation. On the other hand, integrins are also required for propagation of ligand-mediated signalling, leading to a full repertoire of RPTKs activities. In this case integrin and growth factor receptor signalling can act on parallel pathways and synergize to reach a final biological response. As a third possibility, integrins and their immediate target effectors can function as downstream transducers of RPTK signals. Although much less investigated, similar mechanisms are depicted for integrin-dependent activation of cytokine receptors. The molecular mechanisms involved in such events include recruitment of crucial transducing proteins to membrane cytoskeletal complexes, and enhancement of nuclear translocation of transcriptional regulators.

Integrin-Dependent Ligand-Independent Activation of RPTKs

Direct phosphorylation of growth factor receptors by integrin signalling represents a potential mechanism by which integrins can enhance signalling pathways emanating from growth factor receptors. It has been shown, in fact, that integrins stimulates phosphorylation, and at least partial activation, of several RPTKs, including the EGF-R,[77,147] HGF-R,[148,149] PDGFbeta-R,[150] RON kinase[151] and VEGF-R,[152] in the absence of any growth factor ligand. The number of RPTKs involved suggests that this activation scheme can be a broadly used mechanism in adhesion-mediated signalling. In other words, cell-matrix adhesion represents a priming event, and a limiting factor in addition to soluble ligand, for RPTKs activation.

Molecular mechanisms regulating integrin-dependent EGF-R phosphorylation are distinct from the classical activation induced by mitogenic concentrations of soluble EGF. In the early phases of cell adhesion integrins associate with EGF-R on the cell membrane in a macromolecular complex, suggesting that close proximity between integrins and RPTK could locally increase receptor densities and lead to oligomerisation and trans-activation. Similarly, in confluent A431 epithelial cells, the alpha2beta1 integrin associates with phosphorylated EGF-R at cell-cell contacts. The EGFR phosphorylation is abrogated with a function-blocking antibody directed to the integrin alpha2, implying that ligand-independent EGF-R activation relies on clustering with integrins in inter-cellular adhesion.[153] This mechanistic simple explanation is made complex by the fact that integrin-dependent EGF-R phosphorylation requires Src kinase activity, and the assembly of Src and the adaptor protein p130Cas in a multimeric complex with integrins and EGF-R[77] (Fig. 6). The requirement of Src kinase is in line with the mechanism proposed for the integrin-induced activation of RON.[151] In this model in fact integrins activate Src tyrosine kinase, which in turn mediates Ron activation. Cross-communication between integrins and EGF-R may be regulated by ganglioside GM3 the predominant ganglioside of epithelial cells. Recent data show that endogenous accumulation of GM3 disrupts the ligand-independent association of the integrin beta1 subunit with EGFR. Consistently, endogenous depletion of GM3 markedly increases the association of integrin beta1 with EGFR and its phosphorylation. These events depend on Src kinase and PI-3K activity, suggesting that GM3 inhibits integrin-EGFR cross-talk through suppression of Src family kinase and PI-3K signalling.[154]

A second feature of integrin-dependent EGF-R activation is that it leads to phosphorylation of EGF-R on a specific subset of tyrosine residues, only partially overlapping with those phosphorylated by EGF. Upon adhesion EGF-R is phosphorylated on tyrosine 845, 1068, 1086 and 1173, but not on residue 1148, a major site of phosphorylation in response to EGF.[77,155] Interestingly, integrin-dependent phosphorylation of tyrosine residue 845, which is located in the activation loop of the RPTK, could occur through Src kinase activity as a priming event for subsequent activation of the EGF-R kinase activity, which in turn is

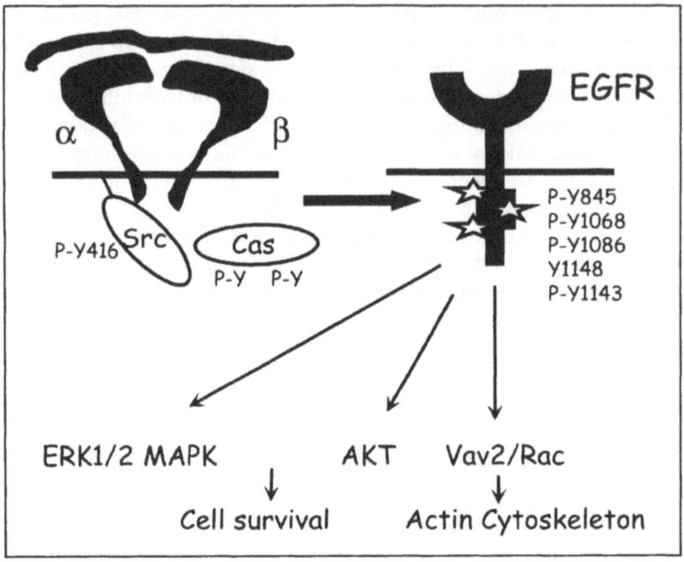

Figure 6. Integrin-dependent activation of EGF receptor. Upon cell-matrix adhesion, integrins associates with the EGF-R, active c-Src kinase (PY416) and p130Cas. As a consequence of cell adhesion, EGF-R is phosphorylated on Tyr 845, 1068, 1086 and 1173, but not on Tyr1148 [16], and transduces signalling on Erk1/2, AKT and Rac, leading to cell survival and actin cytoskeleton organisation.

required for phosphorylation of the other tyrosine residues. A third feature of integrin-dependent EGF-R activation is a lower apparent stoichiometry, which leads to a lower level of tyrosine phosphorylation compared to that observed with mitogenic doses of EGF.[147] Phosphorylation of EGF-R on a specific subset of tyrosine residues could account for this event.

As an alternative mechanism, integrins can cause the recruitment of tyrosine phosphatases such as SHP-2 at the plasma membrane in close proximity to the PDGF-R. SHP-2 de-phosphorylates PDGF-R on specific tyrosine residues involved in Ras-GAP binding, thus decreasing Ras-GAP activation and potentiating Ras signalling,[156] resulting in PDGF-R-downstream events, without involving the activation of the receptor tyrosine kinase.

Few data are available on integrin cross-talk with cytokine receptors. Recent results show that on murine hemopoietic cells, adhesion through alpha4beta1 integrin activates cytokine receptor-associated Jak2 to induce phosphorylation of the erythropoietin receptor and the IL-3 receptor beta subunit.[157] Coimmunoprecipitation studies demonstrate that in endothelial cells the beta common subunit of the IL-3 receptor (IL-3R) associates with the active beta1 integrin. Upon adhesion to fibronectin Jak2 binds to this complex, is activated and leads to phosphorylation of the IL-3R beta common subunit and STAT5A (Defilippi P., Brizzi MF, unpublished observations), giving further insights on the existence of an extensive integrin cross-talk including not only the tyrosine kinase but also the cytokine receptors.

Functional Significance of Integrin-Dependent Receptor Transactivation

The involvement of integrin-dependent receptor transactivation in specific signalling pathways has been delineated by treatment with specific RPTK inhibitors and by expression of dominant-negative mutants. By the use of EGF-R inhibitor AG1478 and of EGF-R/deltaC integrin-dependent EGF-R activation has been shown to regulate adhesion-dependent Erk1/2 MAPK,[147] as well as AKT activation (Fig. 6),[155,158] indicating that integrins can utilize EGF-R as a transducing element in the matrix-induced signalling pathways. To further strengthen the relevance of integrin-induced EGF-R activation, Marcoux and Vuori have recently shown that it is instrumental for integrin-dependent Rac activation, through phosphatidylinositol 3-kinase (PI3-K) activation and GTP loading on Vav2, a known exchange factor for Rac. As a functional consequence, lamellipodia formation, cell spreading and migration have been shown to be consequence of integrin-dependent EGF-R activation.[159]

Integrin-dependent EGF-R activation is sufficient for adhesion-dependent cell survival, by an EGF-R kinase dependent mechanism.[147] It is well defined that cells plated on matrix proteins activates signals that protect cells from anoikis.[160] In cells expressing EGF-R, the ability of cells to survive on fibronectin is blocked by the specific EGF-R kinase inhibitor AG1478 or by the expression of the dominant negative form EGF-R/deltaC. EGF-R-mediated cell survival is blocked by the PI-3 kinase inhibitor wortmannin, suggesting that PI-3K and AKT, well defined players in adhesion-dependent survival to anoikis, are involved.[147] Integrin-EGF-R signalling has also been recently implicated in the regulation of expression of the pro-apoptotic protein Bim, a critical mediator of anoikis in epithelial cells. Bim is strongly induced after cell detachment by concomitant lack of beta1 integrin engagement, down-regulation of EGF-R expression and inhibition of Erk1/2 MAPK signalling.[161] Therefore taken together these data show that upon adhesion, EGF-R trans-activation accounts for a specific repertoire of mechanisms, namely cell survival and actin cytoskeleton organisation involved in cell migration.

According to its lower apparent stoichiometry, integrin-induced EGF-R activation is not sufficient for G1-S cell cycle progression and proliferation,[147] confirming that in un-transformed cells anchorage-induced signalling is a key control step, that cooperates with pathways activated by growth factors to induce cell proliferation. Therefore trans-activation of growth factor receptors by integrins may represent one of the adhesion check-points that modulates growth factor signalling. Interestingly in vivo analysis in a mouse model of inflammatory skin disease indicate that suprabasal integrin expression leads to Erk activation and increased IL-1alpha expression by potentiating activation of the EGF receptor,[162] triggering epidermal hyperproliferation and inflammation. Because of their heterogeneity, there is less understanding of the role of integrin and RTPKs signalling in proliferation of tumour cells in culture. Urokinase plasminogen activator receptor (uPAR), a GPI-linked protease receptor, has been shown to bind and activate alpha5beta1 integrin and ERK signalling, inducing in vivo proliferation of HEp3 human carcinoma. Recent data show that this effect is mediated by adhesion-dependent EGFR trans-activation, unveiling a mechanism whereby uPAR by binding to integrins stimulates EGFR signalling, providing cancer cells with proliferative advantage. In other cell systems, such as an in vivo hepatocyte tumour model, the ability of integrins to transactivate HGF-R is crucial for tumour metastasis, implying that cells could mistake signals form integrins, leading to inappropriate cell growth and tumorigenesis.[163]

Cooperation between Integrins, RPTKs, Cytokine Receptors and Their Ligands: Reciprocal Potentiation

Integrins have been shown to potentiate signalling pathways in response to different growth factors, such as insulin, PDGF, EGF, FGF and VEGF.[164-173] This cooperation begins at the level of the receptors themselves. In fact cell adhesion increases the number of PDGF-R by blocking their degradation by an ubiquitin-dependent pathway.[174] Many integrins form complexes with RPTKs and some appear to have preferred partners. For example, the alphavbeta3 integrin associates and synergizes with the insulin receptor, the PDGF-R and the VEGF-R.[171-173,175] Beta1 integrins associate with the EGF-R[53,153,169] and alpha6beta4 combines with the EGF-R,[176] Erb-B2[177] and the HGF-R.[178] Biochemical analysis revealed that alphavbeta3 integrin occupancy by its matrix ligand is required to get full tyrosine phosphorylation of insulin and PDGFbeta receptors and their binding to signalling molecules such as IRS-1, PLCgamma, Ras-GAP, p85-PI3-K and the tyrosine phosphatases SHP-1 and SHP-2.[171,173,179-181] In endothelial cells, moreover, alphavbeta3 integrin potentiates the activation of VEGF-R and of p85-PI-3K by its ligand.[172] A very recent report sustains the hypothesis that distinct integrin isoforms may selectively modulate RPTKs signalling in response to IGF stimulation. The canonical beta1A integrin associates with the insulin-like growth factor type I receptor (IGF-IR) and insulin receptor substrate-1 (IRS-1) in a complex which support IGF-mediated cell proliferation. In contrast, beta1C, an integrin cytoplasmic variant, completely prevents IGF-mediated cell proliferation and tumor growth by inhibiting IGF-IR auto-phosphorylation in response to IGF-I stimulation. Evidence is provided that the beta1 cytodomain plays an important role in mediating beta1 integrin association with IGF-IR downstream effectors: specifically, beta1A associates with IRS-1 and beta1C with Gab1/Shp2. Therefore this study unravels a novel mechanism mediated by the integrin cytoplasmic domain that differentially regulates cell proliferation in response to IGF.[182]

The molecular bases for specific association between integrins and RPTKs remains to be examined, but there is evidence suggesting that lipid microdomains enhance this association. In the central nervous system interaction between growth factor and integrin signalling pathways provides a mechanism for directing oligodendrocyte progenitor (OP) proliferation and survival. OP proliferation is dependent on the physical coassociation of alphavbeta3 integrin and PDGF-R.[180] At the developmental stage when PDGF ceases to promote proliferation, but is now required for survival, the PDGF-R becomes sequestered in a raft compartment, where it colocalise with the integrin alpha6beta1, generating a signaling environment within the raft for survival-promoting PI3K/Akt activity.[179] The emerging role of lipid rafts in integrin signalling will be also discussed below.

Cross talk between integrins and RPTKs is required for the enhancement of growth factor signals. A canonical pathway which is regulated by the joint action of integrins and RPTKs is that of Ras-Erk. Matrix adhesion causes a weak and transient activation of Erk in cells deprived of growth factors, while soluble growth factors in the absence of matrix adhesion cause a relatively modest activation of Erk. Combined activation of integrins and RPTKs is thus necessary to activate Ras-ERK pathway beyond the threshold necessary for transcription of cyclin D.[183] This relies on the fact that integrin cooperation with RPTKs is required for enhancement of nuclear translocation of transcriptional regulators. Points of regulation have been localized to the level of Raf, MEK (mitogen-activated protein kinase/ERK kinase) and ERK translocation from the cytoplasm to the nucleus. Phosphorylation of the nuclear ERK substrate, Elk-1, is anchorage dependent in response to growth factor treatment of NIH 3T3 fibroblasts.[184] By contrast in malignant melanoma cells SK-MEL-24 and SK-MEL-28, the ERK1/2 pathway adhesion-dependent regulation of ERK1/2 activity is by-passed. In these cells, ERK1/2 translocates to the nucleus and regulates transcription events in an adhesion-independent manner, indicating that alterations in the adhesion requirement for ERK1/2 signalling in melanocytes are associated with the acquisition of malignant cell behaviour.[185]

The third mechanisms by which integrins and RPTKs undergo reciprocal regulation consists in the fact that integrins and their immediate target effectors can function as downstream transducers of RPTK signals to regulate cell migration. The EGF-R and the PDGF-R have been shown to exploit the Fak/SFKs signalling. Upon growth factor stimulation, Fak interacts through its amino terminal FERM domain with the RPTKs and with its carboxy terminal FAT domain with paxillin and talin at integrin attachment sites,[53,186] bridging the two receptor systems. The relevance of Fak in RPTK-dependent regulation is highlighted by the observation that fibroblasts lacking Fak are refractory to PDGF and EGF-induced migration. Reexpression of Fak rescues the migratory defect, indicating that the presence of a Fak molecule able to interact with the RPTKs and regions of integrin clustering is necessary for growth factor-dependent migration. In contrast, in endothelial cells, Fak is constitutively associated with beta1 and beta3 integrins in the presence or absence of growth factors, but VEGF promotes the Src-mediated phosphorylation of Fak and its coupling to the alphavbeta5,[187] further supporting the concept that RPTKs can induce coupling of Fak to integrins. It is likely that RPTKs function by phosphorylating the SFK/Fak targets, thus indicating that SFK and Fak function as a receptor proximal element in response to integrins and RPTKs.

The alpha6beta4 integrin in migrating epithelial and carcinoma cells redistributes with motility structures and mediates cell movement. This peculiar integrin has been shown to associate with the EGF-R and the HGF-R, that upon binding to their ligands, induce phosphorylation of the beta4 cytoplasmic domain and recruitment of Shc and PI-3K.[176,178] This event generates a signalling platform that potentiates EGF/HGF-triggered activation of Ras- and PI3K-dependent pathways, inducing hemidesmosome disassembly and promotion of cell motility and invasion. These observations highlight the fact that in the presence of activated or over-expressed RPTKs, the beta4 integrin can constitute a parallel docking platform that cooperates with RPTKs to amplify the signalling machinery.

Lipid Rafts as Emerging Regulators of Integrin Signalling

Although both ligand binding and lateral aggregation of integrins are essential steps, the mechanisms by which integrin engagement coordinates the activation of distinct signalling pathways is presently unclear. The conventional picture of membrane proteins diffusing freely in the bilayer has evolved in recent years, as a growing body of evidence suggests that there is a selective confinement of lipids and proteins to discrete detergent domains, termed rafts.[188] The lipid raft hypothesis argues that a heterogeneity of membrane microdomains (rafts) provides important locations for cell signalling processes, having distinct concentrations of specific glycosylphosphatidyl inositol (GPI)-linked and signalling molecules.[189,190] These microdomains are rich in cholesterol and glycolipids and therefore can be isolated within membrane fractions exhibiting relative insolubility in detergents by virtue of their increased lipid order. Although initially proposed as platforms for the selective sorting of membrane proteins, it is recognized that lipid rafts may also impose short- and long-range lateral organization of plasma membrane constituents. As rafts may favor protein interactions in specific cell locations, accordingly, a number of signalling molecules, including integrins and integrin-associated proteins, are preferentially activated in rafts (for review see refs. 189, 191). The role of caveolae in integrin function seems to be the opposite of that reported for rafts. Integrins are not generally considered to locate in caveolae,[192] and the interaction between caveolin-1 and beta1 integrins is reported to occur in a Triton X-100-soluble plasma membrane fraction.[193] Other authors suggest a scenario in which beta1 integrin, caveolin, and SFK complexes exist as signalling units in caveolae, but segregate from them in response to ligand-induced clustering and cytoskeletal reorganization.[194]

Lipid rafts are involved in regulation of SFKs. In fact Csk, a negative regulator of SFK, is recruited to a membrane microdomain of lipid rafts by its binding to Cbp, a C-terminal Src kinase (Csk)-binding protein. Interestingly upon cell adhesion onto fibronectin, Cbp becomes transiently phosphorylated and recruits Csk to lipid rafts. These events are completed before the full activation of focal adhesion kinase, indicating that the transient activation and down-regulation of SFK in lipid rafts are earlier events in cell adhesion signalling.[195]

Lipid rafts are also commandeered by integrins to target Rho and Rac GTPases to specific plasma membrane domains and to couple them to their downstream effector molecules.[139] In addition to directly activating Rac, integrins regulate the targeting of Rac to the plasma membrane and its coupling to the downstream effector molecule p21-activated kinase (PAK) in cells adhering to the extracellular matrix.[118,196] Integrins regulate the coupling of Rac to PAK by modulating the plasma membrane to allow Rac targeting rather than by acting directly on either Rac or PAK. The ability of integrins to promote membrane translocation of Rac is due to effects on the membrane binding sites and can be exerted locally so that Rac preferentially translocates to regions of the plasma membrane proximal to sites of integrin-mediated adhesion. Disruption of lipid rafts by cholesterol depletion mimicked loss of cell adhesion, preventing targeting of Rac to the plasma membrane and its coupling to PAK. Therefore cell adhesion via integrins regulates plasma membrane targeting of Rac and its coupling to PAK by preventing internalization of the lipid rafts bound by Rac. Integrins regulate the subcellular distribution of lipid rafts and thus the targeting of Rac and its subsequent coupling to downstream effector molecules by localizing G_{M1}-rich lipid rafts at the leading edge of migrating fibroblasts. The molecular mechanisms involved point to Fak as a critical mediator of integrin signalling in this process.[119]

Conclusions

The picture emerging from these studies is that integrins behave as nodes within webs of signalling, adhesive and cytoskeletal pathways. In conclusion, integrins exert most prominent signalling functions through the activation of cytoplasmic kinases such as SFKs and Fak, cross-talk with RPTKs and cytokine receptors, regulation of -PTPases activity and building up of transducing macromolecular complexes with adaptor molecules. Each of the partners brings a distinctive contribution to downstream signalling and it will be important to elucidate in the next future the mechanisms by which they associate and those by which they respond to extracellular signals to induce and integrate a cellular response.

Note Added in Proof

Defilippi P and Brizzi MF's unpublished results have now been accepted for publication: Defilippi P, Rosso A, Dentelli P et al. Betal Integrin and IL-3 Receptor coordinately regulate STAT5 activation and anchorage-dependent proliferation. J Cell Biol 2005; in press.

Acknowledgements

This work was supported by grants of AIRC, MIUR, Special project "Oncology", Compagnia San Paolo/FIRMS, Torino, Italy, and Progetti Regione Piemonte. Sara Cabodi is a FIRC fellow.

References

1. Maher PA. Activation of phosphotyrosine phosphatase activity by reduction of cell-substrate adhesion. Proc Natl Acad Sci USA 1993; 90:11177-11181.
2. Folkman J, Moscona A. Role of cell shape in growth control. Nature 1978; 273:345-349.
3. Menko AS, Boettiger D. Occupation of the extracellular matrix receptor, integrin, is a control point for myogenic differentiation. Cell 1987; 51:51-57.
4. Solursh M, Jensen KL, Zanetti NC et al. Extracellular matrix mediates epithelial effects on chondrogenesis in vitro. Dev Biol 1984; 105:451-457.
5. Edgar D, Timpl R, Thoenen H. The heparin-binding domain of laminin is responsible for its effects on neurite outgrowth and neuronal survival. EMBO J 1984; 3:1463-1468.
6. Dedhar S. Regulation of expression of the cell adhesion receptors, integrins, by recombinant human interleukin-1 beta in human osteosarcoma cells: Inhibition of cell proliferation and stimulation of alkaline phosphatase activity. J Cell Physiol 1989; 138:291-299.
7. Dike LE, Farmer SR. Cell adhesion induces expression of growth-associated genes in suspension-arrested fibroblasts. Proc Natl Acad Sci USA 1988; 85:6792-6796.
8. Eierman DF, Johnson CE, Haskill JS. Human monocyte inflammatory mediator gene expression is selectively regulated by adherence substrates. J Immunol 1989; 142:1970-1976.
9. Schwartz MA, Ingber DE, Lawrence M et al. Multiple integrins share the ability to induce elevation of intracellular pH. Exp Cell Res 1991; 195:533-535.
10. Pardi R, Bender JR, Dettori C et al. Heterogeneous distribution and transmembrane signaling properties of lymphocyte function-associated antigen (LFA-1) in human lymphocyte subsets. J Immunol 1989; 143:3157-3166.
11. Ferrell Jr JE, Martin GS. Tyrosine-specific protein phosphorylation is regulated by glycoprotein IIb-IIIa in platelets. Proc Natl Acad Sci USA 1989; 86:2234-2238.
12. Nakamura S, Yamamura H. Thrombin and collagen induce rapid phosphorylation of a common set of cellular proteins on tyrosine in human platelets. J Biol Chem 1989; 264:7089-7091.
13. Golden A, Brugge JS, Shattil SJ. Role of platelet membrane glycoprotein IIb-IIIa in agonist-induced tyrosine phosphorylation of platelet proteins. J Cell Biol 1990; 111:3117-3127.
14. Miranti CK, Brugge JS. Sensing the environment: A historical perspective on integrin signal transduction. Nat Cell Biol 2002; 4:E83-90.
15. Schwartz MA, Ginsberg MH. Networks and crosstalk: Integrin signalling spreads. Nat Cell Biol 2002; 4:E65-68.
16. Giancotti FG, Tarone G. Positional control of cell fate through joint integrin/receptor protein kinase signaling. Annu Rev Cell Dev Biol 2003; 19:173-206.

17. Sastry SK, Burridge K. Focal adhesions: A nexus for intracellular signaling and cytoskeletal dynamics. Exp Cell Res 2000; 261:25-36.
18. Burridge K, Chrzanowska-Wodnicka M. Focal adhesions, contractility, and signaling. Annu Rev Cell Dev Biol 1996; 12:463-518.
19. Jockusch BM, Bubeck P, Giehl K et al. The molecular architecture of focal adhesions. Annu Rev Cell Dev Biol 1995; 11:379-416.
20. Critchley DR. Focal adhesions - the cytoskeletal connection. Curr Opin Cell Biol 2000; 12:133-139.
21. Horwitz A, Duggan K, Buck C et al. Interaction of plasma membrane fibronectin receptor with talin—a transmembrane linkage. Nature 1986; 320:531-533.
22. Girault JA, Labesse G, Mornon JP et al. The N-termini of FAK and JAKs contain divergent band 4.1 domains. Trends Biochem Sci 1999; 24:54-57.
23. Chishti AH, Kim AC, Marfatia SM et al. The FERM domain: A unique module involved in the linkage of cytoplasmic proteins to the membrane. Trends Biochem Sci 1998; 23:281-282.
24. Vinogradova O, Velyvis A, Velyviene A et al. A structural mechanism of integrin alpha(IIb)beta(3) „inside-out" activation as regulated by its cytoplasmic face. Cell 2002; 110:587-597.
25. Garcia-Alvarez B, de Pereda JM, Calderwood DA et al. Structural determinants of integrin recognition by talin. Mol Cell 2003; 11:49-58.
26. Tadokoro S, Shattil SJ, Eto K et al. Talin binding to integrin beta tails: A final common step in integrin activation. Science 2003; 302:103-106.
27. Giannone G, Jiang G, Sutton DH et al. Talin1 is critical for force-dependent reinforcement of initial integrin-cytoskeleton bonds but not tyrosine kinase activation. J Cell Biol 2003; 163:409-419.
28. Jiang G, Giannone G, Critchley DR et al. Two-piconewton slip bond between fibronectin and the cytoskeleton depends on talin. Nature 2003; 424:334-337.
29. Priddle H, Hemmings L, Monkley S et al. Disruption of the talin gene compromises focal adhesion assembly in undifferentiated but not differentiated embryonic stem cells. J Cell Biol 1998; 142:1121-1133.
30. Monkley SJ, Zhou XH, Kinston SJ et al. Disruption of the talin gene arrests mouse development at the gastrulation stage. Dev Dyn 2000; 219:560-574.
31. Martel V, Racaud-Sultan C, Dupe S et al. Conformation, localization, and integrin binding of talin depend on its interaction with phosphoinositides. J Biol Chem 2001; 276:21217-21227.
32. Chong LD, Traynor-Kaplan A, Bokoch GM et al. The small GTP-binding protein Rho regulates a phosphatidylinositol 4-phosphate 5-kinase in mammalian cells. Cell 1994; 79:507-513.
33. Di Paolo G, Pellegrini L, Letinic K et al. Recruitment and regulation of phosphatidylinositol phosphate kinase type 1 gamma by the FERM domain of talin. Nature 2002; 420:85-89.
34. Ling K, Doughman RL, Firestone AJ et al. Type I gamma phosphatidylinositol phosphate kinase targets and regulates focal adhesions. Nature 2002; 420:89-93.
35. Ling K, Doughman RL, Iyer VV et al. Tyrosine phosphorylation of type Igamma phosphatidylinositol phosphate kinase by Src regulates an integrin-talin switch. J Cell Biol 2003; 163:1339-1349.
36. Gilmore AP, Burridge K. Cell adhesion. Cryptic sites in vinculin. Nature 1995; 373:197.
37. Brakebusch C, Fassler R. The integrin-actin connection, an eternal love affair. EMBO J 2003; 22:2324-2333.
38. Fukami K, Endo T, Imamura M et al. alpha-Actinin and vinculin are PIP2-binding proteins involved in signaling by tyrosine kinase. J Biol Chem 1994; 269:1518-1522.
39. Shibasaki F, Fukami K, Fukui Y et al. Phosphatidylinositol 3-kinase binds to alpha-actinin through the p85 subunit. Biochem J 1994; 302(Pt 2):551-557.
40. Mukai H, Toshimori M, Shibata H et al. Interaction of PKN with alpha-actinin. J Biol Chem 1997; 272:4740-4746.
41. Greenwood JA, Theibert AB, Prestwich GD et al. Restructuring of focal adhesion plaques by PI 3-kinase. Regulation by PtdIns (3,4,5)-p(3) binding to alpha-actinin. J Cell Biol 2000; 150:627-642.
42. Schaller MD, Parsons JT. pp125FAK-dependent tyrosine phosphorylation of paxillin creates a high-affinity binding site for Crk. Mol Cell Biol 1995; 15:2635-2645.
43. Shen Y, Schaller MD. Focal adhesion targeting: The critical determinant of FAK regulation and substrate phosphorylation. Mol Biol Cell 1999; 10:2507-2518.
44. Chen HC, Appeddu PA, Parsons JT et al. Interaction of focal adhesion kinase with cytoskeletal protein talin. J Biol Chem 1995; 270:16995-16999.
45. Hildebrand JD, Schaller MD, Parsons JT. Paxillin, a tyrosine phosphorylated focal adhesion-associated protein binds to the carboxyl terminal domain of focal adhesion kinase. Mol Biol Cell 1995; 6:637-647.
46. Hanks SK, Calalb MB, Harper MC et al. Focal adhesion protein-tyrosine kinase phosphorylated in response to cell attachment to fibronectin. Proc Natl Acad Sci USA 1992; 89:8487-8491.
47. Schaller MD, Borgman CA, Cobb BS et al. pp125FAK a structurally distinctive protein-tyrosine kinase associated with focal adhesions. Proc Natl Acad Sci USA 1992; 89:5192-5196.

48. Parsons JT. Focal adhesion kinase: the first ten years. J Cell Sci 2003; 116:1409-1416.
49. Schaller MD, Hildebrand JD, Parsons JT. Complex formation with focal adhesion kinase: A mechanism to regulate activity and subcellular localization of Src kinases. Mol Biol Cell 1999; 10:3489-3505.
50. Schaller MD, Schaefer EM. Multiple stimuli induce tyrosine phosphorylation of the Crk-binding sites of paxillin. Biochem J 2001; 360:57-66.
51. Schaller MD, Hildebrand JD, Shannon JD et al. Autophosphorylation of the focal adhesion kinase, pp125FAK, directs SH2- dependent binding of pp60src. Mol Cell Biol 1994; 14:1680-1688.
52. Eide BL, Turck CW, Escobedo JA. Identification of Tyr-397 as the primary site of tyrosine phosphorylation and pp60src association in the focal adhesion kinase, pp125FAK. Mol Cell Biol 1995; 15:2819-2827.
53. Sieg DJ, Hauck CR, Ilic D et al. FAK integrates growth-factor and integrin signals to promote cell migration. Nat Cell Biol 2000; 2:249-256.
54. Poullet P, Gautreau A, Kadare G et al. Ezrin interacts with focal adhesion kinase and induces its activation independently of cell-matrix adhesion. J Biol Chem 2001; 276:37686-37691.
55. Cooper LA, Shen TL, Guan JL. Regulation of focal adhesion kinase by its amino-terminal domain through an autoinhibitory interaction. Mol Cell Biol 2003; 23:8030-8041.
56. Dunty JM, Gabarra-Niecko V, King ML et al. FERM domain interaction promotes FAK signaling. Mol Cell Biol 2004; 24:5353-5368.
57. Thomas JW, Ellis B, Boerner RJ et al. SH2- and SH3-mediated interactions between focal adhesion kinase and Src. J Biol Chem 1998; 273:577-583.
58. Hildebrand JD, Taylor JM, Parsons JT. An SH3 domain-containing GTPase-activating protein for Rho and Cdc42 associates with focal adhesion kinase. Mol Cell Biol 1996; 16:3169-3178.
59. Liu Y, Loijens JC, Martin KH et al. The association of ASAP1, an ADP ribosylation factor-GTPase activating protein, with focal adhesion kinase contributes to the process of focal adhesion assembly. Mol Biol Cell 2002; 13:2147-2156.
60. Bouton AH, Riggins RB, Bruce-Staskal PJ. Functions of the adapter protein Cas: Signal convergence and the determination of cellular responses. Oncogene 2001; 20:6448-6458.
61. O'Neill GM, Fashena SJ, Golemis EA. Integrin signalling: A new Cas(t) of characters enters the stage. Trends Cell Biol 2000; 10:111-119.
62. Toutant M, Costa A, Studler JM et al. Alternative splicing controls the mechanisms of FAK autophosphorylation. Mol Cell Biol 2002; 22:7731-7743.
63. Calalb MB, Polte TR, Hanks SK. Tyrosine phosphorylation of focal adhesion kinase at sites in the catalytic domain regulates kinase activity: A role for Src family kinases. Mol Cell Biol 1995; 15:954-963.
64. Pawson T. Protein modules and signalling networks. Nature 1995; 373:573-580.
65. Chen HC, Guan JL. Association of focal adhesion kinase with its potential substrate phosphatidylinositol 3-kinase. Proc Natl Acad Sci USA 1994; 91:10148-10152.
66. Zhang X, Chattopadhyay A, Ji QS et al. Focal adhesion kinase promotes phospholipase C-gamma1 activity. Proc Natl Acad Sci USA 1999; 96:9021-9026.
67. Han DC, Guan JL. Association of focal adhesion kinase with Grb7 and its role in cell migration. J Biol Chem 1999; 274:24425-24430.
68. Schlaepfer DD, Hanks SK, Hunter T et al. Integrin-mediated signal transduction linked to Ras pathway by GRB2 binding to focal adhesion kinase. Nature 1994; 372:786-791.
69. Schlaepfer DD, Hauck CR, Sieg DJ. Signaling through focal adhesion kinase. Prog Biophys Mol Biol 1999; 71:435-478.
70. Cooper JA, Gould KL, Cartwright CA et al. Tyr527 is phosphorylated in pp60c-src: Implications for regulation. Science 1986; 231:1431-1434.
71. Yeatman TJ. A renaissance for SRC. Nat Rev Cancer 2004; 4:470-480.
72. Yamaguchi H, Hendrickson WA. Structural basis for activation of human lymphocyte kinase Lck upon tyrosine phosphorylation. Nature 1996; 384:484-489.
73. Jung EJ, Kim CW. Interaction between chicken protein tyrosine phosphatase 1 (CPTP1)-like rat protein phosphatase 1 (PTP1) and p60(v-src) in v-src-transformed Rat-1 fibroblasts. Exp Mol Med 2002; 34:476-480.
74. Zheng XM, Wang Y, Pallen CJ. Cell transformation and activation of pp60c-src by overexpression of a protein tyrosine phosphatase. Nature 1992; 359:336-339.
75. Irby RB, Mao W, Coppola D et al. Activating SRC mutation in a subset of advanced human colon cancers. Nat Genet 1999; 21:187-190.
76. Sefton BM, Trowbridge IS, Cooper JA et al. The transforming proteins of Rous sarcoma virus, Harvey sarcoma virus and Abelson virus contain tightly bound lipid. Cell 1982; 31:465-474.

77. Moro L, Dolce L, Cabodi S et al. Integrin-induced epidermal growth factor (EGF) receptor activation requires c-Src and p130Cas and leads to phosphorylation of specific EGF receptor tyrosines. J Biol Chem 2002; 277:9405-9414.
78. Arias-Salgado EG, Lizano S, Sarkar S et al. Src kinase activation by direct interaction with the integrin beta cytoplasmic domain. Proc Natl Acad Sci USA 2003; 100:13298-13302.
79. de Virgilio M, Kiosses WB, Shattil SJ et al. Proximal, selective, and dynamic interactions between integrin alphaIIbbeta3 and protein tyrosine kinases in living cells. J Cell Biol 2004; 165:305-311.
80. Cary LA, Klinghoffer RA, Sachsenmaier C et al. SRC catalytic but not scaffolding function is needed for integrin-regulated tyrosine phosphorylation, cell migration, and cell spreading. Mol Cell Biol 2002; 22:2427-2440.
81. Turner CE. Paxillin and focal adhesion signalling. Nat Cell Biol 2000; 2:E231-236.
82. Defilippi P, Gismondi A, Santoni A et al. Integrins and signal transduction, springer Verlag. 1997.
83. Vuori K, Ruoslahti E. Tyrosine phosphorylation of p130Cas and cortactin accompanies integrin-mediated cell adhesion to extracellular matrix. J Biol Chem 1995; 270:22259-22262.
84. Nojima Y, Morino N, Mimura T et al. Integrin-mediated cell adhesion promotes tyrosine phosphorylation of p130Cas, a Src homology 3-containing molecule having multiple Src homology 2-binding motifs. J Biol Chem 1995; 270:15398-15402.
85. Di Stefano P, Cabodi S, Boeri Erba E et al. P130Cas-associated protein (p140Cap) as a new tyrosine-phosphorylated protein involved in cell spreading. Mol Biol Cell 2004; 15:787-800.
86. Owen JD, Ruest PJ, Fry DW et al. Induced focal adhesion kinase (FAK) expression in FAK-null cells enhances cell spreading and migration requiring both auto- and activation loop phosphorylation sites and inhibits adhesion-dependent tyrosine phosphorylation of Pyk2. Mol Cell Biol 1999; 19:4806-4818.
87. Ruest PJ, Shin NY, Polte TR et al. Mechanisms of CAS substrate domain tyrosine phosphorylation by FAK and Src. Mol Cell Biol 2001; 21:7641-7652.
88. Dolfi F, Garcia-Guzman M, Ojaniemi M et al. The adaptor protein Crk connects multiple cellular stimuli to the JNK signaling pathway. Proc Natl Acad Sci USA 1998; 95:15394-15399.
89. Oktay M, Wary KK, Dans M et al. Integrin-mediated activation of focal adhesion kinase is required for signaling to Jun NH2-terminal kinase and progression through the G1 phase of the cell cycle. J Cell Biol 1999; 145:1461-1469.
90. Gotoh T, Hattori S, Nakamura S et al. Identification of Rap1 as a target for the Crk SH3 domain-binding guanine nucleotide-releasing factor C3G. Mol Cell Biol 1995; 15:6746-6753.
91. Cabodi S, Moro L, Baj G et al. p130Cas interacts with estrogen receptor alpha and modulates non-genomic extrogen signaling in breast cancer cells. J Cell Sci 2004; 117:1603-1611.
92. Burnham MR, Bruce-Staskal PJ, Harte MT et al. Regulation of c-SRC activity and function by the adapter protein CAS. Mol Cell Biol 2000; 20:5865-5878.
93. Brown MT, Cooper JA. Regulation, substrates and functions of src. Biochim Biophys Acta 1996; 1287:121-149.
94. Courtneidge SA. Isolation of novel Src substrates. Biochem Soc Trans 2003; 31:25-28.
95. Ilic D, Furuta Y, Kanazawa S et al. Reduced cell motility and enhanced focal adhesion contact formation in cells from FAK-deficient mice. Nature 1995; 377:539-544.
96. Klinghoffer RA, Sachsenmaier C, Cooper JA et al. Src family kinases are required for integrin but not PDGFR signal transduction. EMBO J 1999; 18:2459-2471.
97. Richardson A, Malik RK, Hildebrand JD et al. Inhibition of cell spreading by expression of the C-terminal domain of focal adhesion kinase (FAK) is rescued by coexpression of Src or catalytically inactive FAK: A role for paxillin tyrosine phosphorylation. Mol Cell Biol 1997; 17:6906-6914.
98. Webb DJ, Donais K, Whitmore LA et al. FAK-Src signalling through paxillin, ERK and MLCK regulates adhesion disassembly. Nat Cell Biol 2004; 6:154-161.
99. Kaplan KB, Swedlow JR, Morgan DO et al. c-Src enhances the spreading of src-/- fibroblasts on fibronectin by a kinase-independent mechanism. Genes Dev 1995; 9:1505-1517.
100. Petit V, Boyer B, Lentz D et al. Phosphorylation of tyrosine residues 31 and 118 on paxillin regulates cell migration through an association with CRK in NBT-II cells. J Cell Biol 2000; 148:957-970.
101. Carragher NO, Frame MC. Calpain: A role in cell transformation and migration. Int J Biochem Cell Biol 2002; 34:1539-1543.
102. Frame MC, Fincham VJ, Carragher NO et al. v-Src's hold over actin and cell adhesions. Nat Rev Mol Cell Biol 2002; 3:233-245.
103. Hsia DA, Mitra SK, Hauck CR et al. Differential regulation of cell motility and invasion by FAK. J Cell Biol 2003; 160:753-767.
104. Honda H, Oda H, Nakamoto T et al. Cardiovascular anomaly, impaired actin bundling and resistance to Src-induced transformation in mice lacking p130Cas. Nat Genet 1998; 19:361-365.

105. Chodniewicz D, Klemke RL. Regulation of integrin-mediated cellular responses through assembly of a CAS/Crk scaffold. Biochim Biophys Acta 2004; 1692:63-76.
106. Shin NY, Dise RS, Schneider-Mergener J et al. Subsets of the major tyrosine phosphorylation sites in Crk-associated substrate (CAS) Are sufficient to promote cell migration. J Biol Chem 2004; 279:38331-38337.
107. Goldberg GS, Alexander DB, Pellicena P et al. Src phosphorylates Cas on tyrosine 253 to promote migration of transformed cells. J Biol Chem 2003; 278:46533-46540.
108. Garton AJ, Tonks NK. Regulation of fibroblast motility by the protein tyrosine phosphatase PTP-PEST. J Biol Chem 1999; 274:3811-3818.
109. Sastry SK, Lyons PD, Schaller MD et al. PTP-PEST controls motility through regulation of Rac1. J Cell Sci 2002; 115:4305-4316.
110. Huang J, Hamasaki H, Nakamoto T et al. Differential regulation of cell migration, actin stress fiber organization, and cell transformation by functional domains of Crk-associated substrate. J Biol Chem 2002; 277:27265-27272.
111. Brabek J, Constancio SS, Shin NY et al. CAS promotes invasiveness of Src-transformed cells. Oncogene 2004.
112. Chang JH, Gill S, Settleman J et al. c-Src regulates the simultaneous rearrangement of actin cytoskeleton, p190RhoGAP, and p120RasGAP following epidermal growth factor stimulation. J Cell Biol 1995; 130:355-368.
113. Tsubouchi A, Sakakura J, Yagi R et al. Localized suppression of RhoA activity by Tyr31/118-phosphorylated paxillin in cell adhesion and migration. J Cell Biol 2002; 159:673-683.
114. Schwartz MA, Assoian RK. Integrins and cell proliferation: regulation of cyclin-dependent kinases via cytoplasmic signaling pathways. J Cell Sci 2001; 114:2553-2560.
115. Barberis L, Wary KK, Fiucci G et al. Distinct roles of the adaptor protein Shc and focal adhesion kinase in integrin signaling to ERK. J Biol Chem 2000; 275:36532-36540.
116. Hirsch E, Barberis L, Brancaccio M et al. Defective Rac-mediated proliferation and survival after targeted mutation of the beta1 integrin cytodomain. J Cell Biol 2002; 157:481-492.
117. Wary KK, Mainiero F, Isakoff SJ et al. The adaptor protein Shc couples a class of integrins to the control of cell cycle progression. Cell 1996; 87:733-743.
118. del Pozo MA, Alderson NB, Kiosses WB et al. Integrins regulate Rac targeting by internalization of membrane domains. Science 2004; 303:839-842.
119. Palazzo AF, Eng CH, Schlaepfer DD et al. Localized stabilization of microtubules by integrin- and FAK-facilitated Rho signaling. Science 2004; 303:836-839.
120. Mainiero F, Murgia C, Wary KK et al. The coupling of alpha6beta4 integrin to Ras-MAP kinase pathways mediated by Shc controls keratinocyte proliferation. EMBO J 1997; 16:2365-2375.
121. Brady-Kalnay SM, Tonks NK. Protein tyrosine phosphatases: From structure to function. Trends Cell Biol 1994; 4:73-76.
122. Retta SF, Barry ST, Critchley DR et al. Focal adhesion and stress fiber formation is regulated by tyrosine phosphatase activity. Exp Cell Res 1996; 229:307-317.
123. Chrzanowska-Wodnicka M, Burridge K. Tyrosine phosphorylation is involved in reorganization of the actin cytoskeleton in response to serum or LPA stimulation. J Cell Sci 1994; 107(Pt 12):3643-3654.
124. Defilippi P, Retta SF, Olivo C et al. p125FAK tyrosine phosphorylation and focal adhesion assembly: Studies with phosphotyrosine phosphatase inhibitors. Exp Cell Res 1995; 221:141-152.
125. den Hertog J, Tracy S, Hunter T. Phosphorylation of receptor protein-tyrosine phosphatase alpha on Tyr789, a binding site for the SH3-SH2-SH3 adaptor protein GRB-2 in vivo. EMBO J 1994; 13:3020-3032.
126. Su J, Muranjan M, Sap J. Receptor protein tyrosine phosphatase alpha activates Src-family kinases and controls integrin-mediated responses in fibroblasts. Curr Biol 1999; 9:505-511.
127. Ponniah S, Wang DZ, Lim KL et al. Targeted disruption of the tyrosine phosphatase PTPalpha leads to constitutive downregulation of the kinases Src and Fyn. Curr Biol 1999; 9:535-538.
128. Pallen CJ. Protein tyrosine phosphatase alpha (PTPalpha): A Src family kinase activator and mediator of multiple biological effects. Curr Top Med Chem 2003; 3:821-835.
129. Lammers R, Lerch MM, Ullrich A. The carboxyl-terminal tyrosine residue of protein-tyrosine phosphatase alpha mediates association with focal adhesion plaques. J Biol Chem 2000; 275:3391-3396.
130. Zeng L, Si X, Yu WP et al. PTP alpha regulates integrin-stimulated FAK autophosphorylation and cytoskeletal rearrangement in cell spreading and migration. J Cell Biol 2003; 160:137-146.
131. von Wichert G, Haimovich B, Feng GS et al. Force-dependent integrin-cytoskeleton linkage formation requires downregulation of focal complex dynamics by Shp2. EMBO J 2003; 22:5023-5035.
132. Serra-Pages C, Kedersha NL, Fazikas L et al. The LAR transmembrane protein tyrosine phosphatase and a coiled-coil LAR-interacting protein colocalize at focal adhesions. EMBO J 1995; 14:2827-2838.

133. O'Grady P, Thai TC, Saito H. The laminin-nidogen complex is a ligand for a specific splice isoform of the transmembrane protein tyrosine phosphatase LAR. J Cell Biol 1998; 141:1675-1684.
134. Garton AJ, Burnham MR, Bouton AH. Association of PTP-PEST with the SH3 domain of p130cas; a novel mechanism of protein tyrosine phosphatase substrate recognition. Oncogene 1997; 15:877-885.
135. Shen Y, Schneider G, Cloutier JF et al. Direct association of protein-tyrosine phosphatase PTP-PEST with paxillin. J Biol Chem 1998; 273:6474-6481.
136. Yu DH, Qu CK, Henegariu O et al. Protein-tyrosine phosphatase Shp-2 regulates cell spreading, migration, and focal adhesion. J Biol Chem 1998; 273:21125-21131.
137. Tsuda M, Matozaki T, Fukunaga K et al. Integrin-mediated tyrosine phosphorylation of SHPS-1 and its association with SHP-2. Roles of Fak and Src family kinases. J Biol Chem 1998; 273:13223-13229.
138. Inagaki K, Yamao T, Noguchi T et al. SHPS-1 regulates integrin-mediated cytoskeletal reorganization and cell motility. EMBO J 2000; 19:6721-6731.
139. Lacalle RA, Mira E, Gomez-Mouton C et al. Specific SHP-2 partitioning in raft domains triggers integrin-mediated signaling via Rho activation. J Cell Biol 2002; 157:277-289.
140. Davies MA, Koul D, Dhesi H et al. Regulation of Akt/PKB activity, cellular growth, and apoptosis in prostate carcinoma cells by MMAC/PTEN. Cancer Res 1999; 59:2551-2556.
141. Tamura M, Gu J, Tran H et al. PTEN gene and integrin signaling in cancer. J Natl Cancer Inst 1999; 91:1820-1828.
142. Liliental J, Moon SY, Lesche R et al. Genetic deletion of the Pten tumor suppressor gene promotes cell motility by activation of Racl and Cdc42 GTPases. Curr Biol 2000; 10:401-404.
143. Gu J, Tamura M, Pankov R et al. Shc and FAK differentially regulate cell motility and directionality modulated by PTEN. J Cell Biol 1999; 146:389-403.
144. Chiarugi P, Pani G, Giannoni E et al. Reactive oxygen species as essential mediators of cell adhesion: The oxidative inhibition of a FAK tyrosine phosphatase is required for cell adhesion. J Cell Biol 2003; 161:933-944.
145. Assoian RK. Anchorage-dependent cell cycle progression. J Cell Biol 1997; 136:1-4.
146. Carrano AC, Pagano M. Role of the F-box protein Skp2 in adhesion-dependent cell cycle progression. J Cell Biol 2001; 153:1381-1390.
147. Moro L, Venturino M, Bozzo C et al. Integrins induce activation of EGF receptor: Role in MAP kinase induction and adhesion-dependent cell survival. EMBO J 1998; 17:6622-6632.
148. Rusciano D, Lorenzoni P, Burger MM. Constitutive activation of c-Met in liver metastatic B16 melanoma cells depends on both substrate adhesion and cell density and is regulated by a cytosolic tyrosine phosphatase activity. J Biol Chem 1996; 271:20763-20769.
149. Wang R, Kobayashi R, Bishop JM. Cellular adherence elicits ligand-independent activation of the Met cell-surface receptor. Proc Natl Acad Sci USA 1996; 93:8425-8430.
150. Sundberg C, Rubin K. Stimulation of betal integrins on fibroblasts induces PDGF independent tyrosine phosphorylation of PDGF beta-receptors. J Cell Biol 1996; 132:741-752.
151. Danilkovitch-Miagkova A, Angeloni D, Skeel A et al. Integrin-mediated RON growth factor receptor phosphorylation requires tyrosine kinase activity of both the receptor and c-Src. J Biol Chem 2000; 275:14783-14786.
152. Wang JF, Zhang XF, Groopman JE. Stimulation of beta 1 integrin induces tyrosine phosphorylation of vascular endothelial growth factor receptor-3 and modulates cell migration. J Biol Chem 2001a; 276:41950-41957.
153. Yu X, Miyamoto S, Mekada E. Integrin alpha 2 beta 1-dependent EGF receptor activation at cell-cell contact sites. J Cell Sci 2000; 113(12):2139-2147.
154. Wang XQ, Sun P, Paller AS. Ganglioside GM3 blocks the activation of epidermal growth factor receptor induced by integrin at specific tyrosine sites. J Biol Chem 2003; 278:48770-48778.
155. Bill HM, Knudsen B, Moores SL et al. Epidermal growth factor receptor-dependent regulation of integrin-mediated signaling and cell cycle entry in epithelial cells. Mol Cell Biol 2004; 24:8586-8599.
156. DeMali KA, Balciunaite E, Kazlauskas A. Integrins enhance platelet-derived growth factor (PDGF)-dependent responses by altering the signal relay enzymes that are recruited to the PDGF beta receptor. J Biol Chem 1999; 274:19551-19558.
157. Kanda E, Jin ZH, Mizuchi D et al. Activation of Rac and tyrosine phosphorylation of cytokine receptors induced by cross-linking of integrin alpha4betal and cell adhesion in hematopoietic cells. Biochem Biophys Res Commun 2003; 301:934-940.
158. Cabodi S, Moro L, Bergatto E et al. Integrin regulation of epidermal growth factor (EGF) receptor and of EGF-dependent responses. Biochem Soc Trans 2004; 32:438-442.
159. Marcoux N, Vuori K. EGF receptor mediates adhesion-dependent activation of the Rac GTPase: A role for phosphatidylinositol 3-kinase and Vav2. Oncogene 2003; 22:6100-6106.

160. Frisch SM, Vuori K, Ruoslahti E et al. Control of adhesion-dependent cell survival by focal adhesion kinase. J Cell Biol 1996; 134:793-799.
161. Reginato MJ, Mills KR, Paulus JK et al. Integrins and EGFR coordinately regulate the pro-apoptotic protein Bim to prevent anoikis. Nat Cell Biol 2003; 5:733-740.
162. Hobbs RM, Watt FM. Regulation of interleukin-1alpha expression by integrins and epidermal growth factor receptor in keratinocytes from a mouse model of inflammatory skin disease. J Biol Chem 2003; 278:19798-19807.
163. Wang R, Ferrell LD, Faouzi S et al. Activation of the Met receptor by cell attachment induces and sustains hepatocellular carcinomas in transgenic mice. J Cell Biol 2001b; 153:1023-1034.
164. Cybulsky AV, McTavish AJ, Cyr MD. Extracellular matrix modulates epidermal growth factor receptor activation in rat glomerular epithelial cells. J Clin Invest 1994; 94:68-78.
165. Guilherme A, Torres K, Czech MP. Cross-talk between insulin receptor and integrin alpha5 beta1 signaling pathways. J Biol Chem 1998; 273:22899-22903.
166. Jones PL, Crack J, Rabinovitch M. Regulation of tenascin-C, a vascular smooth muscle cell survival factor that interacts with the alpha v beta 3 integrin to promote epidermal growth factor receptor phosphorylation and growth. J Cell Biol 1997; 139:279-293.
167. Lee YJ, Streuli CH. Extracellular matrix selectively modulates the response of mammary epithelial cells to different soluble signaling ligands. J Biol Chem 1999; 274:22401-22408.
168. Li J, Lin ML, Wiepz GJ et al. Integrin-mediated migration of murine B82L fibroblasts is dependent on the expression of an intact epidermal growth factor receptor. J Biol Chem 1999; 274:11209-11219.
169. Miyamoto S, Teramoto H, Gutkind JS et al. Integrins can collaborate with growth factors for phosphorylation of receptor tyrosine kinases and MAP kinase activation: Roles of integrin aggregation and occupancy of receptors. J Cell Biol 1996; 135:1633-1642.
170. Rusnati M, Tanghetti E, Dell'Era P et al. alphavbeta3 integrin mediates the cell-adhesive capacity and biological activity of basic fibroblast growth factor (FGF-2) in cultured endothelial cells. Mol Biol Cell 1997; 8:2449-2461.
171. Schneller M, Vuori K, Ruoslahti E. Alphavbeta3 integrin associates with activated insulin and PDGFbeta receptors and potentiates the biological activity of PDGF. EMBO J 1997; 16:5600-5607.
172. Soldi R, Mitola S, Strasly M et al. Role of alphavbeta3 integrin in the activation of vascular endothelial growth factor receptor-2. EMBO J 1999; 18:882-892.
173. Vuori K, Ruoslahti E. Association of insulin receptor substrate-1 with integrins. Science 1994; 266:1576-1578.
174. Baron V, Schwartz M. Cell adhesion regulates ubiquitin-mediated degradation of the platelet-derived growth factor receptor beta. J Biol Chem 2000; 275:39318-39323.
175. Borges E, Jan Y, Ruoslahti E et al. Platelet-derived growth factor receptor beta and vascular endothelial growth factor receptor 2 bind to the beta 3 integrin through its extracellular domain. J Biol Chem 2000; 275:39867-39873.
176. Mariotti A, Kedeshian PA, Dans M et al. EGF-R signaling through Fyn kinase disrupts the function of integrin alpha6beta4 at hemidesmosomes: Role in epithelial cell migration and carcinoma invasion. J Cell Biol 2001; 155:447-458.
177. Falcioni R, Antonini A, Nistico P et al. Alpha 6 beta 4 and alpha 6 beta 1 integrins associate with ErbB-2 in human carcinoma cell lines. Exp Cell Res 1997; 236:76-85.
178. Trusolino L, Bertotti A, Comoglio PM. A signaling adapter function for alpha6beta4 integrin in the control of HGF-dependent invasive growth. Cell 2001; 107:643-654.
179. Baron W, Decker L, Colognato H et al. Regulation of integrin growth factor interactions in oligodendrocytes by lipid raft microdomains. Curr Biol 2003; 13:151-155.
180. Colognato H, Baron W, Avellana-Adalid V et al. CNS integrins switch growth factor signalling to promote target-dependent survival. Nat Cell Biol 2002; 4:833-841.
181. Maile LA, Badley-Clarke J, Clemmons DR. The association between integrin-associated protein and SHPS-1 regulates insulin-like growth factor-I receptor signaling in vascular smooth muscle cells. Mol Biol Cell 2003; 14:3519-3528.
182. Goel HL, Fornaro M, Moro L et al. Selective modulation of type 1 insulin-like growth factor receptor signaling and functions by beta1 integrins. J Cell Biol 2004; 166:407-418.
183. Roovers K, Assoian RK. Integrating the MAP kinase signal into the G1 phase cell cycle machinery. Bioessays 2000; 22:818-826.
184. Aplin AE, Juliano RL. Regulation of nucleocytoplasmic trafficking by cell adhesion receptors and the cytoskeleton. J Cell Biol 2001; 155:187-191.
185. Conner SR, Scott G, Aplin AE. Adhesion-dependent activation of the ERK1/2 cascade is by-passed in melanoma cells. J Biol Chem 2003; 278:34548-34554.

186. Ivankovic-Dikic I, Gronroos E, Blaukat A et al. Pyk2 and FAK regulate neurite outgrowth induced by growth factors and integrins. Nat Cell Biol 2000; 2:574-581.
187. Eliceiri BP, Puente XS, Hood JD et al. Src-mediated coupling of focal adhesion kinase to integrin alpha(v)beta5 in vascular endothelial growth factor signaling. J Cell Biol 2002; 157:149-160.
188. Brown DA, London E. Functions of lipid rafts in biological membranes. Annu Rev Cell Dev Biol 1998; 14:111-136.
189. Brown DA, London E. Structure and function of sphingolipid- and cholesterol-rich membrane rafts. J Biol Chem 2000; 275:17221-17224.
190. Harder T, Simons K. Caveolae, DIGs, and the dynamics of sphingolipid-cholesterol microdomains. Curr Opin Cell Biol 1997; 9:534-542.
191. Pande G. The role of membrane lipids in regulation of integrin functions. Curr Opin Cell Biol 2000; 12:569-574.
192. Lisanti MP, Sargiacomo M, Scherer PE. Purification of caveolae-derived membrane microdomains containing lipid-anchored signaling molecules, such as GPI-anchored proteins, H-Ras, Src-family tyrosine kinases, eNOS, and G-protein alpha-, beta-, and gamma-subunits. Methods Mol Biol 1999; 116:51-60.
193. Wary KK, Mariotti A, Zurzolo C et al. A requirement for caveolin-1 and associated kinase Fyn in integrin signaling and anchorage-dependent cell growth. Cell 1998; 94:625-634.
194. Wei Y, Yang X, Liu Q et al. A role for caveolin and the urokinase receptor in integrin-mediated adhesion and signaling. J Cell Biol 1999; 144:1285-1294.
195. Shima T, Nada S, Okada M. Transmembrane phosphoprotein Cbp senses cell adhesion signaling mediated by Src family kinase in lipid rafts. Proc Natl Acad Sci USA 2003; 100:14897-14902.
196. Del Pozo MA, Kiosses WB, Alderson NB et al. Integrins regulate GTP-Rac localized effector interactions through dissociation of Rho-GDI. Nat Cell Biol 2002; 4:232-239.

Chapter 5

C. elegans Integrins

Michel Labouesse* and Elisabeth Georges-Labouesse

Caenorhabditis elegans has only one beta integrin chain, called PAT-3, and two alpha integrin chains, called PAT-2 and INA-1. *C. elegans* integrins are essential for muscle anchoring to the epidermis, cell migration, axon outgrowth and fasciculation, and epithelial morphogenesis. Genetic analysis of this simple integrin repertoire has led to the identification of protein complexes that interact with integrins during these developmental processes. We discuss how the analysis of muscle anchoring, which is mediated by the PAT-2/PAT-3 heterodimer, has become a model to understand the stepwise assembly of proteins such as ILK, actopaxin, Pinch, Kindlin or vinculin at focal adhesions. Likewise, the analysis of cell migration and axon outgrowth is a model to investigate the signaling events involved in mediating adhesion to the laminin-binding INA-1/PAT-3 heterodimer. In particular, it has revealed that the serine/threonine kinase NIK acts in the same process as integrins during axon outgrowth.

Introduction

C. elegans integrins were identified some ten years ago either through genetic analysis or by using degenerate primers to amplify homologues of vertebrate integrins.[1-3] Mutation analysis defined three integrin genes, one encoding the β chain (*pat-3*) and two encoding α chains (*pat-2* and *ina-1*). Since then several publications have strongly suggested that PAT-3 is indeed the common β chain functioning either with PAT-2 or with INA-1.

With its minimal set of integrins, *C. elegans* represents an ideal model to investigate the in vivo role of integrins and to identify proteins that interact with integrins or modulate their activity. Genetic analysis has so far revealed a requirement for *C. elegans* integrins in anchoring muscles to the epidermis, cell migration, axon outgrowth and fasciculation, and finally in epithelial morphogenesis. The study of mutant phenotypes in this simple multicellular organism allows an in-depth analysis of the cellular mechanisms involving integrins. After briefly outlining the structure of *C. elegans* integrins, we review these processes and discuss proteins interacting with integrins.

Structure of *C. elegans* Integrins

C. elegans *Has a Minimal Set of Integrins*

Sequence comparison shows that PAT-3 is most similar to the β1-integrin chain of vertebrates, that PAT-2 is closer to the group of vertebrate integrins that bind RGD ligands, while INA-1 is closer to a subset of integrins binding to laminin ligands (Fig. 1).[2-5] Note that, in accordance with the *C. elegans* nomenclature, we refer to genes and mutations with small case italicized letters (i.e., *pat-3*), to proteins with capital letters (i.e., PAT-3); in addition, we will use the prefix α and β as a reminder of the type of chain.

*Corresponding Author: Michel Labouesse—Institut de Génétique et de Biologie Moléculaire et Cellulaire, CNRS/INSERM/ULP, 1 rue Laurent Fries, BP 10142 C.U. de Strasbourg, 67404 Illkirch Cedex, France. Email: lmichel@igbmc.u-strasbg.fr

Integrins and Development, edited by Erik H.J. Danen. ©2006 Landes Bioscience.

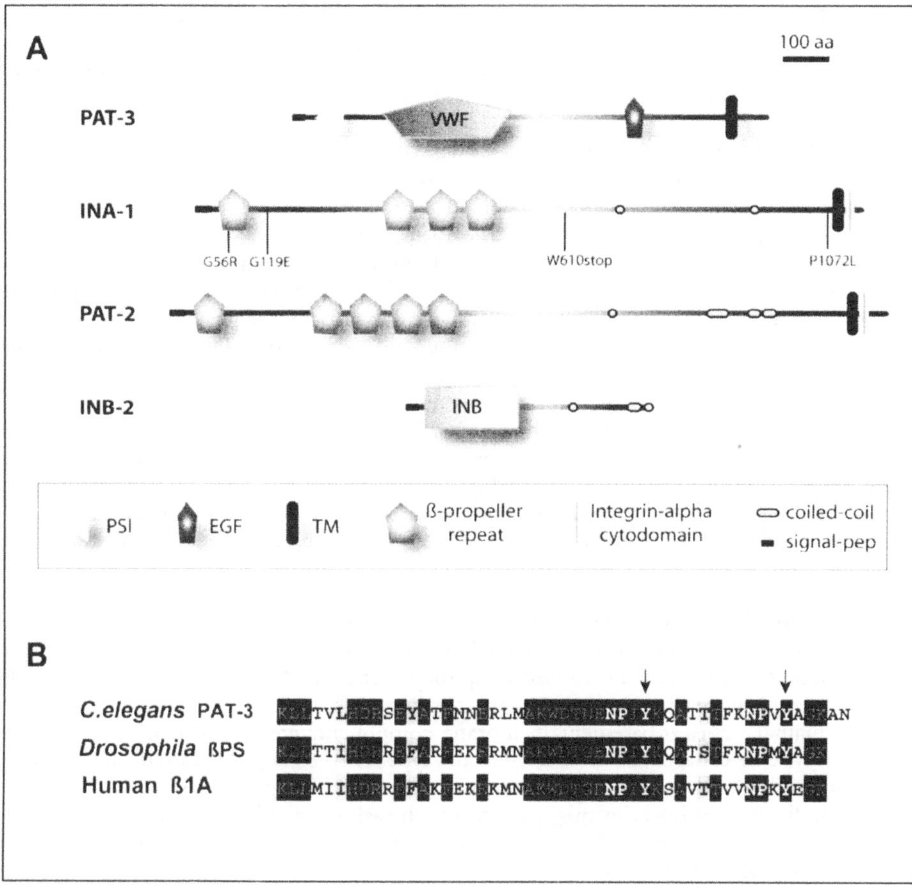

Figure 1. Structure of *C. elegans* integrins. A) Domain organization of *C. elegans* integrins. The extracellular domain of PAT-3 shows strong homology (50%-55% identity) with vertebrate β1- and *Drosophila* βPS integrins; its cytoplasmic domain is even more strongly conserved (63% identity). The extracellular domain of αINA-1 shows 26% identity to the αPS1 integrin chain of *Drosophila*. and is closest to the vertebrate laminin-binding α3, α6, α7-integrins, that of αPAT-2 shows 31% identity to the αPS2 of *Drosophila*, and the vertebrate α8, α5 and αv-integrins. The positions and nature of known *ina-1* mutations are indicated. The INB-2 protein does not have a transmembrane domain, raising questions about its function as an extracellular matrix receptor. Domains are as predicted by the SMART software. B) Alignment of the cytoplasmic domain of βPAT-3, *Drosophila* βPS and human β1A-integrin chains. The positions of the two NPXY motifs are outlined with an arrow.

Consistent with sequence alignments, genetic analysis strongly suggests that αPAT-2/βPAT-3 can interact with two RGD-containing ligands, the proteoglycan perlecan and type-IV collagen, while αINA-1/βPAT-3 can interact with laminins (see below). Co-immunoprecipitation experiments indicate that βPAT-3 can immunoprecipitate at least seven peptides, two of which have an apparent molecular weight in non-reducing conditions consistent with them being αINA-1 or αPAT-2.[3] Conversely, αINA-1 can immunoprecipitate βPAT-3 from worm extracts.[2]

When the *C. elegans* genome sequence was released, a sequence homologous to an incomplete β chain was found (*inb-2*). The corresponding protein is predicted to have a shortened extracellular domain and to lack a transmembrane domain (Fig. 1A), raising doubts about the possibility it acts as a bona fide integrin. Furthermore, a chemically-induced deletion in the gene shows no apparent phenotype, indicating that *inb-2* is unlikely to play a major role.[6] One can thus consider that αPAT-2/βPAT-3 and αINA-1/βPAT-3 represent the ancestral minimal set of integrins necessary to build a multicellular animal.

Important Domains of C. elegans Integrins

The structure of the βPAT-3 extracellular domain is typical of β-integrins (for a recent review, see ref. 7), with a VWF (von Willebrand Factor) domain containing the so-called MIDAS (metal-ion-dependent adhesion site), one EGF type domain, and a recognizable PSI (plexin-semaphorin-integrin) domain at its N-Terminus. The cytoplasmic domain of PAT-3 has a high degree of conservation with the corresponding domain of *Drosophila* βPS and vertebrate β1-integrins (Fig. 1B). In particular, it harbors two NPXY sequences, which have been proposed in vertebrates to mediate talin binding and localisation to focal adhesions.[7] Their replacement by phenylalanines increases the adverse effects of dominant-negative constructs lacking the extracellular domain, thus demonstrating a functional role for these tyrosines.[8]

As their *Drosophila* counterparts, the two *C. elegans* α chains do not carry the I/A domains that are involved in the activation of some vertebrate integrins.[7] Nevertheless, they contain other α-integrin features, such as repeated domains in the extracellular domains predicted to fold as a β-propeller domain, and coiled-coil regions. As all α chains, the cytoplasmic domains of αINA-1 and αPAT-2 contain a K/RxGFFKR sequence lying downstream of the transmembrane domain.

Known and Potential Integrin-Interacting Proteins

The *C. elegans* genome encodes predicted homologues of many proteins known to interact with vertebrate integrins. Prior to discussing the function of *C. elegans* integrins, we briefly outline below which important interactors are present or absent.

The main *C. elegans* basement membrane components are the proteoglycan perlecan, laminins, type-IV collagen, SPARC/osteonectin, type XVIII collagen/endostatin and nidogen/entactin, while the cuticle extracellular matrix is enriched in specific collagens.[9-14] Genetic analysis indicates that integrins might interact with each of the basement membrane ligands, but not with cuticle collagens. Other classical vertebrate integrin ligands, such as fibronectin, have no homologues in *C. elegans*.

Among proteins that interact with integrins on their cytoplasmic face, *C. elegans* contains homologues for most focal adhesion components, some of which were actually discovered in *C. elegans* (UNC-97 and UNC-112). Several homologues have already been characterized by genetic analysis (see below and Table 1), whereas others, such as those for the tyrosine kinase FAK, the signaling protein tensin, the cytoskeletal protein filamin and the adaptor paxillin are in the process of characterization, Finally, *C. elegans* has a plectin homologue called VAB-10A and 20 tetraspanin homologues, among which only TSP-15 has been characterized so far, but they are unlikely to act in the same pathways as integrins.[15,16]

The αPAT-2/βPAT-3 Integrin at Muscle Attachment Sites: A Model for Focal Adhesion Assembly

Muscles and Their Attachments in C. elegans

The discovery of *pat-2* and *pat-3* was the result of genetic screens by Bob Waterston and his colleagues looking for genes required to assemble muscle myofilaments.[1,17] They initially realized that embryos lacking the vinculin homologue DEB-1 are paralyzed and arrest during embryonic morphogenesis at two-fold eggshell length, which prompted them to systematically collect mutations leading to a similar phenotype.[18] The mutations that they identified in this way were named *pat* based on their terminal phenotype characterized by a Paralysis at the two-fold stage (Fig. 2A).[1] Although the tissue distribution and molecular cloning of *pat-2* have not yet

Table 1. C. elegans integrins and proteins acting in the same pathway

Gene Name	Vertebrate Homologue	Phenotype[1,3]	Refs.
Integrins			
pat-3	β1-integrin	Pat, Mig, Com-def, Ste	1,3
pat-2	RGD-binding α-integrin	Pat, Ovu	1,19
ina-1	laminin-binding α-integrin	Mig, Com-def	2
Extracellular Matrix			
unc-52	*Perlecan*	*severe Pat*	13
emb-9	Type-IV α1-collagen	mild Pat	10
let-2	Type-IV α2-collagen	mild Pat	10
ost-1	SPARC/osteonectin	Ste	50
cle-1	Type XVIII collagen/Endostatin	Axon def, Emb	11
nid-1	Nidogen/entactin	Axon def	14
epi-1[2]	α-laminin chain	Mig, Ste	12
Cytoplasmic			
unc-112	Kindlin/kindlerin/Mig-2	*severe Pat*	23
pat-4	ILK	severe Pat	22
pat-6	Actopaxin/CH-ILKBP/parvin	severe Pat	25
unc-97	PINCH	severe Pat	24
deb-1	Vinculin	severe Pat	18
Y71G12B.11	Talin	Muscle def, Ovu, Ste, Mig	19
atn-1	α-actinin	*not known*	51
unc-95	LIM-only	Muscle def	30
unc-98	Zn-fingers (novel)	Muscle def.	31
rnf-5	Ubiquitin E3-ligase	Muscle def.	30
mig-15	NIK ser/thr kinase	Mig, Com-def	36

[1] All phenotypes are inferred from published mutations, except for *ost-1* and talin which have only been analyzed by RNA interference thus far. [2] There is a second laminin α-chain, LAM-3; LAM-3 mutations and integrin mutations result in different phenotypes. The phenotypes of β- and γ-chains have not yet been reported. [3] Abbreviations: Com: commissure; def: defect; Emb: embryonic lethal; Mig: migration defective; Ovu: ovulation defect; Pat: paralyzed at two-fold; Ste: sterile.

been formally reported, it is clear that *pat-2* encodes the alpha-integrin partner of PAT-3 involved in muscle assembly (Ben Williams, personal communication; see also 1,19).

In *C. elegans* as in other invertebrates, movement can occur upon muscle contractions only if muscles are attached to the cuticle exoskeleton. As outlined in Figure 3, myosin and actin filaments are independently anchored to the plasma membrane through structures called the M-line and the dense body, respectively (dense bodies are equivalent to vertebrate Z-discs, see also (Fig. 4A). Muscles are in turn linked to a basal lamina and ultimately to the cuticle through hemidesmosome-like anchoring structures called fibrous organelles, which span the thickness of the epidermis (Fig. 3C). The βPAT-3 integrin chain, and presumably αPAT-2, are found in muscles at the base of dense bodies and M-lines, as well as in plaques between adjacent muscles.[3]

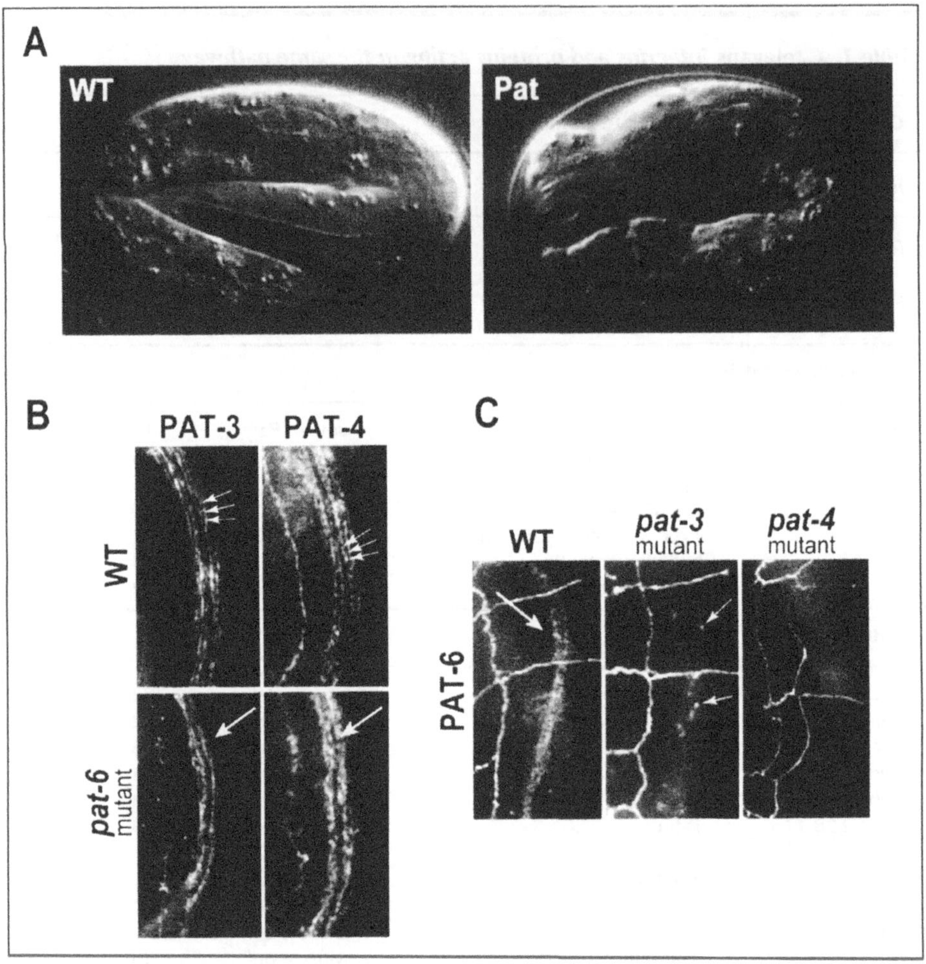

Figure 2. PAT-integrins and interacting proteins at the muscle. A) Illustration of the characteristic Pat phenotype. The wild-type embryo elongates up to 4-fold the length of the eggshell, while severe *pat* mutants arrest at two-fold egg length and are paralyzed (the picture on the right shows a *deb-1*/vinculin mutant). B) Mutations in *pat-6*/actopaxin does not prevent the initial assembly of integrin foci and the recruitment of PAT-4/ILK to those foci (compare distribution of βPAT-3 and PAT-4 in wild-type and *pat-6* embryos). However, the distribution is fuzzier and never evolves into dense bodies and M-lines (small arrows in wild-type, versus large arrow in mutant). C) Conversely, the recruitment of PAT-6/actopaxin to the plasma membrane (long arrow in the left wild-type picture) requires the presence of βPAT-3 and PAT-4/ILK. In *pat-3* or *pat-4* mutants, PAT-6 is barely detectable (small arrows) or has a non-polarized distribution. The thin contiguous lines in B and C correspond to adherens junctions used as positional landmarks. Pictures in B and C are a courtesy of Dr. Ben Williams,[25] with authorization from *Current Biology*.

Muscle attachments also comprise an essential part in the epidermis called fibrous organelles (FOs; see (Fig. 3A,C).[20] Although FOs are functionally and to some extent molecularly analogous to hemidesmosomes since the BPAG1e/plectin-related plakin VAB-10A represents a critical FO component, the main extracellular receptors within FOs are unrelated to integrins, in particular to the typical hemidesmosome β4-integrin.[16,21] Moreover, the distribution of fibrous organelles is not severely altered in *pat-3* mutants.[17]

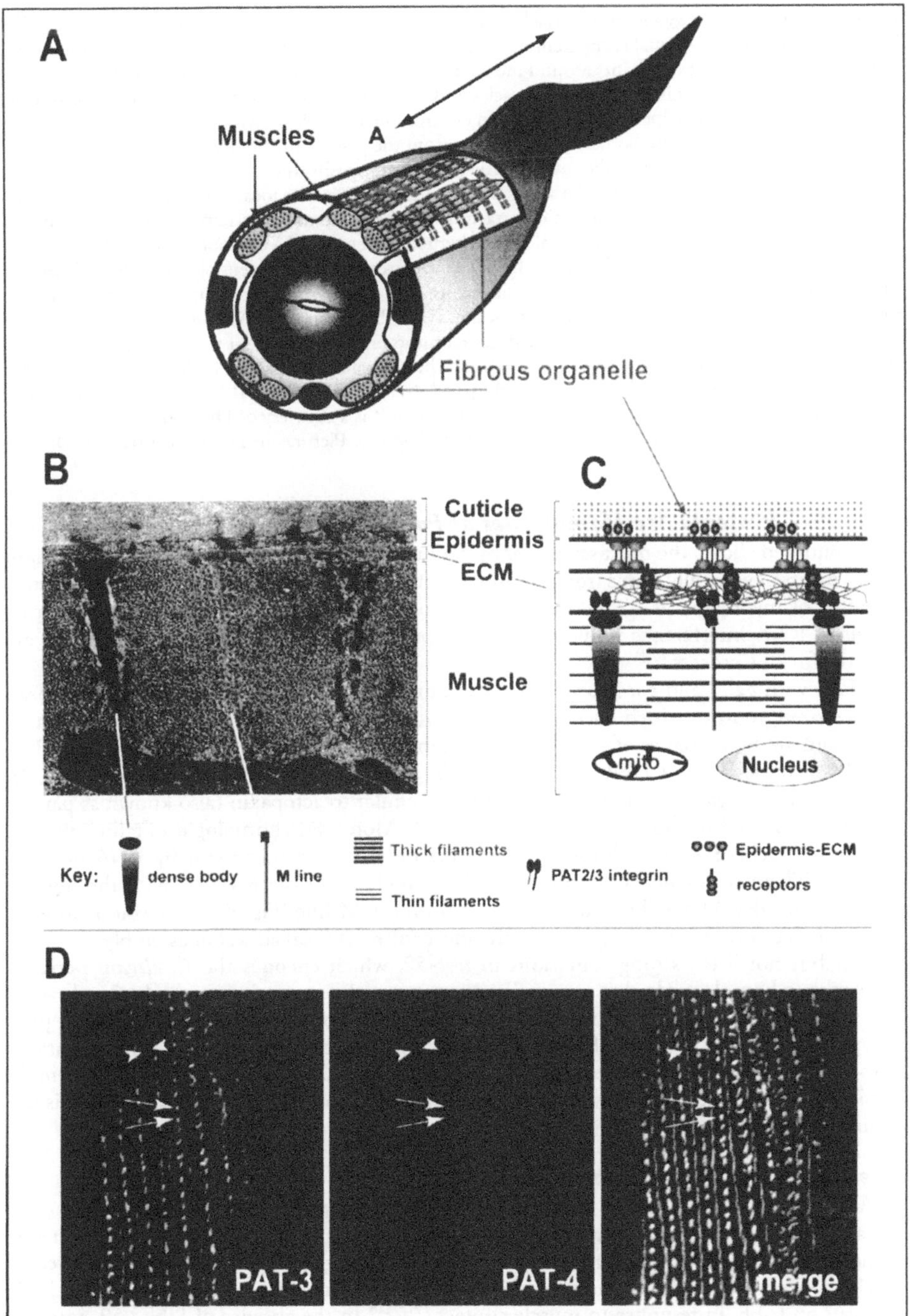

Figure 3. Please see figure legend on following page.

Figure 3. Anatomy of *C. elegans* muscle attachments. A) Cross-section through the *C. elegans* body wall. The internal anatomy of the animal is apparent with the intestine (center), epidermal cells (grey, external) and muscles arranged in four quadrants around the circumference of the animal. A small area of the cuticle is removed to show longitudinally oriented muscles and circumferentially oriented epidermal attachments (fibrous organelles; grey stripes). B) Transmission electron micrograph showing a cross-section through a sarcomere between two dense bodies, which project from the plasma membrane into the cell. Actin thin filaments emanate from dense bodies and interdigitate with myosin thick filaments which attach to the M-line. Note that muscle cells are polarized, such that myofilaments are located on the surface of the cell facing the epidermis. Due to muscle-cuticle attachments, the epidermis is extremely flattened (50 nm in larvae) and not much thicker than the basal lamina located in-between. C) Cartoon showing the respective arrangement of myofilaments and their anchoring complexes in muscles (dense bodies and M-lines; for further details about their molecular composition, see Fig. 4, and in the epidermis (fibrous organelles; therespective thickness of each layer is not to scale). Note that αPAT-2/βPAT-3 integrins are present in muscle cells but not in the epidermis, where other receptors ensure attachment with the basal lamina and the cuticle. (D) High magnification of a small area of the animal coexpressing a βPAT-3::YFP fusion, and a PAT-4/ILK::CFP fusion. As shown in the right panel, βPAT-3 and PAT-4/ILK extensively colocalize at dense bodies (arrows) and M-lines (arrowheads). Picture in B is a courtesy of Dr. Robby Weimer,[49] with authorization from *Journal of Histochemistry and Cytochemistry*. Pictures in D are a courtesy of Dr. Ben Williams,[25] with authorization from *Current Biology*.

Most PAT-Proteins Are Homologues of Focal Adhesion Components

As outlined above, the most severe *pat* mutations cause paralysis of the embryo. The origin of this paralysis generally stems from the absence of dense bodies and M-lines and the defective assembly of myofilaments.[1,17] Likewise, depletion of integrins during larval development by means of RNA interference induces egg-laying problems, as vulval and uterine muscles are essential for this process.[8,19]

Along with *pat-2* and *pat-3*, six more genes causing a severe Pat phenotype have now been characterized, most of which encode proteins that colocalize with integrins at muscle attachments and are homologues of vertebrate focal adhesion components. In addition to DEB-1/vinculin mentioned above, UNC-112 corresponds to kindlin (also known as kindlerin or Mig-2), PAT-4 is the *C. elegans* ILK homologue, PAT-6 is similar to actopaxin (also known as parvin, CH-ILKBP or affixin) and UNC-97 to PINCH.[22-25] Moreover, a homologue of talin is present at muscle attachment sites, although its role has only been characterized by RNA interference.[19,26] These components contribute to dense bodies and the M-lines, with the notable exception of DEB-1/vinculin, which is absent from the M-line (Fig. 4A). An examination of *pat*-mutants can thus bring insights into the mechanism of focal adhesion assembly.

Last but not least, strong mutations in *unc-52*, which encodes the *C. elegans* perlecan proteoglycan homologue lead to a severe Pat phenotype, while mutations in the *let-2* and *emb-9* genes encoding the type-IV alpha1- and alpha2-collagen homologues cause a milder Pat phenotype characterized by a loss of adhesion between muscle cells.[1,10,13] In addition, mutations in the α-laminin chain EPI-1 reduces the number of dense bodies and induces their assembly outside of the muscle-epidermis contact area.[12] It is not known which integrin, if any, is the laminin receptor in muscles.

The Assembly of Muscle Attachments

Analysis of the distribution of PAT-proteins and muscle components in the various *pat* mutants have suggested that sarcomeres in *C. elegans* are built up from the basement membrane inwards (Fig. 4B). During *C. elegans* embryogenesis, gastrulation brings muscle precursors in close contact to dorsal and ventral epidermal cells, which express and secrete UNC-52/perlecan probably in response to muscle contact.[17,27,28] In the absence of UNC-52, integrin foci and muscle attachments fail to assemble.[1,13] Conversely, in the absence of αPAT-2, βPAT-3 or any other dense body/M-line component, UNC-52/perlecan distribution is not affected.[17,22,23,25]

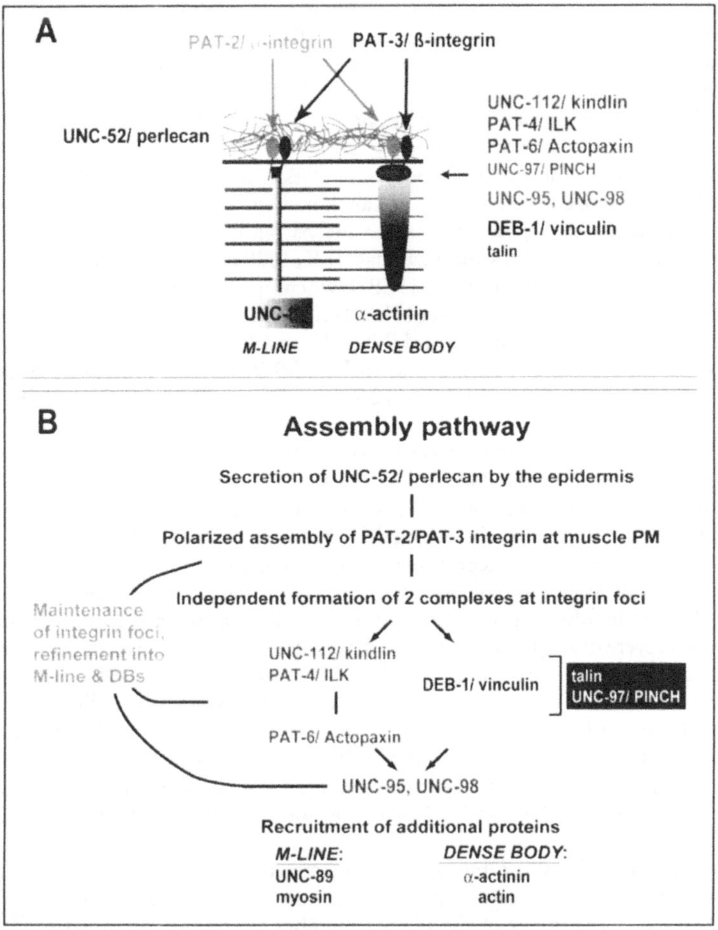

Figure 4. Structure and assembly pathway of dense bodies and M-lines. A) Dense bodies and M-lines contain the same components, except that DEB-1/vinculin and α-actinin are only present in dense bodies, while UNC-89 (a giant titin-like protein) is only present in M-lines. Talin is predominantly found in dense bodies but is also present at lower levels in M-lines. B) Genetic analysis suggests the order of assembly depicted in the drawing (based on ref. 25). The assembly of dense bodies and M-lines starts with deposition of UNC-52/perlecan by the epidermis and progresses inwardly within muscle cells with recruitment of αPAT-2/βPAT-3 integrin to the plasma membrane and then splits into two parallel and partially independent branches. One branch continues with UNC-112/kindlin, PAT-4/ILK and PAT-6/actopaxin, the other branch depends on DEB-1/vinculin. The precise position of UNC-97/PINCH in the assembly process is not known; the strong Pat phenotype observed after RNA interference against *unc-97* and the observation that UNC-97 interacts with PAT-4 argue that UNC-97 might belong to the UNC-112/PAT-4/PAT-6 branch, however in *Drosophila* PINCH is recruited to integrin foci independently of ILK.[33] Likewise, besides the observation that talin recruitment to dense bodies depends on βPAT-3 but not on DEB-1/vinculin,[26] its precise position within the assembly pathway is unclear, in part because the null phenotype of talin defective embryos has not yet been described. The prediction would be that talin acts acts upstream of DEB-1, within the other complex or to activate integrins. Both complexes, together with additional proteins such as the LIM-only protein UNC-95 and the Zn-finger protein UNC-98, are required to maintain muscle attachments. DEB-1 is not required for the formation of the M-line and the further recruitment of UNC-89.

Downstream of integrins, the polarized accumulation of other dense body and M-line components at the plasma membrane is prevented in β*pat-3* mutants. Reciprocally, dense body components are not required for the initial assembly of integrin foci, but they are essential to allow integrin foci to mature into dense bodies and M-lines.[17,18,22,23,25] For instance, the distribution of βPAT-3 integrin deteriorates in *pat-4* and *unc-112* mutants (see Fig. 2B), or after RNA interference against the gene encoding talin.[19,22,23,25]

Interestingly, the assembly of dense bodies involves two distinct complexes, which appear to act in parallel to polarize myofilaments and anchor them at the plasma membrane facing the epidermis (Fig. 4B). On the basis of genetic analysis and yeast two-hybrid interactions, one complex comprises UNC-112/kindlin, PAT-4/ILK, PAT-6/actopaxin, which is recruited last to this complex, while the other involves DEB-1/vinculin.[22,25] In particular, mutations in *pat-4*, *unc-112* or *pat-6* do not strongly affect DEB-1/vinculin distribution, while *deb-1* mutations do not affect the distribution of PAT-4, UNC-112 and PAT-6.

As in vertebrates, PAT-4/ILK can bind to UNC-97/PINCH through its N-terminal domain and to PAT-6/actopaxin and UNC-112/Mig-2 through its kinase domain, but unlike the situation in vertebrates PAT-4/ILK does not interact with the cytoplasmic domain of βPAT-3.[22,25] As in *Drosophila*, the kinase domain of PAT-4/ILK might be nonfunctional, since a mutation in a residue conserved among most kinases does not affect the potential of a PAT-4 transgene to rescue *pat-4* mutations.[22,29] Instead, the kinase domain of PAT-4/ILK must provide a scaffold for protein-protein interaction. How the UNC-112/PAT-4/UNC-97/PAT-6 complex becomes tethered to integrin foci remains to be discovered.

At least two additional proteins, the LIM-domain protein UNC-95 and the Zn-finger protein UNC-98, are involved in the biogenesis of muscle attachments. Mutations in *unc-95* or *unc-98* cause a severe disorganization of myofilaments and compromise the structure of attachments, but do not cause a Pat phenotype.[30,31] In *unc-95* mutants, DEB-1/vinculin is barely polarized indicating that this LIM-only protein distantly related to paxillin might act in conjunction with other LIM-domain proteins to promote the assembly of DEB-1 at dense bodies.[30] Interestingly, the steady-state levels of UNC-95 are limited by the E3- ubiquitin ligase RNF-5, which interacts in yeast and partially colocalizes at dense bodies with UNC-95. Excess or lack of RNF-5 alters UNC-95 levels with the consequent perturbation of muscle attachments.[30]

Interestingly, mutations in *Drosophila* that inactivate PS-integrins or homologues of focal adhesion components also result in muscle attachment defects, and demonstrate that ILK, talin and PINCH recruitment to the plasma membrane depends on integrins.[29,32-34] As the anatomy of fly and worm muscles are different, the defects observed in *Drosophila* and *C. elegans* are also slightly different. Notably, fly integrins can be recruited to the plasma membrane independently of their ECM ligand, and integrin mutations result in more severe muscle defects than mutations in the downstream component ILK.[29,32]

In summary, the analysis of αPAT-2/βPAT-3 in muscles and their interactors has brought the identification of new proteins, and allowed to order them in an assembly pathway. In addition, genetic studies in *C. elegans* and *Drosophila* clarified some conflicting data emerging from in vitro studies in vertebrates. Finally, it has been debated whether the extracellular matrix or its receptors has come first in evolution.[35] Since in *C. elegans* dense bodies are assembled from the ECM inward, it indicates that the ECM can come first, although input from proteins interacting with integrins and from the cytoskeleton is clearly critical for final patterning and the maintenance of integrin adhesion complexes.

The αINA-1/βPAT-3 Integrin in Cell and Axon Migration

αINA-1/PAT-3 Is Required for Cell Migration

The discovery of the other α-integrin chain *ina-1* was the result of genetic screens by G. Garriga and collaborators, who were looking for mutations affecting long range neuronal cell migrations.[2] While α*ina-1* null mutants die soon after hatching with a notched head phenotype, other alleles are viable and revealed a requirement for αINA-1 in cell or axon migration. Similarly, although strong β*pat-3* mutants die too early to examine cell migration, the existence of a weak allele or partial inactivation by RNA interference proved that βPAT-3 is also required for cell and axon migration.[6,8,19,36]

As depicted in Figure 5, the neurons CAN, ALM and HSN born during embryogenesis (Fig. 5A) and post-embryonic neuroblast descendants of QR and QL (not illustrated) migrate a long distance away from their initial birth positions. In partial loss-of-function α*ina-1* mutants those cells migrate poorly or not at all.[2] In addition, the distal tip cells of the gonad (dtc) adopt aberrant migration patterns, thereby affecting gonad morphogenesis.[2] Similar dtc migration and gonad morphogenesis defects are caused by RNA interference during larval development against β*pat-3* or the gene encoding talin, or by expressing a truncated βPAT-3 chain in which its extracellular domain is replaced by an heterologous hemagglutinin (HA) tag that presumably interferes with the activity of endogenous βpat-3 (Fig. 5B).[8,19]

αINA-1/βPAT-3 in Axon Fasciculation and Axon Outgrowth

In addition to cell migration, the αINA-1/βPAT-3 integrin is required for bundling of individual axons in the ventral cord of larvae, and for the proper outgrowth of commissures that ventral neurons extend towards the dorsal midline. In weak α*ina-1* and β*pat-3* mutants, or after RNA interference against β*pat-3*, ventral cord axons fail to bundle together into a fascicle, and commissures stop prematurely when they meet an obstacle (Fig. 6B).[2,6,36] However, αINA-1 appears dispensable for axon extension along the anterior-posterior axis.[2]

Both αINA-1 and βPAT-3 are expressed in cells that are affected by *ina-1* mutations, and furthermore genetic mosaic experiments prove that *ina-1* acts cell-autonomously in migrating neurons.[2,3] The extracellular matrix ligand recognized by αINA-1/βPAT-3 during cell and axon migration is likely to include laminins, since mutants for the α laminin chain EPI-1 share many features of αINA-1 mutants, including the neuronal migration, commissural and some gonad defects.[12,37,38] Mutations inactivating type XVIII collagen CLE-1 also induce phenotypes quite similar to those observed in αINA-1/βPAT-3 defective animals.[11] Besides talin, which as outlined above is certainly required for distal tip cell migration, many additional proteins are known to be required for cell migration and/or axon outgrowth.[39,40] However, none of them has been rigorously proven to act in the same pathway as αINA-1/βPAT-3, except, as further discussed below, the NIK protein kinase homologue MIG-15, and possibly the Rac GTPases CED-10 and MIG-2.[36]

Yeast two-hybrid experiments identified the C-terminal domain of NIK as a partner of the cytoplasmic domain of mouse β1-integrin and established that the *C. elegans* homologues βPAT-3 and MIG-15 can also interact.[36] Genetic experiments in *C. elegans* validated the biological significance of this interaction, since *mig-15* mutations show premature commissure outgrowth arrest as α*ina-1* mutations and synergize with them.[36] The physical interaction between MIG-15 and βPAT-*3* and the genetic synergistic interactions between *mig-15* and α*ina-1* strongly suggest that MIG-15 and αINA-1/βPAT-3 act in the same pathway. It remains unclear whether MIG-15 acts upstream of αINA-1/βPAT-3 to activate this integrin, or as a downstream effector.

In *Drosophila*, disruption of the MIG-15 orthologue misshapen or of the βPS-integrin leads to axonal targeting defects.[41-43] In that case, no evidence for interaction between βPS and misshapen has been brought as yet. Interestingly, misshapen appears to act upstream of the kinase JNK.[44] It remains to be seen whether this is also the case in *C. elegans*.

αINA-1/βPAT-3 in Epithelia

In vertebrates and *Drosophila*, integrins play a major role in epithelial morphogenesis.[34,45-48] In *C. elegans*, although much remains to be done to understand their cellular function, it is clear that integrins are also important in epithelia. Mutations in *epi-1* or α*ina-1*, and RNA

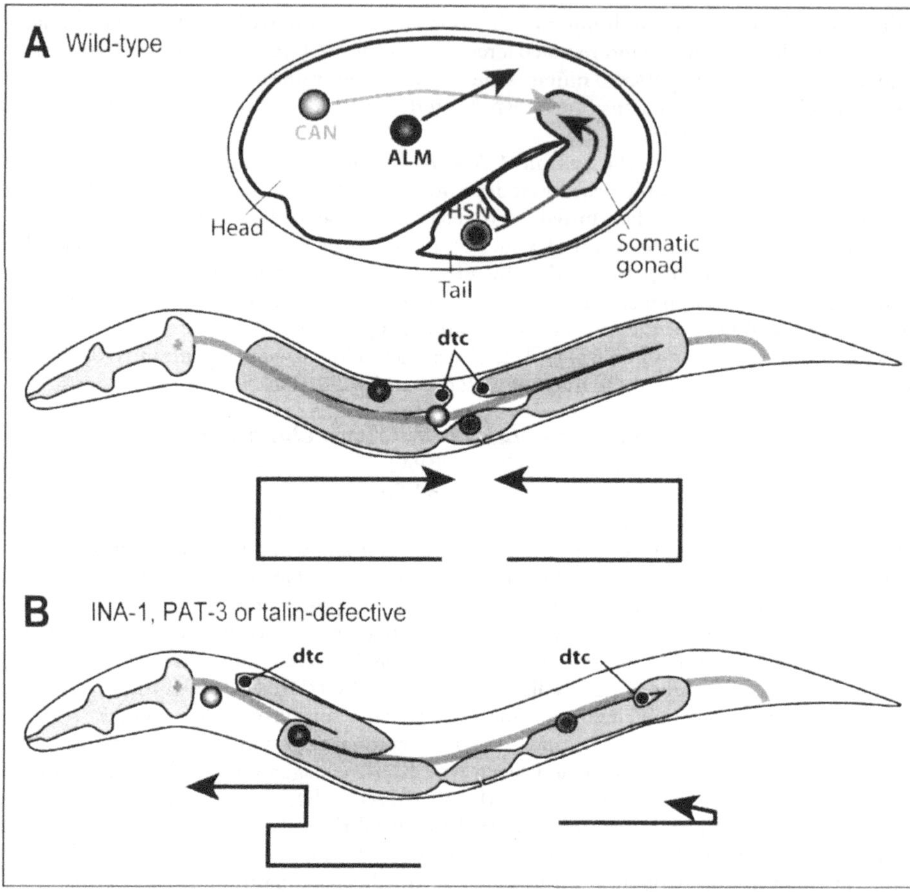

Figure 5. Cell migration defects in αINA-1/βPAT-3 and talin defective animals. A) The neurons called CAN, ALM and HSN (one on each side of the animal) are born in the head, the anterior part of the body and the tail of the embryo, respectively; they migrate towards the gonad (their final positions are shown in the image of the adult below; gonad, grey bag). Additional neuroblasts require integrin activity for their migration, but are not represented for the sake of clarity. Morphogenesis of the somatic gonad starting from the primordium comprising only two cells in the embryo to the complex gonad observed in the adult depends on the migration of the two distal tip cells (dtc; the embryo is not drawn to scale with the adult). The dtcs follow the path symbolized by the two arrows below the adult, starting migration at the end of the second larval stage. B) In α*ina-1* mutants, the migration of CAN, ALM and HSN neurons is defective, causing them to stop prematurely (the migration behavior of these neurons in βPAT-3 or talin defective animals has not yet been examined). The migration path of dtcs in viable α*ina-1* mutants, in βPAT-3 or in talin defective animals is also abnormal. In this hypothetical example, the anterior dtc made an extra anterior turn, while the posterior dtc failed to migrate dorsally and stopped prematurely.[19]

interference against talin or β*pat-3* cause variable ovulation defects and uterus organization problems.[8,12,19] Those defects, as well as the notched head phenotype of α*ina-1* null mutants indicate that αINA-1/βPAT-3, laminins and talin might promote epithelial morphogenesis in some tissues. In talin-defective animals, it is known that the actin cytoskeleton is disorganized in the myoepithelial sheath cells of the somatic gonad that control ovulation in response to signals from the oocyte and sperm.[19]

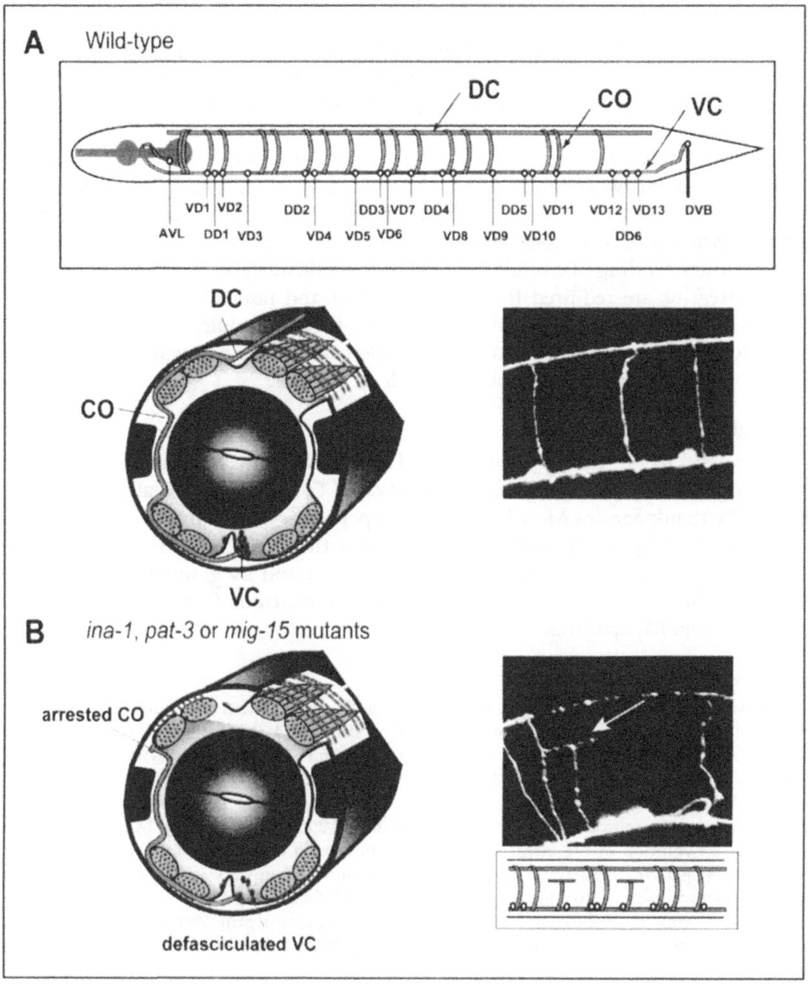

Figure 6. Axonal defects in αINA-1/βPAT-3 and MIG-15 defective animals. A) Anatomy of the entire GABAergic nervous system. Neuron cell bodies are located in the ventral cord (VC). GABAergic neurons in the ventral cord send a short projection anteriorly, which branches to form a commissure (CO) that reaches the dorsal cord (DC) where it further branches to make a short anterior-posterior extension. Most commissures travel along the right side of the animal on a basal lamina deposited by the epidermis and/or the intestine, as further illustrated in the lower cartoon on the left. On their way to the dorsal cord, commissures meet obstacles; in particular, they have to navigate between muscles and the epidermis where muscles make attachments with the epidermis (see Fig. 2). In the ventral cord, most axons form a bundle on the left side of the epidermal ridge, while three axons bundle together on the right side. As illustrated in the right picture, GABAergic axons and cell bodies can be visualized with a GFP expressed under the control of a GABAergic-specific promoter (only a small area of an adult body is shown). B) In viable α*ina-1*, β*pat-3* or *mig-15* mutants, commissures often fail to reach the dorsal cord and stop when they meet an obstacle (such as a muscle), in which case they extend a short anterior-posterior branch as they normally do in the dorsal cord. In addition, ventral cord axons form distinct bundles instead of a unique fascicle and commissures navigate more frequently along the left side of the animal. A typical premature arrest in a *mig-15* defective animal, as visualized with the GFP expressed in GABAergic neurons, is illustrated in the right image (arrow) and further symbolized in the cartoon below. With authorization from *Current Biology* for pictures.[36]

Thus, αINA-1/βPAT-3 appears required for most cell migrations, for axon navigation along the dorso/ventral axis, for axon fasciculation and for some aspects of epithelial morphogenesis. Since these processes involve the localized reorganization of the cytoskeleton, it suggests more generally a role for αINA-1/βPAT-3 in cell polarity events leading to modulation of the cytoskeleton.

In conclusion, with its set of three chains representing both the RGD-binding and the laminin-binding integrins, and the existence of most classical integrin-associated molecules, *C. elegans* represents a model of choice to dissect integrin related pathways. Most notably, it has brought the discovery of some focal adhesion proteins, a dissection of focal adhesion assembly and a better understanding of muscle biogenesis. Genetic analysis in *C. elegans* has also established that integrins are required for axon migration and fasciculation. Parallel work in *C. elegans* and *Drosophila* should keep illuminating the basic principles of integrin function and signaling. Indeed, the collection of mutations in elements of the pathways as well as the RNAi approach are powerful and unique tools for such an analysis.

Acknowledgements

We are grateful to Dr. Robby Weimer and Dr. Ben Williams for providing some pictures, *Current Biology* and the *Journal of Histochemistry and Cytochemistry* for allowing us to reproduce them. We thank Sandra Metz for help in preparing figures. Work in the authors' laboratories is supported by institutional funding from the CNRS, INSERM and Hôpitaux Universitaires de Strasbourg the European Community, and by grants from the Association pour la Recherche sur le Cancer, the Ligue Régionale contre le Cancer and the Association Française contre les Myopathies.

References

1. Williams BD, Waterston RH. Genes critical for muscle development and function in Caenorhabditis elegans identified through lethal mutations. J Cell Biol 1994; 124(4):475-490.
2. Baum PD, Garriga G. Neuronal migrations and axon fasciculation are disrupted in ina-1 integrin mutants. Neuron 1997; 19(1):51-62.
3. Gettner SN, Kenyon C, Reichardt LF. Characterization of beta pat-3 heterodimers, a family of essential integrin receptors in C. elegans. J Cell Biol 1995; 129(4):1127-1141.
4. Hughes AL. Evolution of the integrin alpha and beta protein families. J Mol Evol 2001; 52(1):63-72.
5. Hutter H, Vogel BE, Plenefisch JD et al. Conservation and novelty in the evolution of cell adhesion and extracellular matrix genes. Science 2000; 287(5455):989-994.
6. Huang X, Huang P, Robinson MK et al. UNC-71, a disintegrin and metalloprotease (ADAM) protein, regulates motor axon guidance and sex myoblast migration in C. elegans. Development 2003; 130(14):3147-3161.
7. Hynes RO. Integrins: Bidirectional, allosteric signaling machines. Cell 2002; 110(6):673-687.
8. Lee M, Cram EJ, Shen B et al. Roles for beta(pat-3) integrins in development and function of Caenorhabditis elegans muscles and gonads. J Biol Chem 2001; 276(39):36404-36410.
9. Johnstone IL. Cuticle collagen genes. Expression in Caenorhabditis elegans. Trends Genet 2000; 16(1):21-27.
10. Gupta MC, Graham PL, Kramer JM. Characterization of alpha1(IV) collagen mutations in Caenorhabditis elegans and the effects of alpha1 and alpha2(IV) mutations on type IV collagen distribution. J Cell Biol 1997; 137(5):1185-1196.
11. Ackley BD, Crew JR, Elamaa H et al. The NC1/endostatin domain of Caenorhabditis elegans type XVIII collagen affects cell migration and axon guidance. J Cell Biol 2001; 152(6):1219-1232.
12. Huang CC, Hall DH, Hedgecock EM et al. Laminin alpha subunits and their role in C. elegans development. Development 2003; 130(14):3343-3358.
13. Rogalski TM, Williams BD, Mullen GP et al. Products of the unc-52 gene in Caenorhabditis elegans are homologous to the core protein of the mammalian basement membrane heparan sulfate proteoglycan. Genes Dev 1993; 7(8):1471-1484.
14. Kim S, Wadsworth WG. Positioning of longitudinal nerves in C. elegans by nidogen. Science 2000; 288(5463):150-154.
15. Moribe H, Yochem J, Yamada H et al. Tetraspanin protein (TSP-15) is required for epidermal integrity in Caenorhabditis elegans. J Cell Sci 2004; 117(Pt 22):5209-5220.

16. Bosher JM, Hahn BS, Legouis R et al. The Caenorhabditis elegans vab-10 spectraplakin isoforms protect the epidermis against internal and external forces. J Cell Biol 2003; 161(4):757-768.
17. Hresko MC, Williams BD, Waterston RH. Assembly of body wall muscle and muscle cell attachment structures in Caenorhabditis elegans. J Cell Biol 1994; 124(4):491-506.
18. Barstead RJ, Waterston RH. Vinculin is essential for muscle function in the nematode. J Cell Biol 1991; 114(4):715-724.
19. Cram EJ, Clark SG, Schwarzbauer JE. Talin loss-of-function uncovers roles in cell contractility and migration in C. elegans. J Cell Sci 2003; 116(Pt 19):3871-3878.
20. Francis R, Waterston RH. Muscle cell attachment in Caenorhabditis elegans. J Cell Biol 1991; 114(3):465-479.
21. Hahn BS, Labouesse M. Tissue integrity: Hemidesmosomes and resistance to stress. Curr Biol 2001; 11(21):R858-861.
22. Mackinnon AC, Qadota H, Norman KR et al. C. elegans PAT-4/ILK functions as an adaptor protein within integrin adhesion complexes. Curr Biol 2002; 12(10):787-797.
23. Rogalski TM, Mullen GP, Gilbert MM et al. The UNC-112 gene in Caenorhabditis elegans encodes a novel component of cell-matrix adhesion structures required for integrin localization in the muscle cell membrane. J Cell Biol 2000; 150(1):253-264.
24. Hobert O, Moerman DG, Clark KA et al. A conserved LIM protein that affects muscular adherens junction integrity and mechanosensory function in Caenorhabditis elegans. J Cell Biol 1999; 144(1):45-57.
25. Lin X, Qadota H, Moerman DG et al. C. elegans PAT-6/actopaxin plays a critical role in the assembly of integrin adhesion complexes in vivo. Curr Biol 2003; in press.
26. Moulder GL, Huang MM, Waterston RH et al. Talin requires beta-integrin, but not vinculin, for its assembly into focal adhesion-like structures in the nematode Caenorhabditis elegans. Mol Biol Cell 1996; 7(8):1181-1193.
27. Hresko MC, Schriefer LA, Shrimankar P et al. Myotactin, a novel hypodermal protein involved in muscle-cell adhesion in Caenorhabditis elegans. Journal of Cell Biology 1999; 146(3):659-672.
28. Spike CA, Davies AG, Shaw JE et al. MEC-8 regulates alternative splicing of unc-52 transcripts in C. elegans hypodermal cells. Development 2002; 129(21):4999-5008.
29. Zervas CG, Gregory SL, Brown NH. Drosophila integrin-linked kinase is required at sites of integrin adhesion to link the cytoskeleton to the plasma membrane. J Cell Biol 2001; 152(5):1007-1018.
30. Broday L, Kolotuev I, Didier C et al. The LIM domain protein UNC-95 is required for the assembly of muscle attachment structures and is regulated by the RING finger protein RNF-5 in C. elegans. J Cell Biol 2004; 165(6):857-867.
31. Mercer KB, Flaherty DB, Miller RK et al. Caenorhabditis elegans UNC-98, a C2H2 Zn finger protein, is a novel partner of UNC-97/PINCH in muscle adhesion complexes. Mol Biol Cell 2003; 14(6):2492-2507.
32. Bokel C, Brown NH. Integrins in development: Moving on, responding to, and sticking to the extracellular matrix. Dev Cell 2002; 3(3):311-321.
33. Clark KA, McGrail M, Beckerle MC. Analysis of PINCH function in Drosophila demonstrates its requirement in integrin-dependent cellular processes. Development 2003; 130(12):2611-2621.
34. Brown NH, Gregory SL, Rickoll WL et al. Talin is essential for integrin function in Drosophila. Dev Cell 2002; 3(4):569-579.
35. Brown NH. An integrin chicken and egg problem: Which comes first, the extracellular matrix or the cytoskeleton? Curr Opin Cell Biol 2000; 12(5):629-633.
36. Poinat P, De Arcangelis A, Sookhareea S et al. A conserved interaction between beta1 integrin/PAT-3 and Nck-interacting kinase/MIG-15 that mediates commissural axon navigation in C. elegans. Curr Biol 2002; 12(8):622-631.
37. Forrester WC, Perens E, Zallen JA et al. Identification of Caenorhabditis elegans genes required for neuronal differentiation and migration. Genetics 1998; 148(1):151-165.
38. Forrester WC, Garriga G. Genes necessary for C. elegans cell and growth cone migrations. Development 1997; 124(9):1831-1843.
39. Lundquist EA, Reddien PW, Hartwieg E et al. Three C. elegans Rac proteins and several alternative Rac regulators control axon guidance, cell migration and apoptotic cell phagocytosis. Development 2001; 128(22):4475-4488.
40. Merz DC, Culotti JG. Genetic analysis of growth cone migrations in Caenorhabditis elegans. J Neurobiol 2000; 44(2):281-288.
41. Su YC, Maurel-Zaffran C, Treisman JE et al. The Ste20 kinase misshapen regulates both photoreceptor axon targeting and dorsal closure, acting downstream of distinct signals. Mol Cell Biol 2000; 20(13):4736-4744.

42. Ruan W, Pang P, Rao Y. The SH2/SH3 adaptor protein dock interacts with the Ste20-like kinase misshapen in controlling growth cone motility. Neuron 1999; 24(3):595-605.
43. Hoang B, Chiba A. Genetic analysis on the role of integrin during axon guidance in Drosophila. J Neurosci 1998; 18(19):7847-7855.
44. Su YC, Treisman JE, Skolnik EY. The Drosophila Ste20-related kinase misshapen is required for embryonic dorsal closure and acts through a JNK MAPK module on an evolutionarily conserved signaling pathway. Genes Dev 1998; 12(15):2371-2380.
45. Devenport D, Brown NH. Morphogenesis in the absence of integrins: Mutation of both Drosophila {beta} subunits prevents midgut migration. Development 2004; 131(21):5405-5415.
46. Torgler CN, Narasimha M, Knox AL et al. Tensin stabilizes integrin adhesive contacts in Drosophila. Dev Cell 2004; 6(3):357-369.
47. Narasimha M, Brown NH. Novel functions for integrins in epithelial morphogenesis. Curr Biol 2004; 14(5):381-385.
48. De Arcangelis A, Georges-Labouesse E. Integrin and ECM functions: Roles in vertebrate development. Trends Genet 2000; 16(9):389-395.
49. Rostaing P, Weimer RM, Jorgensen EM et al. Preservation of immunoreactivity and fine structure of adult C. elegans tissues using high-pressure freezing. J Histochem Cytochem 2004; 52(1):1-12.
50. Fitzgerald MC, Schwarzbauer JE. Importance of the basement membrane protein SPARC for viability and fertility in Caenorhabditis elegans. Curr Biol 1998; 8(23):1285-1288.
51. Barstead RJ, Kleiman L, Waterston RH. Cloning, sequencing, and mapping of an alpha-actinin gene from the nematode Caenorhabditis elegans. Cell Motil Cytoskeleton 1991; 20(1):69-78.

CHAPTER 6

Integrins and Associated Proteins in *Drosophila* Development

Maithreyi Narasimha and Nicholas H. Brown*

Abstract

We discuss insights into integrin function obtained through phenotypic and structure function analysis of integrins and associated proteins in the fruitfly, *Drosophila* melanogaster. Integrins contribute to fly development by mediating both dynamic and stable adhesive interactions. The minimal redundancy of proteins involved and the ease with which they can be manipulated in the context of the whole organism and studied at the cellular level make it an ideal model organism to study the "core" complexes that mediate diverse integrin functions. Such an analysis has revealed a hierarchy of molecular interactions that mediate stable adhesion between two layers of cells through an intervening extracellular matrix.

Introduction

The fruitfly *Drosophila melanogaster* has proved to be a valuable model organism for the study of integrin biology. Apart from the obvious advantage of being genetically tractable, integrin biology has benefited from the fact that integrins and their associated proteins are well conserved in the fly and encoded by smaller numbers of genes. This makes the analysis of function of a particular type of protein (by removing it from an otherwise normal animal) much easier in the fly than in vertebrate organisms. Thus, it is possible to test what a particular protein is actually required for in the intact organism, which sometimes differs significantly from what it can do when perturbed or overexpressed in cells in culture.

The wealth of knowledge on the mechanisms of development in *Drosophila* has made it possible to put integrin function within the framework of developmental decisions and events. Of course the relatively simple body plan of *Drosophila* precludes the analysis of some integrin dependant functions that are of great interest in humans and take place in specialized cell types that flies lack (for example, lymphocyte activation, platelet aggregation and neural crest migration). Nonetheless, many different cell types in *Drosophila*, including haemocytes (the blood cells of the fly), are migratory and some of these depend on integrins for their migration. We view *Drosophila* as a means to identify a "core" integrin complex that was present in the metazoan ancestor that gave rise to both flies and humans. By identifying this core complex, we anticipate that we will understand how the basic integrin machinery has become modified during evolution to perform an increased variety of functions.

Here we highlight the insights into integrin function obtained through studies in *Drosophila*. The reader is also referred to previous reviews.[1-3] In our view, we have learned two main

*Corresponding Author: Nicholas H. Brown—The Gurdon Institute and Department of Anatomy, University of Cambridge, Tennis Court Road, Cambridge CB2 1QN, U.K. Email: n.brown@gurdon.cam.ac.uk

Integrins and Development, edited by Erik H.J. Danen. ©2006 Landes Bioscience.

Table 1. Drosophila integrin genes and their mutant phenotypes

Protein/Gene	Features	Phenotype	Function
Integrin βPS *myospheroid*	β integrin	Embryonic lethal, with muscle detachment, dorsal holes, delayed gut migration, blisters in the wing	Major β integrin subunit in heterodimers with αPS1,2,3
Integrin βυ *βυ*	β integrin	Viable, fertile, blocks gut migration in the additional absence of βPS	Dispensible and divergent β subunit in heterodimers with αPS3
Integrin αPS1 *multiple edematous wings*	α integrin	Larval lethal with gut defects, blisters in the wing	In heterodimers with βPS
Integrin αPS2 *inflated*	α integrin	Embryonic lethal with muscle detachment, delayed gut migration, blisters in the wing	In heterodimers with βPS
Integrin αPS3 *scab/volado*	α integrin	Embryonic lethal, with dorsal holes, delayed gut migration. Mutant may also affect αPS4	In heterodimers with βPS and βυ
Integrin αPS4 *scab*	α integrin	Unclear whether some *scab* alleles affect this subunit too	?
Integrin αPS5	α integrin	No mutant	?

lessons. The first is that some integrin-associated molecules are more important than others and that there is a molecular hierarchy among integrin associated proteins. The second is that almost all integrin dependent processes during development involve two groups of cells that interact via integrins and an intervening extracellular matrix (ECM). Thus, rather than thinking of integrins as mediating cell adhesion to a substrate, it is more accurate to think of them as mediating cell-ECM-cell interactions.

Integrins and Associated Proteins: Simplicity Is a Good Thing

The completion of the fly genome sequence allowed a clear tabulation of how many integrin subunits flies have, and how many of the integrin-associated proteins identified in vertebrates are conserved in *Drosophila*[4] (Tables 1-6). The fly has 5 α integrin subunits (αPS1-5) and 2 β integrin subunits (βPS and βν [beta-nu]). The PS is historical, referring to the fact that the distribution of integrins was found to be position specific rather than cell type specific.[5] It remains a useful part of the name, since for example αPS1 is not closely related to vertebrate α1, but instead is equally similar to α3, α6 and α7. In almost all cases there is a single orthologue in flies of each cytoplasmic integrin-associated protein, compared to up to 4 closely related proteins in vertebrates. The ECM proteins that make basement membranes are well conserved, but many other ECM proteins found in vertebrates, notably fibrillar collagens and fibronectin, are not present in the fly.[4] Flies also lack cytoplasmic intermediate filaments and the specialised integrin subunit that connects with them, β4. As *C. elegans* has both fibrillar collagens and cytoplasmic intermediate filaments (although not a β4 integrin subunit), these appear to have been lost in the fly, rather than not present in the common ancestor of flies and humans.

Table 2. Integrin associated proteins whose loss of function phenotypes includes defects in integrin-mediated adhesion in muscles or the wings

Protein/Gene	Features	Phenotype	Function
Laminin α1,2 *wing blister*	EGF, Laminin repeats	Lethal, germ band retraction defects, hypomorphic alleles have wing blisters	Ligand for αPS2βPS, αPS2βPS heterodimer
Tiggrin *tiggrin*	Novel	Larval lethal, weak muscle detachment	Ligand for PS2 heterodimer
Talin *rhea*	FERM, actin binding domains	Embryonic lethal, muscle detachment, dorsal holes, blisters in the wing	Cytoskeletal linker binds integrin and actin
PINCH *steamer duck*	LIM domains	Embryonic lethal, actin detachment from the muscle membrane, blisters in wing clones	Scaffold protein
Integrin linked kinase *ilk*	Atypical(non) kinase, ankyrin repeats	Embryonic lethal, actin detachment from the muscle membrane, blisters in wing clones	Scaffold protein
Short stop *short stop/ kakapo*	CH, plakin rpts, spectrin rpts, GAS2	Axons fail to reach target, dendritic branching defects, tendon cells rip, tracheal defects, oogenesis defects, blisters in wing clones	Cytoskeletal linker as actin and microtublule binding protein
Tensin *blistery*	SH2, PTB	Wing blisters	Actin binding protein
Trio *trio*	Dbl, PH domains, spectrin repeats	Muscle detachment, Axon guidance	RhoGEF

Genetics of Integrins and Integrin-Associated Proteins

Mutations have been isolated within many but not all of the genes encoding *Drosophila* integrins and their associated proteins. There are three sources of these mutations.

The first source is from mutations that were identified because they caused a phenotype of interest and were kept within the fly community. Subsequent molecular characterisation of these mutant genes revealed that some encode integrins or integrin-associated proteins. For example, the first reported allele in an integrin gene was described in 1918 as a mutation that caused blisters in the wings.[6] The gene was therefore named *inflated* and was subsequently found to encode the integrin αPS2 subunit. Other genes of this type are *myospheroid* (βPS),[7] *wing blister* (one of the two laminin α subunits)[8] and *blistery* (tensin).[9]

Based on the hypothesis that the wing blister phenotype, which results in the separation of the two normally closely apposed wing epithelial layers, would be diagnostic of mutations in genes encoding integrins and their associated proteins, new genetic screens were performed. Some of these were designed to isolate mutations specifically in the αPS1 subunit gene,[10] while other more exhaustive screens were aimed at identifying the majority of genes in the genome that can cause this phenotype when mutated).[11,12] These screens made use of a newly developed genetic trick, mitotic recombination using FLP-FRT, to screen mosaic flies that were mostly heterozygous for the newly generated mutations, but had clones of cells within the wing that were homozygous mutant. In this way genes needed for adhesion in the wing that are also required at earlier stages of development, such as the development of the embryo, could still be

Table 3. Drosophila *integrin associated proteins whose mutant phenotype includes dorsal closure defects, but not defects in strong integrin-mediated adhesion at muscle attachment sites*

Protein/Gene	Features	Phenotype	Function
DOCK 180 *myoblast city*	SH3 domain	Myoblast fusion and dorsal closure defects	GTPase interacting protein
Laminin α3,5 *LamininA*	EGF, Laminin repeats	Laminin A	Lethal, defects in muscle, gut and heart
Mig 15 *misshapen*	Protein S/T kinase	Dorsal closure	Signaling/ Adaptor protein
PAK Dm PAK	S/T kinase	Axon guidance, dorsal closure	Signaling
Rac *Rac1, Rac2, mtl* triple mutant	Small GTPase	Dorsal closure	Signaling
Src FK Src42, Src64, Btk/Tec29	Non receptor tyrosine kinase	Redundancy between members; dorsal closure, cellularisation and oogenesis defects	Signaling

recovered. All of the genes recovered in these screens have not been characterised yet, but this approach has so far yielded mutations in three genes encoding integrin-associated proteins: a spectraplakin cytoskeletal linker (*short stop,* previously known as *kakapo,*[13,14]), which is an orthologue of vertebrate ACF7/MACF1 and BPAG1; the cytoskeletal linker talin (*rhea,*[15]); and the LIM domain adaptor protein PINCH (*steamer duck,*[16]). Although the hope was that these screens would identify proteins essential for integrin function that had not been previously identified, there is yet to be an example of this.

The third source is from reverse genetic approaches aimed at isolating mutations in particular genes of interest. A tool that has greatly aided these efforts is the generation of large collections of insertions of modified transposable elements and the mapping of their site of insertion.[17] The transposon may or may not inactivate the gene when it inserts nearby (the P-element transposon, which is the basis for the majority of these lines, preferentially inserts close to the promoter). However, the ability to induce the transposon to jump out has been used very effectively to cause the deletion of sequences adjacent to the site of insertion (in approximately 2% of the cases). In the ideal case, the generated deletion will remove enough of the gene of interest to generate a molecular null, while not affecting the adjacent genes. Mutations in the genes encoding the integrin ECM ligand tiggrin,[44] the integrin βν subunit,[19] focal adhesion kinase (FAK)[20] and definitive null mutations in the gene encoding tensin[21,22] were recovered this way. A mutation in integrin-linked-kinase (ILK) was identified within a collection of lethal mutations that had been generated by more standard classical genetic methods.[23]

Some of the sophisticated molecular genetic tools available in *Drosophila* that have been applied to the study of integrins include (i) insertion of single copies of modified genes into the genome, allowing the rigorous testing of the importance of functional domains within proteins; (ii) the elimination of the maternal contribution of a gene, enabling the complete removal of a gene product from the developing embryo, and (iii) advanced techniques for generating mosaic animals that contain mutant cells within an otherwise normal animal, facilitating a direct comparison of wild type and mutant cells side by side in the intact animal (as is achieved in the mouse with the Cre/loxP system). These tools have made it feasible to study the precise roles and requirements of particular proteins in the context of the whole organism.

Table 4.* Drosophila *integrin associated proteins whose reported mutant phenotypes do not have any overlap with integrin mutant phenotypes

Protein/Gene	Features	Phenotype	Function
α-actinin *Actinin*	CH domain, spectrin repeats	Larval lethal, sarcomeric architecture is defective	Actin binding
Collagen IV *Cgc25*	Collagen repeats	Lethal	Component of basement membrane
Collagen IV *viking*	Collagen repeats	Lethal	Component of basement membrane
Csk		?Lethal, Proliferation defects	Negative regulator of Src kinases
Filamin 1 *cheerio*	CH domain, Filamin repeats	Defects in actin structures in the oocyte	Actin crosslinking
Focal adhesion kinase *FAK56*	FERM domain, tyrosine kinase	Viable, fertile with no visible defects	Signalling
Nck *dreadlocks*	SH2, SH3 domains	Axon guidance defects	Signaling / Adaptor protein
Perlecan *terribly reduced optic lobes*	Proteoglycan	Larval lethal, defective neuroblast stem cell divisions	Component of basement membrane
M-Spondin	Thrombospondin Type 1	Viable and fertile, with no visible defects	ECM ligand
Syndecan *Syndecan*	Transmembrane heparan sulphate proteoglycan	Semilethal, axon and muscle guidance defects	
VASP/MENA *enabled*	PH domain	Axon guidance	Cytoskeletal protein
Vinculin *Vinculin*		Viable and fertile with no visible defects	Cytoskeletal linker

We will first describe the genetics of integrins and their associated proteins and then go on to describe what these mutations have told us about the role of integrins in development and the mechanisms of integrin function.

Integrin Genes

As mentioned, the fly has a small repertoire of integrin subunits (5α and 2β).[3,4] The βPS subunit appears to be the orthologue of vertebrate β1, but this cannot be deduced on the basis of sequence conservation alone, since βPS has similar levels of sequence conservation with more than one vertebrate β subunit (Fig. 1). A very comprehensive analysis of the embryonic mutant phenotype caused by the absence of βPS was performed by Ted Wright in 1960, and impressively, he predicted that the gene product would have a role in the formation of the ECM. He named the gene *myospheroid* (*mys*) based on the most prominent aspect of the phenotype, the rounding up of the muscles.[7]

We suspect that βPS forms heterodimers with all five α subunits, but this has only been demonstrated for αPS1-3. The sequence similarity of the αPS1 subunit with the vertebrate α3, α6 and α7 extends to ligand preference, as αPS1βPS binds to laminin.[24] The gene

Table 5.* Drosophila *integrin associated proteins for which mutations do not yet exist

Protein/Gene	Features	Function
Crk	SH2, SH3 domains	Signaling/ Adaptor protein
Filamin 2 CG30092	CH domain, Filamin repeats	Actin crosslinking
Graf	Gap, PH domain	GTPase activator
Laminin β	EGF repeats	ECM component
Laminin γ	EGF repeats	ECM component
Nidogen	EGF repeats	ECM component
Parvin CG35258	CH domains	Actin binding
Papilin	Kunitz, Ig domains, endopeptidase activity	Component of basement membrane
Paxillin DmPax/ PDLP	LIM domains	Scaffold protein
p130Cas CG1212	SH3 domain	Cytoskeletal protein
Tenascin	EGF, FN III repeats	ECM ligand
TES CG6522	LIM domains	Cytoskeletal protein
UNC112 CG14991	FERM, PH domain	Scaffold protein
UNC112 CG7729	FERM, PH domain	Scaffold protein
Zyxin Zyxin102	LIM domains	Cytoskeletal protein

Table 6. Integrin associated proteins for which no orthologues exist in the fly

Protein/Gene
Migfilin
Layalin
Polycystin
PSTPIP
PST PEST
Vimentin
SHPS1
uPAR

encoding αPS1 was named *multiple edematous wings* (*mew*),[10] due to the wing blister phenotype. The αPS2 subunit is equally similar to vertebrate α5, α8, αV and αIIb, and αPS2βPS shares with these integrins the ability to bind ligands containing the tripeptide sequence RGD.[25] The three remaining integrin subunits, αPS3-5, are very closely related but do not have a clear orthologue in vertebrates (all 5 α subunits lack an I domain). The αPS3βPS heterodimer binds

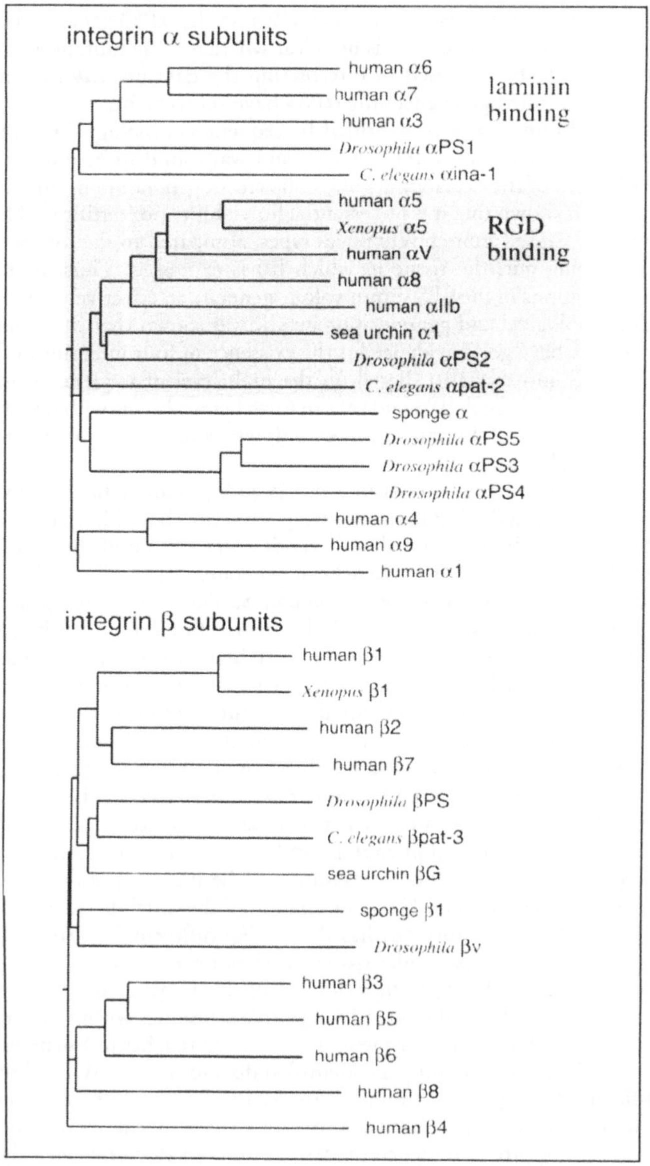

Figure 1. Phylogenetic tree of selected integrin α and β subunits.

to both forms of laminin.[26,27] The αPS3-5 subunits appear to be the result of recent duplication events, because the *Drosophila pseudoobscura* genome contains only one, in addititon to αPS1 and αPS2 (our observations). In *Drosophila melanogaster*, the transcription units encoding αPS3 and αPS4 are adjacent, suggesting that they could be controlled by the same cis-regulatory elements, which may explain some of the illogical aspects of the molecular genetics of this locus. The locus is named *scab* because of the scab in the embryonic epidermis that results from a failure in dorsal closure,[28] and also *volado* (forgetful), because of defects in short term memory.[29] Mutations in *scab* clearly affect αPS3, so the αPS3 gene is often thought of as equivalent to the *scab* gene. However, it is not clear whether these mutations also affect αPS4, so some caution should be exercised in interpreting the data on *scab* mutants until this is cleared up. Mutations in the gene encoding αPS5 have yet to be described.

The second β subunit, βν, was identified by sequence homology, but is not an obvious orthologue of any vertebrate β subunit (Fig. 1), and was found to be only expressed at high levels in the endoderm of the developing gut.[30] The recent generation of null mutations in the gene encoding βν has shown that it is not essential for viability nor fertility.[19] However, the loss of both βν and βPS caused more severe phenotypes, compared to the loss of βPS alone, but only in the developing gut, the tissue in which βν is expressed. Thus, in all tissues tested besides the gut, mutations in the βPS subunit alone generate an effectively null integrin phenotype. Using a cell biological and genetic approach, it was shown that βν forms heterodimers with αPS3. Thus, we have good evidence for the existence of four integrins in flies: αPS1βPS, αPS2βPS, αPS3βPS and αPS3βν. Based on the high level of sequence similarity between αPS3-5, it seems likely that both βPS and βν can form heterodimers with αPS4 and αPS5. We suspect that αPS1 and αPS2 do not form heterodimers with βν, but such heterodimers have not been absolutely ruled out.

Since integrins must form a heterodimer in the endoplasmic reticulum to become transported to the cell surface, a fly that lacks the two β subunits should have no integrin function at all. Thus, by removing both maternal and zygotic contribution of these subunits, embryos that completely lack integrin function have been generated for the first time;[19] a comparable experiment has yet to be done in worm or mouse. Similarly, it is now possible to produce clones of cells in the adult that completely lack integrin function. Surprisingly, certain processes such as the regulation of proliferation, that vertebrate studies have suggested are integrin dependent, occurred normally in the complete absence of integrins in the fly.

As one might expect, the loss of individual α subunits causes subsets of the phenotype of the complete loss of integrins. For some developmental events the loss of a single α subunit produces a phenotype that is just as strong but does not perturb other developmental processes mediated by other integrin heterodimers. In other cases there seems to be redundancy between the different α subunits, most commonly between αPS1 and αPS3, so that the loss of each individual subunit produces a mild phenotype and only the double mutant is as strong as the absence of the β subunits (see below for examples). The phenotypes fit with the expression patterns of the different α subunits. Thus, the analysis of the distribution and consequences of the loss of function of integrin subunits has shown that different integrin heterodimers have functions that are restricted to particular tissues, as elaborated below.

With a genetically tractable organism it can be valuable to generate a large variety of mutations in a given locus with the goal of identifying specific residues within the protein that have key functions. The use of a variety of genetic screening approaches to isolate mutant alleles in the gene encoding the αPS2 subunit gene identified distinct subfunctions of this protein that operated in different developmental events.[31] An extensive analysis of strong mutations in the βPS subunit did not prove so informative, as the majority were frameshifts or stop codons,[32] but an elegant genetic approach designed to identify mutants that only partially impaired βPS function produced a large variety of mutations of interest.[33] Particularly surprising is the finding that mutations in some very highly conserved residues caused relatively weak phenotypes.

Genes Encoding Integrin-Associated Proteins

At this point in time, we can divide the genes encoding integrin-associated proteins into the following three groups: (i) those for which mutations exist and the mutant phenotypes are wholly or partially similar to those produced by mutations in the genes encoding integrins, suggesting that they function with integrins (Tables 2 -3), (ii) those for which mutations exist, but the phenotypes are distinct from integrin gene mutations, suggesting that they are not essential for integrin function (Table 4), and (iii) those for which mutations have yet to be isolated (Table 5). Some proteins associated with integrins in vertebrates do not have orthologues in the fly (Table 6).

Extracellular Matrix and Integrin Ligands

One of the complexities of integrins and their ligands relates to the question of "who comes first?". Does the ECM get assembled first and then recruit and cluster integrins, or do the integrins lead to the assembly of the ECM.[34] There is evidence to suggest both scenarios occur.

So far only three integrin ligands have been identified in *Drosophila*. Two of these are laminins. In flies there are two laminin α subunit genes, 1 β subunit and 1 γ subunit encoded in the genome. The α subunit gene encoded by the *lamininA* locus is most similar to vertebrate α3 and α5 laminin subunits, and contributes to basement membranes throughout the animal.[35-37] The second α subunit is encoded by the *wing blister* gene, is equally similar to vertebrate α1 and α2, and has a much more restricted distribution.[38] We shall refer to the two trimers formed by these two α subunits as lamininA and lamininW. LamininA has been shown to be a substrate for both αPS1βPS and αPS3βPS,[24,27] but in general the distribution of the lamininA containing basement membranes is not altered in embryos lacking integrins.[39] Recently, an extracellular matrix containing lamininA was identified that requires integrins for its assembly, which lies between two extraembryonic cell layers in the embryo, the amnioserosa and yolk cell,[40] and is important for integrin function in the process of dorsal closure. This ECM does not seem to be a full basement membrane, as it is not visible as such by electron microscopy,[89] and it lacks collagen IV.[40] The Wing blister α subunit contains an RGD site, and a fragment containing this site is an effective ligand for αPS2βPS and αPS3βPS.[27,42] Mutations in these two laminins cause phenotypes that mimic certain aspects of the integrin phenotypes, but they certainly do not completely recapitulate the integrin mutant phenotype, indicating that there must be other ligands. Other components of the basement membrane are also conserved in *Drosophila*, including collagen IV, perlecan, and nidogen/entactin, but the mutants available do not produce clear integrin-like phenotypes.[43]

The novel ECM protein Tiggrin highlights the potential problem in identifying the missing integrin ligands in *Drosophila*. It is a novel secreted protein, containing an RGD site,[18] but so far a similar protein cannot be found in organisms other than *Drosophila* species, not even in other insects such as the mosquito(our observations). This suggests that completely novel proteins can adopt a function in the ECM, and therefore that predicting other ECM proteins from the genome sequence may prove difficult. Tiggrin was isolated from the media conditioned by the *Drosophila* Kc cell line, and functions as a ligand for αPS2βPS.[18,44] The distribution of Tiggrin in the embryo is highly restricted, where it is present in stripes at the points where the muscles attach to the epidermis-the major site of integrin adhesion in the embryo (Fig. 2). These muscle attachment sites contain the largest ECM in the fly, called the tendon matrix, which consists of an aggregate 1-2 μm across, containing fibrils of unknown composition (given the absence of genes encoding fibrillar collagen).[45,39] Tiggrin can be captured from the extracellular fluid at these sites, as demonstrated by the normal distribution of Tiggrin when its site of synthesis was changed from the mesodermal cells to cells of the nervous system (*tiggrin^X; elav::Gal4/UAS::tiggrin*).[44] Surprisingly, integrins are not required for the capture of Tiggrin at the muscle attachment site, and the tendon matrix, as visualised by EM, still assembles.[18,39,45,46] A null *tiggrin* mutation causes a weakening of the attachment site, but does not fully mimic loss of integrins, indicating the presence of additional ligands.[44] A protein M-spondin, related to vertebrate F-spondin is found at the muscle attachment site, but its mutation does not disrupt integrin adhesion.[47] Mutation in the gene encoding the muscle-specific transcription factor Muscleblind causes a great reduction in the tendon matrix without altering tiggrin distribution or strongly affecting muscle attachment, further highlighting the redundancy of the components the tendon matrix.[48]

In summary, three integrin ECM ligands have been identified and genetic evidence suggests other unidentified ligands must exist. The combination of genetic redundancy and novelty in sequence of Tiggrin suggests that identification of additional ligands will most likely succeed by a proteomics approach.

Intracellular Integrin-Associated Proteins

A large number of intracellular proteins are enriched at sites of integrin adhesion in vertebrate cells.[49] The *Drosophila* genome sequence has revealed that the majority of these proteins are conserved[4] (Tables 2-5). Mutations are available in many of them, making it possible to examine whether the absence of the protein produces a phenotype indicative of an essential role in contributing to integrin adhesion. Furthermore, this analysis will reveal whether the protein has functions independent of integrins, as revealed by phenotypes that are not shared with integrin mutants. One of the advantages of *Drosophila* is that different signaling pathways appear to have clearly independent functions,[50] suggesting that some of the crosstalk between pathways observed in vertebrate cells in culture is not critical for the function of these pathways within the intact organism. To give an example, components of PI3 kinase and MAP kinase pathways have been implicated in integrin function, but mutations in PI3 kinase, MAP kinase and integrins produce clearly distinct phenotypes in the developing wing, with PI3 kinase regulating cell size,[51] MAP kinase regulating proliferation,[52] and integrins mediating adhesion between the two layers of the wing.[103] Model organisms such as *Drosophila* can therefore be used to advantage to separate components into their core pathways.

The list of intracellular proteins confirmed via genetics to participate in integrin dependant adhesion in Drosophila is fairly small and includes the spectraplakin Short stop (shot), talin, integrin-linked-kinase (ILK), PINCH, and tensin (Table 2). Short stop appears to have a specific role in linking integrins to microtubules in the embryonic epidermis and the wing epithelia,[13,14,86] and has other functions including axon extension (the failure in this led to the name) and oocyte specification which are independent of integrins.[53,54,109] Talin appears to be required for almost all integrin functions, with the exception of cell migration mediated by αPS3βv[19] and possibly signalling to regulate gene expression (for caveats see the section on signaling).[15] PINCH is required for integrin function in strong adhesion and does affect one dynamic morphogenetic process, apparently through the regulation of signaling.[96] ILK appears to be only required for integrin function in strong adhesion and not in more dynamic events, while a function for tensin is only apparent during adhesion in the wing, as the null mutant flies are viable and fertile with a wing blister phenotype.[16,22,23]

Several proteins including vinculin and focal adhesion kinase (FAK) are not required for integrin function; both can be removed without obvious detriment to the fly.[20,55] This is unexpected because mutations in these genes in other organisms cause lethality.[56,57] Overexpression of FAK causes defects that mimic loss of integrin function. This was previously thought to suggest that a dominant negative effect on endogenous FAK, but the fact that removing FAK does not cause this phenotype shows that this is not the explanation. Instead FAK must be acting to inhibit other proteins. In the muscles, overexpression of FAK causes the detachment of integrins from the ECM, suggesting that FAK downregulates integrin affinity for the ECM.[20] Several other proteins have essential functions that are distinct from those mediated by integrins, including one of the two filamins (encoded by *cheerio*) and α-actinin[58-61] while others, such as talin have additional functions that are apparently integrin independent (ref. 15 and unpublished observations).

The roles of a further set of proteins in integrin function have not been easy to resolve. For example, cells lacking Rho cannot divide, and thus it is not possible to test whether clones of cells completely lacking Rho have a failure in integrin adhesion. Weaker (hypomorphic) mutations in Rho can proliferate and have defects in cell polarity while integrin adhesion appears normal. Thus, one can only conclude that if Rho is required for integrin adhesion, this requires lower levels of the protein than for polarity, as is true for its role in cytokinesis.[62] For other proteins, genetic redundancy combined with diverse roles preclude a simple conclusion. This is true for the Src kinases, where there is redundancy between the two Src kinases (Src42A and

Src 64B), and the related Tec (Btk) kinase. Src64 and Tec have a role in regulating the actin cytoskeleton of the ring canals of the ovary, which is completely independent of integrins.[63] Combinations of mutants in these kinases show defects in dorsal closure, which is an integrin dependent process, but also involves many other pathways.[64] The recent identification of mutations in Csk has confirmed its role as a negative regulator of the Src family of kinases in vivo but the presence of integrin-like phenotypes has not yet been examined.[65,66]

Several proteins of interest await genetic characterisation, including the second filamin, the two orthologues of the Kindler syndrome protein/UNC-112, p130Cas, Crk, and the LIM domain-containing proteins paxillin, zyxin and TES/testin.

Integrin Functions in Fly Development: Sculpting Tissues through Adhesion and Migration

The three most prominent defects in development caused by the loss of βPS integrin function in the fly are: detachment and rounding up of the embryonic muscles, a hole in the dorsal epidermis of the embryo, and blisters in the adult wing. It should be noted that null mutants in the gene encoding βPS cause embryonic lethality due to the first two defects and the wing blister phenotype of the adult fly can only be observed when chimeric flies are generated with patches of tissues that are mutant, or with unusual mutant alleles that particularly affect function in the adult. These prominent defects can be largely explained by the loss of an integrin adhesive function in keeping different layers of cells attached to each other through an intervening ECM, and it is still not clear whether integrin signalling is involved.

Integrin Signalling: Still Elusive

Although vertebrate cell culture has provided a wealth of literature supporting the existence of, requirements for and pathways mediating integrin signalling,[67,68] studies in the fly have not been very forthcoming. We should clarify how we define integrin signalling. The assembly of a link between the integrin cytoplasmic tail and the cytoskeleton by a series of protein-protein interactions does not, in our view, constitute integrin signalling. However, if integrins trigger a cascade of protein modification, such as phosphorylation, that is required for the assembly of this link, then we would describe that as integrin signalling. Similarly, integrin regulation of gene expression must involve a signalling pathway that leads into the nucleus. There is an uncomfortable grey area that is harder to define: if the assembly of a link between integrins and the cytoskeleton requires conformational changes in each of the linker components, then does this constitute a signalling pathway? We leave that for the reader to decide.

Despite the identification of genes that are regulated by intracellular events downstream of the integrin βPS subunit cytoplasmic tail,[69] so far of all the mutations tested, only mutations in the αPS1 subunit affect this signalling, (refs. 15, 20, unpublished results). Part of the problem is that when one produces embryos completely lacking some of the candidate proteins, then the embryo does not develop far enough to assay the known target genes (this is true for both βPS and talin). In other cases integrin mediated adhesion is essential for the correct transmission of signals, for example in the transmission of an EGF signal from the muscles to the tendon cells[70] where it regulates MAPK activity through the localisation of the EGFR ligand. Potential functions of integrin signalling synergising with TGFβ and EGF signalling have also been suggested during wing development, but the mechanisms are not yet clear.[71,72]

More extensive analysis of the flies lacking integrin function has revealed defects in migration, maintenance of epithelial polarity, dynamic regulation of synapses and regulation of gene expression. These developmental defects will be described in more depth below, but it is also worth mentioning some of the things that integrins do not do in fly development. Perhaps most importantly in the context of development, there is no evidence that integrins are involved in cell fate choice. This is in contrast to transmembrane receptors of several well known signalling pathways, including Notch and the receptors for a variety of secreted ligands: EGF, FGF, Wnt, Hedgehog, and TGF-β. The cell-cell adhesion molecules cadherins also present an interesting comparison. Cadherins do not appear to directly influence cell fate, but it may be more than a coincidence that the linker molecule β-catenin/Armadillo, which links the cytoplasmic domain of cadherins to the actin-binding protein α-catenin, is also a key component of Wnt signalling. In addition, cadherin-mediated adhesion plays a crucial role in the establishment of the body axes.[73,74] In *Drosophila*, integrin-mediated adhesion is not essential for proliferation,[19] so anchorage-dependent growth does not appear to operate in this organism. Integrins are not generally required for differentiation, as many cell type-specific proteins are produced normally in embryos lacking integrins. However, a screen through a small set of genes expressed at late stages of gut development identified two that were regulated by integrins.[69] This suggested that integrins are part of a mechanism for coordinating differentiation and morphogenesis of a tissue.

Migration

Integrins are essential for some, but not all, cellular migrations during *Drosophila* development. In the embryo, integrins are essential for the movement of three groups of cells over a substrate formed by the visceral mesoderm. The primordial midgut endoderm arises in two populations, one at each end of the embryo, that undergo an epithelial to mesenchymal transition, crawl over the visceral mesoderm in an integrin-dependent fashion to meet in the middle and then reform an epithelium. The development of the endoderm is the only tissue where the integrin βν subunit plays a role, which appears to be redundant with βPS. In the absence of βPS alone, the migration is severely delayed, but eventually occurs,[75,76] while in the absence of both β subunits, the cells never move.[19] The trachea, the "lungs" of the fly, is a set of branching tubes that migrate to reach a variety of target tissues. One of the branches in every segment - the branch that migrates over the visceral mesoderm to reach the endoderm- requires integrins for its migration.[77] Similarly, the tip of the salivary gland, which is an invaginated epithelial sac, migrates in a posterior direction over the visceral mesoderm using integrins.[78] In each case, integrins are required both in the visceral mesoderm (αPS2βPS) and in the migrating tissue, which employs one or both of αPS1βPS and αPS3βPS. The current model is that αPS2βPS is required in the visceral mesoderm to assemble an ECM that serves as a substrate for the migration of these different tissues. The composition of this predicted ECM is not known. A further example of integrin-dependent migration is when a portion of the extraembryonic epithelial layer called the amnioserosa extends over the epidermis during the process of germ band retraction (Fig. 4). This involves αPS3, αPS1 and the second laminin isoform, lamininW.[27]

These cases show that integrins play a crucial role in the positioning of tissues within the embryo. In the absence of integrins, the cells that lead these movements lack their characteristic lamellipodial and filopodial extensions.[27,76,78] Integrins are not required for all cell movements in the embryo. As mentioned, other tracheal branches reach their target without integrins, and the germ cells move normally in the absence of integrins from their original position outside the embryo into the gonad. Thus, the mechanism of cell movement in development can be separated from the use of particular receptors to specify traction.

Integrin Function in Muscle Attachment; A Focal Adhesion-Like Structure within the Embryo

The role of integrins in muscle attachment has proven to be a useful model for integrin function in stable adhesion within the embryo. The eponymous "rounded up" muscle phenotype of mutations in the βPS integrin subunit (encoded by the gene *myospheroid*,[7,79]) is easily visualised. The major sites of integrin accumulation in the developing embryo are muscle attachments (see Fig. 2). Integrins are required for the maintenance of these structures, not their formation. The muscle attachment sites are points of contact between the ends of the muscles and specialised epidermal cells, the tendon cells. The tendon cells are specified independently of the existence of the muscles, and appear to send signals to attract the muscles.[80] The muscle

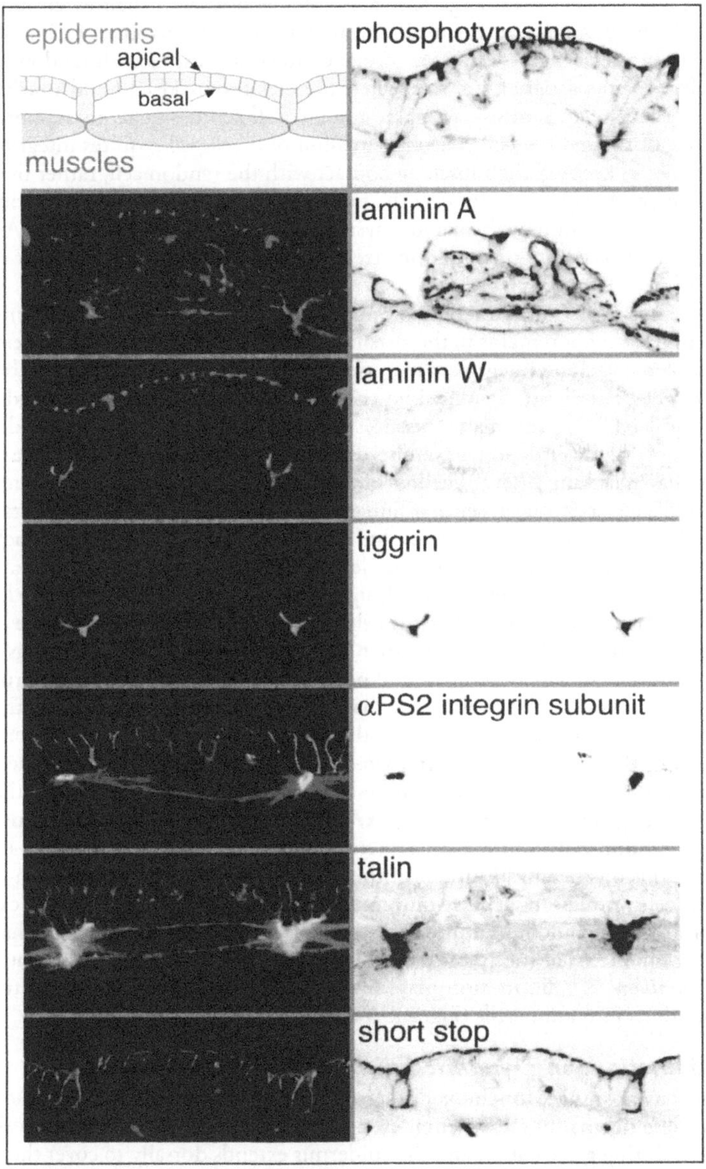

Figure 2. Localisation of integrins and associated proteins at muscle attachment sites. The muscle attachment site is a Y-shaped structure formed by the apposition of the basal surface of the specialised epidermal cell, the tendon cell and the two muscle cells through an intervening matrix (see schematic). Single fluorescent channels are shown inverted in the right column. An antibody that recognises proteins phosphorylated on tyrosine outlines all cells but is enriched at apical epidermal adherens junctions and the muscle attachment site. The integrin associated protein talin (merged image on the left shows talin in green, actin in red and Fasciclin 3 in blue) and the RGD ligand tiggrin (laminin W in red and tiggrin in green in merged image) localise to the Y-shaped structure. The αPS2 integrin (green in merged image with actin in red and Fasciclin 3 in blue) is enriched at points of muscle cell contact. The ECM ligand laminin A (red in merged image, Fasciclin 3 in blue) localises to the basement membranes around the muscle while laminin W (red in merged image, phosphotyrosine is blue) localises to the V but not to the stem of the Y. The BPAG like protein encoded by short stop (red; Fasciclin 3 is blue) localises to the basal surface only in the tendon cell.

founder cells and myoblasts undergo fusion simultaneously with their extension toward and adhesion to the tendon cells. Integrins do not to contribute to myoblast fusion in *Drosophila*, as they can do in vertebrates.[81] After the initial contact of muscles with tendon cells, the tendon matrix builds up, particularly at sites where muscles also make end-end attachment to each other. The contact of the muscle with the tendon cell sends signals to support the further differentiation of the tendon cell. The transmission of the signal requires integrins, but this is due to their role in keeping the muscle in contact with the tendon cell, rather than being part of the signalling pathway.[70] Shortly after the muscles are attached they begin to contract, which they do prior to innervation by the nervous system or the formation of striations. As the muscles develop further the contractions build in strength, and eventually the resultant larva hatches and crawls away.

This gradual increase in the strength of muscle contraction provides a useful measure of the strength of adhesion that remains in the absence of a particular component. For example, in an embryo completely lacking integrins, the muscles detach soon after contractions begin.[7,19,75,82] If just the αPS2βPS integrin, the only known integrin in the muscles, is removed, the muscles detach slightly later,[39,82] presumably because the αPS1βPS integrin in the tendon cell holds the structure together a little longer; embryos lacking both αPS1 and αPS2 have equivalent defects as embryos lacking βPS.[39] The loss of talin results in the muscles detaching at the same developmental stage as in the absence of integrins,[15] while the loss of ILK results in a detachment some 8 hours later[23] (out of the 22 hrs of embryogenesis). This later detachment looks different, with less rounding up of the muscles. Closer examination revealed that whereas the early defect results from the muscles detaching from the tendon matrix,[15] in the later cases the point of detachment occurs first inside the cell, separating the integrins from the cytoskeleton, resulting in an "actin detachment" rather than muscle detachment[23] (Fig. 3). This suggests that there is some redundancy between ILK and other components, such that the structure is only weakened in the absence of ILK alone, so that stronger contractions can be withstood before the attachment structure fails. Tensin is tightly localised to the muscle attachment, as is phosphorylated FAK, but removal of these proteins does not cause visible defects in the muscles.[20,22]

The final muscle attachment structure consists of the ends of the contractile machinery connected via integrins to the tendon matrix, which is in turn bound by integrins on the basal surface of the tendon cells. In these cells microtubule bundles connect the basal integrin junctions to junctions on the apical surface of the cell that binds to the exoskeleton or cuticle. In this way, muscle contractions achieve movement of the exoskeleton. The cuticle can be considered an apical ECM, but the receptors for this ECM are not integrins and have yet to be identified. Components of the apical ECM were identified as the wing blister mutants *dumpy*, *piopio* and *papillote*.[83-85] Short stop plays a key role in connecting the microtubules to the integrin junctions in tendon cells (Fig. 2).[14,86]

Integrin Functions in Dynamic Epithelial Morphogenesis

Integrins have complex functions during the processes of germband retraction and dorsal closure. In these normal developmental events, the embryo retracts back from the extension that occurred during gastrulation and the epidermis extends dorsally to cover the "hole" in the surface of the embryo that has been covered by the extraembryonic epithelial layer called the amnioserosa (Fig. 4). At the end of dorsal closure the amnioserosa invaginates and undergoes apoptosis, while the dorsal edges of the epidermis seal up along the dorsal midline. The retraction of the embryo and extension of the epidermis dorsally are driven by cell shape changes; there is no cell proliferation in the epidermis at this stage. Dorsal closure bears similarities in both appearance and molecular mechanism with embryonic wound healing.[41]

Recently a third player in these epithelial transformations was identified, in addition to the epidermis and the amnioserosa, which is the giant yolk cell that lies in the centre of the embryo.[40] This cell is a product of the unusual early development of insect embryos. After fertilisation, rather than undergoing cell cleavages *Drosophila* undergoes 14 rounds of nuclear

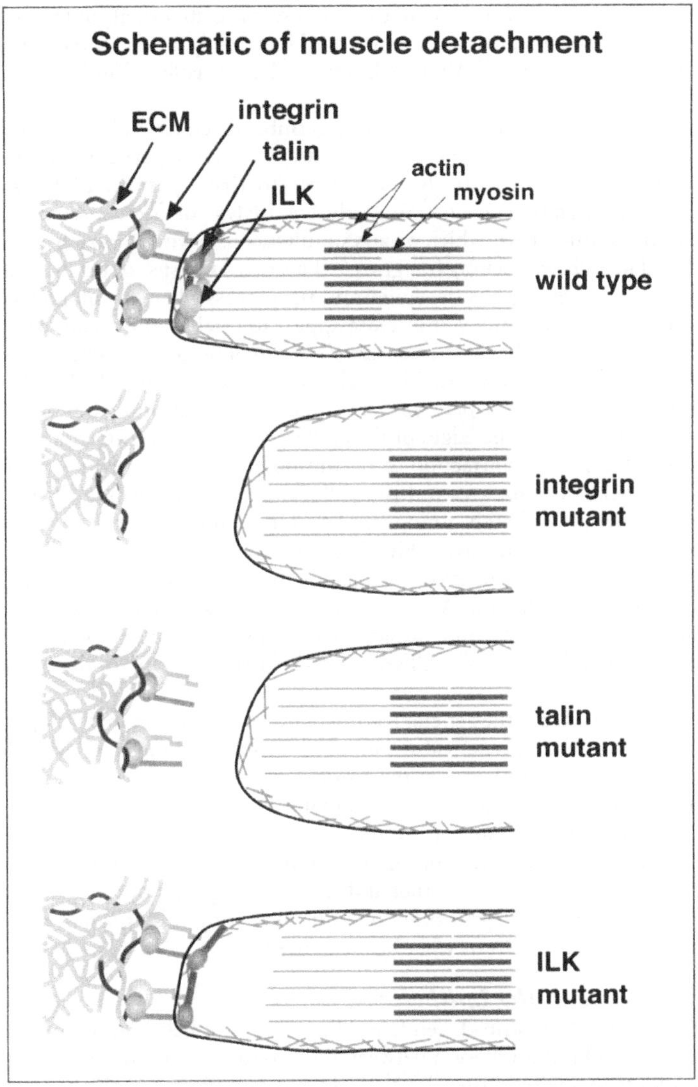

Figure 3. Distinct phenotypes generated at the muscle attachment sites by mutations in integrins and associated proteins. In the absence of integrins, the muscle cell membrane detaches from the extracellular matrix. In talin mutant embryos, a similar detachment results with integrins left behind with the extracellular matrix. In embryos mutant for the integrin linked kinase, the membrane remains attached to the extracellular matrix but the sarcomeric actin detaches from the membrane.

division, without any cell division. The nuclei then migrate to the surface of the egg, and the plasma membrane invaginates around each nucleus. When the invaginating membranes meet underneath each nucleus this generates a monolayer of blastoderm cells and a single large internal cell, the yolk cell, which contains a few nuclei.

Integrins have been known to contribute to germ band retraction and dorsal closure for some time, but how they contribute has not been clear, and recent work has suggested that they have multiple roles. At the end of embryogenesis, many embryos lacking integrin function contain a dorsal hole.[7] The contraction of the amnioserosa is abnormal, and tears appear within the amnioserosa.[40] If the integrins are also removed maternally, a more severe defect is observed were germ band retraction fails.[75] All three layers, epidermis, yolk cell and amnioserosa, express integrins (Fig. 4).[40]

Recent work using live imaging to examine germband retraction has shown that both the amnioserosa and the yolk cell are very active cells.[27,87] One hypothesis has the amnioserosa sending membrane ruffles over the epidermis to drive germband retraction via an integrin-dependent migration event.[27] Similarly, live analysis of dorsal closure showed that the contraction of the amnioserosa, which occurs simultaneously with the extension of the epidermis, was delayed in embryos lacking integrins.[88] This was suggested to be due to a defect in the zippering up of the dorsal edges of the epidermis, the process where the filopodia of the epidermal cells at the interface with the amnioserosa appear to reach over the amnioserosa and make contact with the opposing edge of the epiderms, and then by contraction, draw the two edges together.[89] Our own observations provide an alternative explanation, which is that the integrin defect (and hence requirement) preceeds the zippering, in that the amnioserosa cells fail to contract sufficiently to allow the edges of the epidermis to come within zippering range.

The integrin dependant contraction of the amnioserosa is mediated in part by facilitating adhesion between the amnioserosa and the yolk cell via an intervening ECM.[40] This ECM contains the basement membrane component lamininA, and in contrast to the general assembly of basement membranes, the assembly of lamininA between the amnioserosa and yolk cell is integrin dependent. Loss of integrin mediated adhesion by mutations in the integrin subunits αPS3 or βPS, or the lamininA α subunit resulted in a separation between the yolk cell and the amnioserosa and a delay in contraction and closure. Examination of the mutant phenotypes also suggested that the yolk cell serves as a cellular substrate for the contraction of the amnioserosa.

These results showed that the general feature of integrin function in development, of mediating adhesion between cell layers, is also employed during dynamic epithelial rearrangements. However, the loss of adhesion between the amnioserosa and the yolk cell is not in itself sufficient to cause a dorsal hole; in contrast to removing integrins, the *lamininA* mutant embryos do not end up with a dorsal hole. This indicates that integrins have additional roles that are independent of laminin. One such role appears to be mediating adhesion between the lateral sides of the amnioserosa cell to each other and to the leading edge of the dorsal epidermis.[40] It is not clear yet how direct integrin function is in mediating this apparent cell-cell adhesion between the lateral sides of cells or what ligand is used.

Integrin Functions in the Epithelia

Further examples of integrin-dependent adhesion of epithelial layers are in the adult fly. The best characterised is the adhesion between the two surfaces of the developing wing; loss of this adhesion leads to a wing blister.[103,108,115] This is reminiscent of the human skin diseases epidermolysis bullosa and bullous pemphigoid. Since the former is caused by loss of the hemidesmosome integrin α6β4, which is absent in flies, one might not think that the fly would provide a good model for this disease. However, the wing blister mutant *short stop/kakapo* turns out to encode a fly orthologue of the protein bullous pemphigoid antigen 1, BPAG1, and more recently the human orthologue of a protein required for integrin function in *C. elegans* has been shown to be linked to the skin blistering disease Kindler's syndrome.[14,90,116] This suggests that an understanding of basic mechanisms of integrin function in invertebrates may well provide a useful model for these human epithelial diseases.

The wing arises from an epithelial sac, the imaginal disc, that grows during larval life. When the larva pupariates, the larval epidermis degenerates and a new epidermis is produced from the imaginal discs and histoblasts. At the start of pupal development the wing imaginal disc evaginates, generating a two-layered epithelial wing. Integrins are essential to hold these two layers together. Certainly part of the function is to generate strong adhesive junctions, but

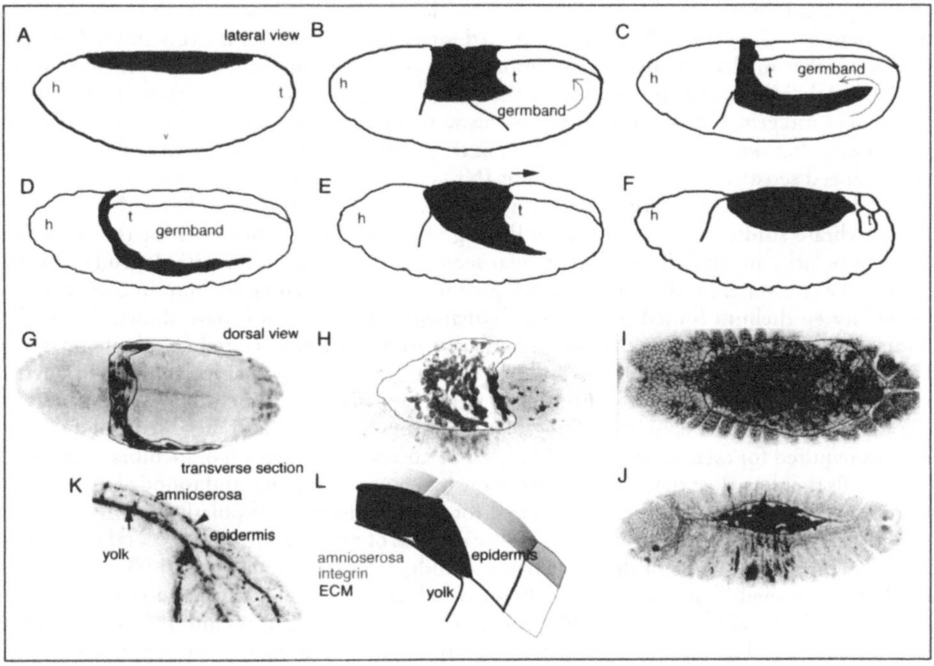

Figure 4. Germ band retraction and dorsal closure are dependant on integrin ECM interactions. A-F) schematic representations of Drosophila embryogenesis (lateral views) highlighting the morphogenetic movements of germband extension (A-D), and germband retraction (D-F, adapted from ref. 122). During germband extension, the germband extends in the direction of the blue arrow in (B,C) such that it forms a U-shaped structure with the tail (t) of the embryo situated progressively closer to the head (h) of the embryo (C,D). During germband retraction, the tail retracts (E), progressively (in the direction of the arrow in E) to the pole opposite the head. G-I) dorsal views of embryos at the beginning, midway through and at the end of germ band retraction corresponding to the schematics in D-F respectively. The amnioserosa (filled/outlined in green in A-J) becomes folded during germband extension and then stretches out during retraction such that at the end, it forms a flat sheet covering most of the dorsal surface of the embryo (I). During dorsal closure, contraction of the amnioserosa results in the reduction of the surface area of the amnioserosa visible on the dorsal surface (compare I and J) and together with the collective movement of the epidermis to cover the dorsal surface of the embryo (F, J). A section through an embryo during dorsal closure (K) reveals βPS integrin expression at the interface between the epidermis and the amnioserosa (arrowhead) and between the amnioserosa and the yolk cell (arrow). At both sites, they mediate adhesion through an intervening laminin containing extracellular matrix (schematised in L).

signaling may also be involved.[117] The integrins on the two sides of the wing are different; αPS1βPS is on the dorsal side, while αPS2βPS is on the ventral side.[91] The laminin α subunit encoded by the gene *wing blister* is essential, while mutations in the other laminin α subunit or tiggrin do not give a wing blister phenotype.[38,44,118] Integrins become strongly localised into tight discs where the basal side of each epithelial cell attaches to a cell on the other surface of the wing.[92] A similar function for integrins is in the attachment of the retinal cells to the underlying basment membrane in the developing eye.[93]

Thus, the defects caused by the absence of integrins in the epithelial layers have revealed that the primary role of integrins is to hold cell layers together. However, other potential roles for integrins in the epithelia have been explored. While integrins do not seem to mediate anchorage dependant growth, there is correlative evidence for a role for integrins in anchorage dependent cell survival. Premature degeneration of the amnioserosa seen in a small proportion of integrin mutants has been attributed to the loss of adhesion between the yolk and the amnioserosa.[87] Genetic studies have uncovered interactions between integrins and ecdysteroid hormones that mediate many metamorphic cell death events during fly development that include dorsal closure and the shaping of the wing.[94,95] It has also been suggested that tensin transduces integrin signals to the JNK pathway to mediate apoptosis in the wing. It seems more likely that tensin acts in parallel to the JNK pathway and that the loss of tensin uncovers an enhanced sensitivity to the reduction of JNK signals.[21] More recent studies have also suggested a role for PINCH in the regulation of JNK activity during dorsal closure.[96]

Vertebrate studies have provided conflicting views on the requirements for the establishment of polarity in epithelia, with particular reference to the role of integrin dependant adhesion to the substrate. Studies on the development of the follicular epithelium in *Drosophila*, a secondary epithelium fomed by a mesenchymal-epithelial transition have shown that while contact with the basement membrane is sufficient to exclude apicolateral membrane markers from the "basal" membrane domain and prevent the lateral spread of apicolateral components in the absence of apical cues, it is insufficient to fully polarise the epithelium.[97] On the basis of the analysis of secondary epithelia lacking all integrin function it has been suggested that integrins are not required for establishment of polarity. The aberrant distribution of membrane markers in the cells that have lost their attachment to the basement membrane and rounded up has lead to the suggestion that integrins are required for the maintenance of polarity.[19] It is not been established whether the cells round up because of a loss of polarity. However, secondary epithelia are atypical, since some of them, such as the midgut epithelium in which integrin function has been examined,[19] are not completely polarised until long after their establishment. It is therefore still plausible that the establishment of polarity is very robust and that integrins play a redundant role. This would be consistent with the finding that many molecules that contribute to polarity act in parallel and result in only modest defects when absent.[98]

Integrin Function in the Nervous System

A modest defect in axon pathfinding has been described for integrin mutants,[99] which contrasts with the severe phenotype caused by loss of Short stop.[100] Although the formation of the synapses is not itself affected, changes in the levels of integrins results in dramatic alterations in their growth and sprouting.[29,39,101] The potential to regulate functional synaptic plasticity processes may account for the role of integrins in the formation of memory.[29] This function was uncovered through the analysis of hypomorphic mutants in βPS and αPS3 subunit. Studies have revealed that integrin effects on the synapses may be mediated through its association with discs large and the resulting modulation of Fasciclin II through CamII kinase activity.[102]

Complementary and Two-Sided Requirements

In all the above-mentioned integrin-dependent developmental events, βPS containing integrins are required on both faces to mediate adhesion or facilitate migration. This is best inferred from the deleterious effects on adhesion of disrupting integrin function on one side (e.g., a mutant clone on one side of the wing, or driving the expression of a dominant negative integrin construct in just the yolk or the amnioserosa during dorsal closure[40,103]). A recurring theme in the deployment of *Drosophila* integrins is therefore that they mediate adhesion between cell layers rather than of a cell layer to a basement membrane (muscles and tendon cell at muscle attachment sites, yolk and amnioserosa during dorsal closure, dorsal and ventral epithelial layers in the wing).

The α subunits often demonstrate complementarity of distribution in the apposing layers (but not necessarily mutual exclusion, with more than one α expressed in a given layer in some instances) and show requirements predominantly in that layer. This is evident from the altered distribution and poor transport of the βPS subunit in the absence of the α subunit exclusively in that layer.[104] Overexpression of the wrong heterodimer in the wing epithelial layer produces a blister analogous to that seen in the loss of function clones. The reason for this is not yet clear, but genetic studies have suggested that it is a gain of function phenotype.[71,117] One mechanism that may explain complementarity is that adhesion is facilitated by the bridging of a ligand that binds to sites in two α subunits. It has indeed been shown that the specificity that each α subunit brings to integrin function resides in the extracellular domain; swapping the cytoplasmic domains between αPS1 and αPS2 had no detectable effects on their function.[104]

Insights into the Mechanism of Integrin Function

Establishing a Hierarchy: Not All Integrin-Associated Proteins Are Equally Important

As progress is made toward isolating mutations in every protein required for integrin function, the mutations isolated so far can be used to address two fundamental problems of integrin biology relating to the diversity and complexity of the intracellular complexes recruited by integrins. Our discussion will focus on the proteins characterised both molecularly and genetically in flies: talin, ILK, PINCH, tensin and Short stop. First a brief mention about the structure of these proteins. Talin is large, (>2500 residues) and has two well defined domains, an N-terminal FERM domain, which generally seems to be used for interacting with membrane proteins, and a C-terminal I/LWEQ domain, shared with the yeast actin binding protein Sla2.[105] It has been found to bind to several proteins of interest, including the integrin β subunit cytoplasmic tail, FAK, vinculin, PI4-5 kinase, and actin. Integrin-linked-kinase (ILK) has an N-terminus containing ankyrin repeats and a kinase domain that is diverged at several residues normally highly conserved in active kinases, making it unclear whether it has substantial kinase activity.[106,107] The ankyrin repeats bind to PINCH, and this has been confirmed in vertebrate cells, worms and flies. The kinase domain binds to integrins (but this could not be verified in flies or worms), the protein parvin (also known as actopaxin, affixin, PAT-6), paxillin, and to UNC-112. PINCH consists of 5 LIM domains, and binds to ILK and the adaptor Nck2. The former interaction has been confirmed in flies and worms, but not the latter. The fly tensin appears to have lost the N-terminal phosphatase domain, although it retains actin binding activity. At the C-terminus, tensin contains two phosphotryosine binding domains, SH2 and PTB.[22] Short stop is a giant protein of the spectrin and plakin families, containing an N-terminal actin binding domain and a C-terminal microtubule binding domain.[109]

An outstanding question in integrin biology is how the same integrin performs different functions at different times and places during development. For example, we wish to understand what changes an integrin from one that mediates migration to one that forms stable adhesive junctions. The simplest explanation is that the differences result from a change in the intracellular proteins available to interact with integrins and therefore it seems likely that the intracellular machinery downstream of integrins is different for the two integrin functions. The genetics of the integrin-associated proteins has supported this view, by showing that the components are only essential for some integrin functions.

Talin is required for almost all integrin functions, including migration, muscle attachment, germband retraction, dorsal closure and wing adhesion, and so can be viewed as part of the "core" of integrin intracellular machinery. This view has been supported by recent work in vertebrate cells,[110] despite the fact that the analysis is hampered by the existence of two talin homologues in mouse and in humans. For example, the knockout of the single fly talin gene results in the inability of the integrins to form to focal adhesion-like structures in the wing primordium,[15] while mouse cells lacking talin1 are still able to form focal adhesions, although at a reduced rate and delayed in time.[111,112] There are two known integrin-dependent processes in the fly that do not require talin function: signalling to regulate gene expression (or FAK phosphorylation in the mouse[112]) and βν-dependent migration, as previously discussed. The talin mutant phenotype in the midgut is the same as βPS, with a delay and eventual recovery of migration and unlike the double β subunit mutation in which migration is blocked.[19] These results show that the βν-mediated migration, which can only be observed as the completion of migration in the absence of βPS, does not use talin. This is further supported by the fact that the well characterised binding site for talin on β subunit cytoplasmic domains is not conserved in βν. The protein that substitutes for talin in βν-mediated migration is not known.

Other integrin-associated proteins characterised to date have even more limited functions. Mutations in the gene encoding the interacting protein ILK shows only a subset of the defects found in animals lacking integrins or talin, namely defects in muscle attachment and wing adhesion[23] while PINCH mutants have dorsal closure defects in addition to these.[16,96] Flies lacking tensin only have a wing blister phenotype, which is distinct from the blisters produced by other mutations in that it is caused by the normal mechanical abrasion produced by stroking the wings with the legs to help flatten the wing.[22] Thus, it is possible to rescue the tensin mutant wing blister phenotype by immobilising the legs of the mutant flies. The distribution of these proteins does not fully correlate with function: ILK colocalises with integrins and talin during dorsal closure, but is not required there, while tensin is neither associated with these proteins during dorsal closure nor required there, but it is found with them at the muscle ends even though it is not required. It is possible that the activity of these proteins is regulated differently for different functions. The distribution of Short stop more accurately reflects its function. In the embryo, Short stop is only expressed in the tendon cells, not the muscles, and its function with integrins appears to be limited to cases where integrin junctions are connected to microtubules, a junction that appears to be analogous to the hemidesmosome of vertebrate epithelial cells.[13,14] Short stop has roles in the development of the trachea, nervous system and oocyte that are independent of integrins.[54,109] In the converse situation, several proteins are known to be required for dorsal closure but not other integrin events, e.g., jun kinase kinase.[121] Proteins that are specifically required for integrin-dependent migration have not yet been identified.

The second related problem is to discover why the intracellular machinery connected to integrins is so complex; over 60 proteins have been documented that may play a role in this connection.[49] Several of them have the potential to bind directly to integrins and actin. Even if we take the rigorous line of requiring mutations to give phenotypes similar to integrin mutants, we still have several intracellular proteins required, and there are many remaining to be tested. By comparing the detailed cell biology of the different mutant phenotypes it has been possible to build up a hierarchy of function and obtain clues towards the different roles of the proteins. The large attachment sites of embryonic muscle have been most amenable for this analysis, due to the ease of antibody staining compared to the attachments within the pupal wing. It has been informative to document the timing of the defect and the position of the failure in the ECM-integrin/plasma membrane-actin cytoskeleton link (Fig. 3). Thus, in embryos lacking integrins the detachment occurs right after the muscles become contractile, and occurs between the ECM and plasma membrane, with some plasma membrane-actin connection still formed.[7,23,39,45] In the absence of talin, the detachment occurs at the same early stage, with a detachment between integrins and actin. Muscles can be seen without any integrin on the cell surface, and it appears as if the integrin has been left behind, still attached to the ECM. Whether the integrins are pulled out of the plasma membrane or whether a small "vesicle" of plasma membrane and cytoskeleton remains associated is not known. In embryos lacking either ILK and PINCH the detachment occurs at the very end of embryogenesis, about 8 hours later than in the absence of integrins or talin.[16,23] Our hypothesis is that this indicates that the connection between integrins and actin has been weakened, rather than eliminated, and it takes building up the muscle structure and the production of stronger contractions to generate a force strong enough to break the connection. The failure in the connection also occurs at a different point, between the integrins/plasma membrane and actin cytoskeleton, an "actin detachment" phenotype rather than "muscle detachment".[23] Thus, components of the integrin complex function at different levels in the link between the membrane and the cytoskeleton.

Examination of the distribution of the components in the different mutants also supports this hierarchical view. Integrins are required for the recruitment of talin, PINCH and tensin, and reciprocally none of the intracellular proteins are required for the localisation of integrins to the ends of the muscles.[15,16,22] In summary, the chronology of the onset of the phenotypes and the dependence of components on others for their localisation at the muscle attachment sites allows the ordering of the molecules into a hierarchy:

integrins → talin → ILK → tensin
→ PINCH

that is not strictly linear. This strengthens the notion that each component performs a crucial function, separable in time and dependant on interactions that precede it. Some are crucial for the establishment of the junction and show early phenotypes (integrin and talin). Others (ILK complex) play crucial roles linking the junctions, once established, to the actin cytoskeleton[23] and still others (e.g., tensin) have been hypothesised to strengthen this link.[22]

Structure function analyses facilitated by engineering mutations and deletions in the genomic context have also provided insights into how the molecules function in the context of the whole organism. Such studies have revealed that ILK functions as a scaffold for protein-protein interactions and not as a kinase.[23] In worms as in flies, mutations that are predicted to eliminate kinase activity do not impair function while in the mouse the results of similar experiments are less clear cut.[113,114,120] Similar studies on tensin have identified a novel mechanism of recruitment implicating a crucial role for the PTB domain in targeting the protein to the muscle end where it can interact with phosphoproteins through its SH2 domain.[22]

Future Prospects

We anticipate that the analysis of all integrins and associated proteins in *Drosophila* will shed light on the hierarchical organisation and molecular architecture of the "core" complex that mediates stable adhesion. A detailed analysis of morphogenetic movements that involve "dynamic adhesion" in mutants in integrin associated proteins should help understand how different the composition of the complex is from the one that mediates stable adhesion. As mentioned earlier in this chapter, expression does not imply an essential function and it is likely that certain components are differentially activated in the two types of adhesion. How activity is modulated will be important to understand. From a cell biological standpoint, it will be interesting to understand the molecular interactions that facilitate the translation of integrin-substrate interactions into changes in cell morphology and behaviour.

Acknowledgements

We thank members of the Brown lab 'Writing Workshop' for comments and suggestions.

References

1. Bokel C, Brown NH. Integrins in development: Moving on, responding to, and sticking to the extracellular matrix. Dev Cell 2002; 3(3):311-321.
2. Brower DL. Platelets with wings: The maturation of Drosophila integrin biology. Curr Opin Cell Biol 2003; 15(5):607-613.
3. Brown NH, Gregory SL, Martin-Bermudo MD. Integrins as mediators of morphogenesis in Drosophila. Dev Biol 2000; 223(1):1-16.
4. Hynes RO, Zhao Q. The evolution of cell adhesion. J Cell Biol 2000; 150(2):F89-96.
5. Wilcox M, Brower DL, Smith RJ. A position-specific cell surface antigen in the Drososphila wing imaginal disc. Cell 1981; 25:159-164.
6. Weinstein A. Coincidence of crossing over in Drosophila melanogaster (Amelophila). Genetics 1918; 3:135-172.
7. Wright TRF. The phenogenetics of the embryonic mutant, lethal myospheroid, in Drosophila melanogaster. Development 1960; 118:845-858.
8. Woodruff RC, Ashburner M. The genetics of a small autosomal region of Drosophila melanogaster containing the structural gene for alcohol dehydrogenase. I. Characterization of deficiencies and mapping of ADH and visible mutations. Genetics 1979; 92(1):117-132.
9. Glass HB. New mutants report. D.I.S. 1937; 2:7-8.
10. Brower DL, Bunch TA, Mukai L et al. Nonequivalent requirements for PS1 and PS2 integrin at cell attachments in Drosophila: Genetic analysis of the alpha PS1 integrin subunit. Development 1995; 121(5):1311-1320.

11. Prout M, Damania Z, Soong J et al. Autosomal mutations affecting adhesion between wing surfaces in Drosophila melanogaster. Genetics 1997; 146(1):275-285.
12. Walsh EP, Brown NH. A screen to identify Drosophila genes required for integrin-mediated adhesion. Genetics 1998; 150(2):791-805.
13. Strumpf D, Volk T. Kakapo, a novel cytoskeletal-associated protein is essential for the restricted localization of the neuregulin-like factor, vein, at the muscle-tendon junction site. J Cell Biol 1998; 143(5):1259-1270.
14. Gregory SL, Brown NH. kakapo, a gene required for adhesion between and within cell layers in Drosophila, encodes a large cytoskeletal linker protein related to plectin and dystrophin. J Cell Biol 1998; 143(5):1271-1282.
15. Brown NH, Gregory SL, Rickoll WL et al. Talin is essential for integrin function in Drosophila. Dev Cell 2002; 3(4):569-579.
16. Clark KA, McGrail M, Beckerle MC. Analysis of PINCH function in Drosophila demonstrates its requirement in integrin-dependent cellular processes. Development 2003; 130(12):2611-2621.
17. Bellen HJ, Levis RW, Liao G et al. The BDGP gene disruption project: Single transposon insertions associated with 40% of Drosophila genes. Genetics 2004; 167(2):761-781.
18. Fogerty FJ, Fessler LI, Bunch TA et al. Tiggrin, a novel Drosophila extracellular matrix protein that functions as a ligand for Drosophila alpha PS2 beta PS integrins. Development 1994; 120(7):1747-1758.
19. Devenport D, Brown NH. Morphogenesis in the absence of integrins: Mutation of both Drosophila beta subunits prevents midgut migration. Development 2004; 131(21):5405-5415.
20. Grabbe C, Zervas CG, Hunter T et al. Focal adhesion kinase is not required for integrin function or viability in Drosophila. Development 2004; 131(23):5795-5805.
21. Lee SB, Cho KS, Kim E et al. blistery encodes Drosophila tensin protein and interacts with integrin and the JNK signaling pathway during wing development. Development 2003; 130(17):4001-4010.
22. Torgler CN, Narasimha M, Knox AL et al. Tensin stabilizes integrin adhesive contacts in Drosophila. Dev Cell 2004; 6(3):357-369.
23. Zervas CG, Gregory SL, Brown NH. Drosophila integrin-linked kinase is required at sites of integrin adhesion to link the cytoskeleton to the plasma membrane. J Cell Biol 2001; 152(5):1007-1018.
24. Gotwals PJ, Fessler LI, Wehrli M et al. Drosophila PS1 integrin is a laminin receptor and differs in ligand specificity from PS2. Proc Natl Acad Sci USA 1994; 91(24):11447-11451.
25. Bunch TA, Brower DL. Drosophila PS2 integrin mediates RGD-dependent cell-matrix interactions. Development 1992; 116(1):239-247.
26. Stark KA, Yee GH, Roote CE et al. A novel alpha integrin subunit associates with betaPS and functions in tissue morphogenesis and movement during Drosophila development. Development 1997; 124(22):4583-4594.
27. Schock F, Perrimon N. Retraction of the Drosophila germ band requires cell-matrix interaction. Genes Dev 2003; 17(5):597-602.
28. Nusslein-Volhard C, Wieschaus E, Kluding H. Mutations affecting the pattern of the larval cuticle in Drosophila melanogaster I. Zygotic loci on the second chromosome. Roux's Arch Dev Biol 1984; 193:267-2892.
29. Grotewiel MS, Beck CD, Wu KH et al. Integrin-mediated short-term memory in Drosophila. Nature 1998; 391(6666):455-460.
30. Yee GH, Hynes RO. A novel, tissue-specific integrin subunit, beta nu, expressed in the midgut of Drosophila melanogaster. Development 1993; 118(3):845-858.
31. Bloor JW, Brown NH. Genetic analysis of the Drosophila alphaPS2 integrin subunit reveals discrete adhesive, morphogenetic and sarcomeric functions. Genetics 1998; 148(3):1127-1142.
32. Jannuzi AL, Bunch TA, Brabant MC et al. Disruption of C-terminal cytoplasmic domain of betaPS integrin subunit has dominant negative properties in developing Drosophila. Mol Biol Cell 2002; 13(4):1352-1365.
33. Jannuzi AL, Bunch TA, West RF et al. Identification of integrin beta subunit mutations that alter heterodimer function in situ. Mol Biol Cell 2004; 15(8):3829-3840.
34. Brown NH. An integrin chicken and egg problem: Which comes first, the extracellular matrix or the cytoskeleton? Curr Opin Cell Biol 2000; 12(5):629-633.
35. Montell DJ, Goodman CS. Drosophila laminin: Sequence of B2 subunit and expression of all three subunits during embryogenesis. J Cell Biol 1989; 109(5):2441-2453.
36. Kusche-Gullberg M, Garrison K, MacKrell AJ et al. Laminin A chain: Expression during Drosophila development and genomic sequence. Embo J 1992; 11(12):4519-4527.
37. Fessler LI, Campbell AG, Duncan KG et al. Drosophila laminin: Characterization and localization. J Cell Biol 1987; 105(5):2383-2391.

38. Martin D, Zusman S, Li X et al. Wing blister, a new Drosophila laminin alpha chain required for cell adhesion and migration during embryonic and imaginal development. J Cell Biol 1999; 145(1):191-201.
39. Prokop A, Martin-Bermudo MD, Bate M et al. Absence of PS integrins or laminin A affects extracellular adhesion, but not intracellular assembly, of hemiadherens and neuromuscular junctions in Drosophila embryos. Dev Biol 1998; 196(1):58-76.
40. Narasimha M, Brown NH. Novel functions for integrins in epithelial morphogenesis. Curr Biol 2004; 14(5):381-385.
41. Jacinto A, Woolner S, Martin P. Dynamic analysis of dorsal closure in Drosophila: From genetics to cell biology. Dev Cell 2002; 3(1):9-19.
42. Zavortink M, Bunch TA, Brower DL. Functional properties of alternatively spliced forms of the Drosophila PS2 integrin alpha subunit. Cell Adhes Commun 1993; 1(3):251-264.
43. Voigt A, Pflanz R, Schafer U et al. Perlecan participates in proliferation activation of quiescent Drosophila neuroblasts. Dev Dyn 2002; 224(4):403-412.
44. Bunch TA, Graner MW, Fessler LI et al. The PS2 integrin ligand tiggrin is required for proper muscle function in Drosophila. Development 1998; 125(9):1679-1689.
45. Newman Jr SM, Wright TRF. A histological and ultrastructural analysis of developmental defects produced by the mutation, lethal(1) myospheroid, in Drosophila melanogaster. Developmental Biology 1981; 86:393-402.
46. Martin-Bermudo MD, Brown NH. The localized assembly of extracellular matrix integrin ligands requires cell-cell contact. J Cell Sci 2000; 113(Pt 21):3715-3723.
47. Umemiya T, Takeichi M, Nose A. M-spondin, a novel ECM protein highly homologous to vertebrate F-spondin, is localized at the muscle attachment sites in the Drosophila embryo. Dev Biol 1997; 186(2):165-176.
48. Artero R, Prokop A, Paricio N et al. The muscleblind gene participates in the organization of Z-bands and epidermal attachments of Drosophila muscles and is regulated by Dmef2. Dev Biol 1998; 195(2):131-143.
49. Zamir E, Geiger B. Molecular complexity and dynamics of cell-matrix adhesions. J Cell Sci 2001; 114(Pt 20):3583-3590.
50. Noselli S, Perrimon N. Signal transduction. Are there close encounters between signaling pathways? Science 2000; 290(5489):68-69.
51. Weinkove D, Neufeld TP, Twardzik T et al. Regulation of imaginal disc cell size, cell number and organ size by Drosophila class I(A) phosphoinositide 3-kinase and its adaptor. Curr Biol 1999; 9(18):1019-1029.
52. Diaz-Benjumea FJ, Hafen E. The sevenless signalling cassette mediates Drosophila EGF receptor function during epidermal development. Development 1994; 120(3):569-578.
53. Lee S, Kolodziej PA. Short Stop provides an essential link between F-actin and microtubules during axon extension. Development 2002; 129(5):1195-1204.
54. Roper K, Brown NH. A spectraplakin is enriched on the fusome and organizes microtubules during oocyte specification in Drosophila. Curr Biol 2004; 14(2):99-110.
55. Alatortsev VE, Kramerova IA, Frolov MV et al. Vinculin gene is nonessential in Drosophila melanogaster. FEBS Lett 1997; 413(2):197-201.
56. Xu W, Baribault H, Adamson ED. Vinculin knockout results in heart and brain defects during embryonic development. Development 1998; 125(2):327-337.
57. Ilic D, Furuta Y, Kanazawa S et al. Reduced cell motility and enhanced focal adhesion contact formation in cells from FAK-deficient mice. Nature 1995; 377(6549):539-544.
58. Fyrberg C, Ketchum A, Ball E et al. Characterization of lethal Drosophila melanogaster alpha-actinin mutants. Biochem Genet 1998; 36(9-10):299-310.
59. Roulier EM, Fyrberg C, Fyrberg E. Perturbations of Drosophila alpha-actinin cause muscle paralysis, weakness, and atrophy but do not confer obvious nonmuscle phenotypes. J Cell Biol 1992; 116(4):911-922.
60. Sokol NS, Cooley L. Drosophila filamin encoded by the cheerio locus is a component of ovarian ring canals. Curr Biol 1999; 9(21):1221-1230.
61. Robinson DN, Smith-Leiker TA, Sokol NS et al. Formation of the Drosophila ovarian ring canal inner rim depends on cheerio. Genetics 1997; 145(4):1063-1072.
62. Strutt D, Johnson R, Cooper K et al. Asymmetric localization of frizzled and the determination of notch-dependent cell fate in the Drosophila eye. Curr Biol 2002; 12(10):813-824.
63. Roulier EM, Panzer S, Beckendorf SK. The Tec29 tyrosine kinase is required during Drosophila embryogenesis and interacts with Src64 in ring canal development. Mol Cell 1998; 1(6):819-829.
64. Tateno M, Nishida Y, Adachi-Yamada T. Regulation of JNK by Src during Drosophila development. Science 2000; 287(5451):324-327.

65. Read RD, Bach EA, Cagan RL. Drosophila C-terminal Src kinase negatively regulates organ growth and cell proliferation through inhibition of the Src, Jun N-terminal kinase, and STAT pathways. Mol Cell Biol 2004; 24(15):6676-6689.
66. Pedraza LG, Stewart RA, Li DM et al. Drosophila Src-family kinases function with Csk to regulate cell proliferation and apoptosis. Oncogene 2004; 23(27):4754-4762.
67. Damsky CH, Ilic D. Integrin signaling: It's where the action is. Curr Opin Cell Biol 2002; 14(5):594-602.
68. DeMali KA, Wennerberg K, Burridge K. Integrin signaling to the actin cytoskeleton. Curr Opin Cell Biol 2003; 15(5):572-582.
69. Martin-Bermudo MD, Brown NH. Uncoupling integrin adhesion and signaling: The betaPS cytoplasmic domain is sufficient to regulate gene expression in the Drosophila embryo. Genes Dev 1999; 13(6):729-739.
70. Martin-Bermudo MD. Integrins modulate the Egfr signaling pathway to regulate tendon cell differentiation in the Drosophila embryo. Development 2000; 127(12):2607-2615.
71. Baker SE, Lorenzen JA, Miller SW et al. Genetic interaction between integrins and moleskin, a gene encoding a Drosophila homolog of importin-7. Genetics 2002; 162(1):285-296.
72. Araujo H, Negreiros E, Bier E. Integrins modulate Sog activity in the Drosophila wing. Development 2003; 130(16):3851-3864.
73. Gonzalez-Reyes A, St Johnston D. The Drosophila AP axis is polarised by the cadherin-mediated positioning of the oocyte. Development 1998; 125(18):3635-3644.
74. Godt D, Tepass U. Drosophila oocyte localization is mediated by differential cadherin-based adhesion. Nature 1998; 395(6700):387-391.
75. Roote CE, Zusman S. Functions for PS integrins in tissue adhesion, migration, and shape changes during early embryonic development in Drosophila. Dev Biol 1995; 169(1):322-336.
76. Martin-Bermudo MD, Alvarez-Garcia I, Brown NH. Migration of the Drosophila primordial midgut cells requires coordination of diverse PS integrin functions. Development 1999; 126(22):5161-5169.
77. Boube M, Martin-Bermudo MD, Brown NH et al. Specific tracheal migration is mediated by complementary expression of cell surface proteins. Genes Dev 2001; 15(12):1554-1562.
78. Bradley PL, Myat MM, Comeaux CA et al. Posterior migration of the salivary gland requires an intact visceral mesoderm and integrin function. Dev Biol 2003; 257(2):249-262.
79. Leptin M, Bogaert T, Lehmann R et al. The function of PS integrins during Drosophila embryogenesis. Cell 1989; 56(3):401-408.
80. Schnorrer F, Dickson BJ. Muscle building; mechanisms of myotube guidance and attachment site selection. Dev Cell 2004; 7(1):9-20.
81. Schwander M, Leu M, Stumm M et al. Beta1 integrins regulate myoblast fusion and sarcomere assembly. Dev Cell 2003; 4(5):673-685.
82. Brown NH. Null mutations in the alpha PS2 and beta PS integrin subunit genes have distinct phenotypes. Development 1994; 120(5):1221-1231.
83. Wilkin MB, Becker MN, Mulvey D et al. Drosophila dumpy is a gigantic extracellular protein required to maintain tension at epidermal-cuticle attachment sites. Curr Biol 2000; 10(10):559-567.
84. Jazwinska A, Ribeiro C, Affolter M. Epithelial tube morphogenesis during Drosophila tracheal development requires Piopio, a luminal ZP protein. Nat Cell Biol 2003; 5(10):895-901.
85. Bokel C, Prokop A, Brown NH. Papillote and Piopio, Drosophila ZP-domain proteins required for cell adhesion to the apical extracellular matrix and microtubule organisation. J Cell Sci 2004; In Press.
86. Prokop A, Uhler J, Roote J et al. The kakapo mutation affects terminal arborization and central dendritic sprouting of Drosophila motorneurons. J Cell Biol 1998; 143(5):1283-1294.
87. Reed BH, Wilk R, Schock F et al. Integrin-dependent apposition of Drosophila extraembryonic membranes promotes morphogenesis and prevents anoikis. Curr Biol 2004; 14(5):372-380.
88. Hutson MS, Tokutake Y, Chang MS et al. Forces for morphogenesis investigated with laser microsurgery and quantitative modeling. Science 2003; 300(5616):145-149.
89. Jacinto A, Wood W, Balayo T et al. Dynamic actin-based epithelial adhesion and cell matching during Drosophila dorsal closure. Curr Biol 2000; 10(22):1420-1426.
90. Ashton GH, McLean WH, South AP et al. Recurrent mutations in kindlin-1, a novel keratinocyte focal contact protein, in the autosomal recessive skin fragility and photosensitivity disorder, Kindler syndrome. J Invest Dermatol 2004; 122(1):78-83.
91. Brabant MC, Fristrom D, Bunch TA et al. Distinct spatial and temporal functions for PS integrins during Drosophila wing morphogenesis. Development 1996; 122(10):3307-3317.
92. Fristrom D, Wilcox M, Fristrom J. The distribution of PS integrins, laminin A and F-actin during key stages in Drosophila wing development. Development 1993; 117(2):509-523.
93. Longley Jr RL, Ready DF. Integrins and the development of three-dimensional structure in the Drosophila compound eye. Dev Biol 1995; 171(2):415-433.

94. D'Avino PP, Thummel CS. The ecdysone regulatory pathway controls wing morphogenesis and integrin expression during Drosophila metamorphosis. Dev Biol 2000; 220(2):211-224.
95. Kozlova T, Thummel CS. Essential roles for ecdysone signaling during Drosophila mid-embryonic development. Science 2003; 301(5641):1911-1914.
96. Kadrmas JL, Smith MA, Clark KA et al. The integrin effector PINCH regulates JNK activity and epithelial migration in concert with Ras suppressor 1. J Cell Biol 2004; 167(6):1019-1024.
97. Tanentzapf G, Smith C, McGlade J et al. Apical, lateral, and basal polarization cues contribute to the development of the follicular epithelium during Drosophila oogenesis. J Cell Biol 2000; 151(4):891-904.
98. Bilder D. Epithelial polarity and proliferation control: Links from the Drosophila neoplastic tumor suppressors. Genes Dev 2004; 18(16):1909-1925.
99. Hoang B, Chiba A. Genetic analysis on the role of integrin during axon guidance in Drosophila. J Neurosci 1998; 18(19):7847-7855.
100. Lee S, Harris KL, Whitington PM et al. Short stop is allelic to kakapo, and encodes rod-like cytoskeletal-associated proteins required for axon extension. J Neurosci 2000; 20(3):1096-1108.
101. Rohrbough J, Grotewiel MS, Davis RL et al. Integrin-mediated regulation of synaptic morphology, transmission, and plasticity. J Neurosci 2000; 20(18):6868-6878.
102. Beumer K, Matthies HJ, Bradshaw A et al. Integrins regulate DLG/FAS2 via a CaM kinase II-dependent pathway to mediate synapse elaboration and stabilization during postembryonic development. Development 2002; 129(14):3381-3391.
103. Brower DL, Jaffe SM. Requirement for integrins during Drosophila wing development. Nature 1989; 342(6247):285-287.
104. Martin-Bermudo MD, Dunin-Borkowski OM, Brown NH. Specificity of PS integrin function during embryogenesis resides in the alpha subunit extracellular domain. Embo J 1997; 16(14):4184-4193.
105. Nayal A, Webb DJ, Horwitz AF. Talin: An emerging focal point of adhesion dynamics. Curr Opin Cell Biol 2004; 16(1):94-98.
106. Brakebusch C, Fassler R. The integrin-actin connection, an eternal love affair. Embo J 2003; 22(10):2324-2333.
107. Zervas CG, Brown NH. Integrin adhesion: When is a kinase a kinase? Curr Biol 2002; 12(10):R350-351.
108. Wilcox M, DiAntonio A, Leptin M. The function of PS integrins in Drosophila wing morphogenesis. Development 1989; 107(4):891-897.
109. Roper K, Gregory SL, Brown NH. The 'spectraplakins': Cytoskeletal giants with characteristics of both spectrin and plakin families. J Cell Sci 2002; 115(Pt 22):4215-4225.
110. Tadokoro S, Shattil SJ, Eto K et al. Talin binding to integrin beta tails: A final common step in integrin activation. Science 2003; 302(5642):103-106.
111. Monkley SJ, Zhou XH, Kinston SJ et al. Disruption of the talin gene arrests mouse development at the gastrulation stage. Dev Dyn 2000; 219(4):560-574.
112. Giannone G, Jiang G, Sutton DH et al. Talin1 is critical for force-dependent reinforcement of initial integrin-cytoskeleton bonds but not tyrosine kinase activation. J Cell Biol 2003; 163(2):409-419.
113. Mackinnon AC, Qadota H, Norman KR et al. C. elegans PAT-4/ILK functions as an adaptor protein within integrin adhesion complexes. Curr Biol 2002; 12(10):787-797.
114. Troussard AA, Mawji NM, Ong C et al. Conditional knock-out of integrin-linked kinase demonstrates an essential role in protein kinase B/Akt activation. J Biol Chem 2003; 278(25):22374-22378.
115. Zusman S, Patel-King RS, Ffrench-Constant C et al. Requirements for integrins during Drosophila development. Development 1990; 108(3):391-402.
116. Jobard F, Bouadjar B, Caux F et al. Identification of mutations in a new gene encoding a FERM family protein with a pleckstrin homology domain in Kindler syndrome. Hum Mol Genet 2003; 12(8):925-935.
117. Brabant MC, Fristrom D, Bunch TA et al. The PS integrins are required for a regulatory event during Drosophila wing morphogenesis. Ann NY Acad Sci 1998; 857:99-109.
118. Henchcliffe C, Garcia-Alonso L, Tang J et al. Genetic analysis of laminin A reveals diverse functions during morphogenesis in Drosophila. Development 1993; 118(2):325-337.
119. Wu C, Dedhar S. Integrin-linked kinase (ILK) and its interactors: A new paradigm for the coupling of extracellular matrix to actin cytoskeleton and signaling complexes. J Cell Biol 2001; 155(4):505-510.
120. Sakai T, Li S, Docheva D et al. Integrin-linked kinase (ILK) is required for polarizing the epiblast, cell adhesion, and controlling actin accumulation. Genes Dev 2003; 17(7):926-940.
121. Stronach BE, Perrimon N. Stress signaling in Drosophila. Oncogene 1999; 18(45):6172-6182.
122. Hartenstein V. Atlas of Drosophila development. New York: Cold Spring Harbor Laboratory Press, 1993.

CHAPTER 7

Integrins in the Skin

Cord Brakebusch*

Introduction

The skin protects the body of mammalians from mechanical and chemical damage, is a barrier against infections and prevents dehydration. It can be divided into an upper layer, called epidermis, and a lower layer, called dermis (Fig. 1). The epidermis consists mainly of keratinocytes, while the dermis is a loose connective tissue containing dermal fibroblasts, macrophages, neurons and other cells. These layers are separated by a basement membrane (BM), a sheet like structure made from different extracellular matrix molecules including laminin 5, collagen IV and nidogen, that mechanically connects epidermis and dermis and provides signals for cell proliferation and polarisation.[4] At the epidermal side, basal keratinocytes are attached to the BM. In healthy conditions only these keratinocytes are proliferating. When further differentiating, keratinocytes detach from the BM, move upward and become suprabasal keratinocytes. Suprabasal keratinocytes later terminally differentiate, to form a layer of dead cells called stratum corneum. It is mainly this layer, which provides the protective barrier function to the skin. During terminal differentiation keratinocytes fill up with keratin filaments, seal the space between the cell bodies by lipids and degrade the nucleus.[1] This process has many similarities to apoptosis, but is much slower and requires the production of many proteins.[20]

Hair follicles are skin appendages that arise from skin stem cells (Fig. 2). Hair follicle formation is induced in the epidermis during embryogenesis by signals from the dermis involving classical Wnt signaling.[1,2,32] They consist of different layers which are grouped into outer root sheath (ORS), inner root sheath (IRS) and hair shaft.[43] The ORS cells at the outside are contiguous with the basal keratinocytes. In this layer, stem cells migrate down to the hair bulb region where they first expand as transit amplifying matrix cells and then differentiate to the companion layer, IRS (Henle layer, Huxley layer, cuticle) and hair shaft (cuticle, cortex, medulla). During the transition from stem cells to transit amplifying matrix cells and the differentiation into IRS and hair shaft cells signal from the dermis, provided by the fibroblasts of the dermal papilla, are crucial. Vice versa, soluble signals from the keratinocytes are influencing the dermal papilla fibroblasts resulting in a bidirectional information exchange involving Shh, Bmps, Fgfs and Wnts.[1,2,32]

Hair follicles continuously cycle between periods of growth, called anagen, and periods of involution and resting, called catagen and telogen, respectively.[43] During embryogenesis, stem cells from the epidermis form the hair follicle. Later, stem cells seem to reside mainly in the bulge region of the hair follicle close to the sebaceous glands. Elegant transfer experiments indicated that bulge derived stem cells are not only important for hair follicle growth, but also migrate to the interfollicular epidermis where they can contribute to transit-amplifying basal keratinocytes.[34]

In unwounded, healthy skin integrins are expressed only in the basal keratinocyte layer of healthy animals, with the exception of $\alpha v\beta 8$, which is restricted to suprabasal layers

*Cord Brakebusch—Max Planck Institute for Biochemistry, Department of Molecular Medicine, Am Klopferspitz 18a, 82152 Martinsried, Germany. Email: brakebus@biochem.mpg.de

Integrins and Development, edited by Erik H.J. Danen. ©2006 Landes Bioscience.

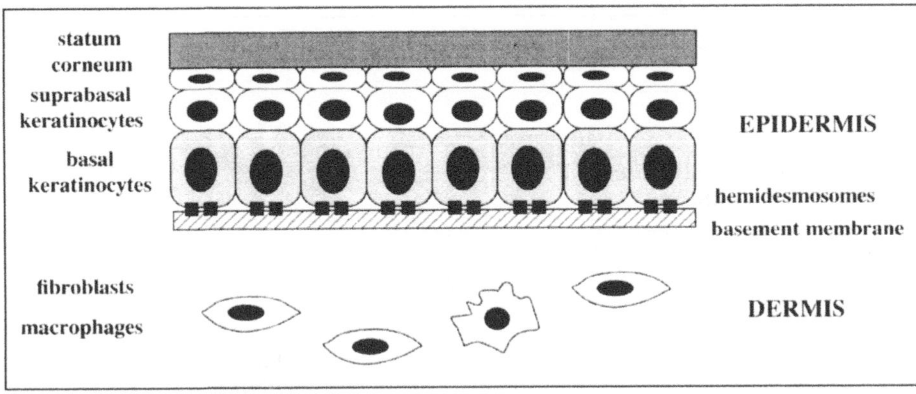

Figure 1. Skin epidermis. Epidermis and dermis are separated by a basement membrane (BM). The epidermis consists primarily of keratinocytes. Mediated via hemidesmosomes, α3β1 integrin and other adhesion receptors the basal keratinocytes are attached to the BM. Basal keratinocytes proliferate or differentiate further into suprabasal keratinocytes. Terminal differentiation results in formation of the stratum corneum, which seals the skin towards the environment. The dermis contains fibroblasts, macrophages and loose connective tissue.

(Table 1[50]). The laminin 5 receptor, α6β4 integrin is concentrated at the basal side where it is part of hemidesmosomes, specialized junctions that mediate the attachment of basal keratinocytes to the underlying BM. Integrin α3β1 also binds to laminin 5, but is found at basal, lateral and apical sides of the basal keratinocytes. At the basal side, it strengthens the adhesion to the BM. The collagen and laminin binding α2β1 integrin is present at the lateral sides. To a lower degree, tenascin C binding α9β1 and the vitronectin receptor αvβ5 are found in normal skin. During wound healing, αvβ5 and α9β1 are upregulated and α5β1 and αvβ6 are newly expressed in basal keratinocytes. α5β1 binds to fibronectin while αvβ6 mediates attachment to fibronectin, vitronectin, tenascin C and latent TGFβ, which promotes infiltration of inflammatory cells.

Development

The analysis of mice with targeted deletion of integrin genes provided profound insight into the function of integrins during skin development (Table 2). Ablation of either α6 or β4 integrin resulted in complete loss of hemidesmosomes, detachment of keratinocytes from the BM and huge blister formation.[15,48] All mutant animals died shortly after birth. A similar

Table 1. Integrins in skin

Integrin	Ligands	Expression
α2β1	collagen, laminin 5	basal keratinocytes, constitutive
α3β1	laminin 5	basal keratinocytes, constitutive
α6β4	laminin 5	basal keratinocytes,constitutive
αvβ5	vitronectin	basal keratinocytes,weak, upregulated during wounding
α9β1	vitronectin	basal keratinocytes,weak, upregulated during wounding
αvβ8	vitronectin	suprabasal keratinocytes, constitutive
α5β1	fibronectin	induced during wounding and in pathological conditions
αvβ6	fibronectin, tenascin, vitronectin, latent TGFβ	induced during wounding and in pathological conditions

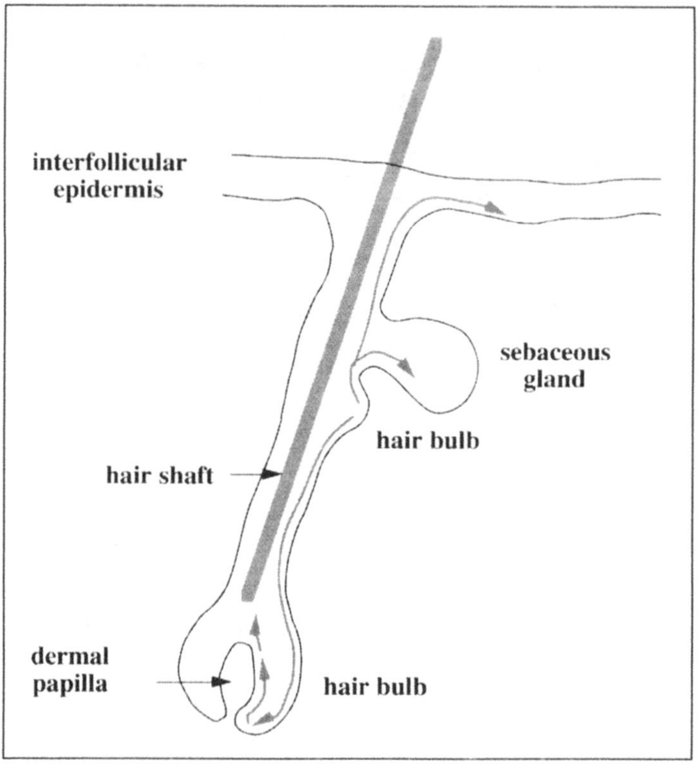

Figure 2. Hair follicles. Hair follicles are skin appendages derived from the same stem cells as the interfollicular epidermis. During the growth phase of the hair follicle, stem cells migrate from the bulge region through the outer root sheath (ORS) layer towards the hair bulb region. There they receive signals from the fibroblasts of the dermal papilla, which induce differentiation to the transit amplifying cells of the hair matrix. Further signals from the dermal papilla result in differentiation into various lineages of the inner root sheath (IRS) layer and the hair shaft. Stem cells from the bulge regions also contribute to sebaceous glands and to interfollicular epidermis.

phenotype is known from patients suffering from a subtype of epidermolysis bullosa (EB) called junctional EB with pyloric atresia.[37] Indeed, these patients carry mutations in the α6 or β4 integrin gene. Depending on the mutation, the severity of the disease can vary ranging from neonatal lethality to adult patients with blister related wounds over large parts of the body.[3] The BM itself seems to stay intact in the absence of α6β4 and also the differentiation of keratinocytes is unperturbed.[15,48]

Mice lacking α3 integrin die at around birth due to impaired kidney function.[30] At that time, already striking defects are present in the skin.[13] In some regions the BM seems to split in the middle giving rise to blisters between epidermis and dermis that in the foot pad region are visible macroscopically. A similar, but more severe phenotype was observed in mice that lack all β1 integrins restricted to keratinocytes.[8,40] Also here the BM splits in the middle and blisters are forming. After birth, the BM aberrations became more severe leading to loss of ultrastructurally detectable BM over large regions of the dermal-epidermal junction.[8] Furthermore, BM material was incompletely processed and aberrantly deposited in the dermis. To what degree the defective processing of the laminin γ2 subunit contributes to the defect in BM maintenance is not clear, since a noncleavable form of laminin γ2 can be introduced into the BM in vivo.[18] It could be that in the absence of β1 integrin also the processing of laminin α3 is

Table 2. Integrin function in skin

Proliferation	
	β1 (secondary due to effect on stem cell maintenance or migration?)
Apoptosis	
	no clear role of β1 and α6b4
Basement membrane	
	α3β1 (BM attachment, BM maintenance)
	α6β4 (BM attachment, hemidesmosome formation)
Cell migration	
	β1 (crucial in vitro and in vivo)
	α2β1 (essential in vitro, dispensable in vivo)
	αvβ6 (important in vitro, dispensable in vivo)
Tumor	
	α6β4 (tumor formation, metastasis)
	αvβ6 (metastasis?)
	β1 (metastasis?)
Stem cells	
	β1 (partial suppression of differentiation? suppression of stem cell migration?)

defective and that this is the reason for the impairment of BM integrity. Alternatively, α3β1 integrin binding to BM molecules could be crucial for the efficient integration of the proteins into the BM. β1 integrins crosstalk with α6β4, since loss of β1 integrin decreased the expression of α6β4 integrin on basal keratinocytes, which conceivably contributes to the decreased adhesion to the BM.

β1 integrin expression on keratinocytes correlated with high proliferation potential in vitro. In addition, dominant negative inhibition of integrin signaling induced keratinocyte differentiation in vitro[28,51] It was therefore suggested that β1 integrins have a crucial functional role in the maintenance of skin stem cells. However, mice with a keratinocyte specific ablation of the β1 integrin gene did not confirm an essential role of β1 integrin, since no premature differentiation of basal keratinocytes was observed.[8,40] Also terminal differentiation was unperturbed. However, mutant keratinocytes showed an aberrant expression of keratin 6, which is also seen in hyperproliferative situations such as wound healing or tumor formation.

Proliferation of β1 integrin-deficient basal keratinocytes was reduced by 70%, suggesting that β1 integrin mediated binding to extracellular matrix molecules is regulating cell cycle progression. β1 integrins might affect cell proliferation directly by synergizing with growth factor receptor signaling on Erk activation and cyclin D1 expression.[7] Alternatively or in addition, they could be crucial for the migration of bulge-derived stem cells to the interfollicular epidermis.

Integrin mediated attachment of epithelial cells to BM is often important to prevent apoptosis.[17] In vitro, loss of α3β1-mediated adhesion to laminin 5 impairs activation of FAK and Erk and causes apoptosis.[31] In vivo, however, β1 integrin dependent signals are not essential for cell survival since the frequency of apoptotic cells outside of the blisters was not increased in the absence of β1.[8,40] Mice with a combined knockout of α3 and α6 integrin die at birth and show a simple additive phenotype, indicating no obvious redundant functions between α3β1 and α6β4.[14] Interestingly, in some parts of the epidermis of E15.5 mutant embryos the epidermis was still attached to the BM without any signs of apoptosis indicating the presence of alternative adhesion and survival mechanisms.

In the absence of β1 integrin hair follicle morphogenesis was severely impaired.[8] The ORS layer thickened, the IRS layers were partially folded and the hair shaft became aberrantly pigmented and amorphous. Proliferating cells were lost in the hair matrix within 2 weeks. Since no apoptotic cells were observed, the reduced proliferation is not caused by a premature catagen when the hair follicle partially involutes (see above). Whether it is due to a lack of mitogenic integrin signaling or to an impaired migration of skin stem cells to the hair matrix region is not clear. Infiltrating macrophages and granulocytes accumulate then around the deformed hair follicles and remove them completely so that in older mice no hair follicles are left. These inflammatory cells also induce cytokines such as TGF-β1 CTGF and activin, resulting in a scleroderma-like fibrosis of older mutant mice.

Deletion of α2 or α9 integrin did not result in any skin defects.[11,25] However, since the β1 knockout is much more severe than the knockout of α3 integrin alone it is clear that α2β1 and/or α9β1 have at least partially redundant functions with α3β1 integrin. Mice lacking β5 or β6 integrin have no primary defect in skin development.[24,26]

Stem Cells

Multipotent skin stem cells exist in the basal keratinocytes layer and in the bulge region of the hair follicle. Stem cells from the bulge region can migrate down to the hair matrix or upward to the interfollicular epidermis. Whether additional stem cell populations exist in other regions of the hair follicle is debated.[36]

It was suggested that β1 integrin is not only highly expressed on keratinocyte stem cells, but that it is also essential for their maintenance (Jones and Watt, 1993; Jones et al, 1995).[28,51] However, since also in the complete absence of β1 integrin basal keratinocytes continue to proliferate for at least a year, this is clearly not the case.[8] Nevertheless, β1 integrins might contribute to stem cell maintenance, since in the absence of β1 integrin the frequency of proliferating keratinocytes was reduced. It was further speculated, that all skin stem cells in adult tissues are derived from the bulge region of the hair follicle. However, also in the absence of hair follicles epidermal keratinocytes proliferate indicating that keratinocyte stem cells exist outside the bulge region in adult mice.[8]

More recent studies questioned, whether β1 integrin expression is indeed increased on keratinocyte stem cells. First, no increased immunofluorescent staining for β1 integrin was detected on label retaining cells in the bulge region and other parts of the epidermis.[10] Secondly, an array expression study of skin stem cells defined as slow dividing cells expressing keratin 5 indicated no increased mRNA levels of β1 integrin as compared to average cells of the basal keratinocyte or ORS layer.[47] Interestingly, β6 integrin was highly upregulated in stem cells suggesting a possible role of αvβ6 integrin in stem cell function. However, mice lacking β6 integrin show normal hair growth.[24]

Wound Healing

Skin wounds are first closed by a blot clot and a provisional wound matrix. Later keratinocytes from the wound edge epidermis, including also hair follicles, proliferate and migrate to reepithelialize and close the wound. During wound healing the expression of αvβ5, αvβ6, α5β1 and α9β1 integrin are upregulated on keratinocytes.

β1 integrins play an important role in wound healing in vivo.[21] In the absence of β1 integrin wound edge keratinocytes initially do not migrate because they cannot efficiently adhere to extracellular matrix molecules of the wound matrix such as fibronectin and collagen I. In vitro, β1 integrin mediated adhesion to collagen I and laminin 5 is crucial for migration. After a few days, also β1-null keratinocytes begin to migrate, conceivably due to the increased expression of αvβ5 and αvβ6, which mediate binding to fibronectin, vitronectin and tenascin C. In vitro, αvβ6 is important, but not essential for keratinocyte migration.[27] In vivo, αvβ5 and αvβ6 are dispensable for wound healing in the presence of β1 integrin.[24,26] In contrast to unwounded epidermis, β1-deficient keratinocytes in the wound area proliferate normally, quite likely due to proliferation-promoting cytokines released during the wound response.[21]

In the absence of β1 integrin, cell-cell contacts between wound edge keratinocytes are tighter than in normal mice indicating a crosstalk between β1 integrins and cell-cell adhesion receptors.[21] Defective cell-cell interactions were also observed later during wound healing, since the keratinocyte sheets meeting in the wound center have problems to fuse in mutant mice.[21] Instead they often continue to grow and push each other up above the granulation tissue.

In vitro, various integrin receptors are crucial for cell migration in a scratch assay. First, cooperation between α6β4 and the tyrosine kinase receptor Ron is required for the activation of α3β1-mediated cell spreading and binding to laminin 5.[41] α3β1 activation, on the other hand, activates α2β1, which then mediates migration on collagen.[33] In vivo however, mice lacking the only known Ron ligand, macrophage stimulating protein (MSP), or integrin α2β1 show normal wound healing,[6,11] indicating that the requirements for in vivo migration are rather different from those in the scratch assay in vitro.

Tumor Formation

Most tumors are derived from a single cell with about five mutations in humans and two to three mutations in mice.[22,23] Since terminal differentiation of keratinocytes takes about 2-3 weeks the chances are rather low to acquire several transforming mutations during that time. Stem cells, however, have a long life span and can accumulate mutations that lead to transformation. Since integrins are expressed on keratinocyte stem cells and regulate keratinocyte proliferation in vitro and in vivo it is conceivable that changes in integrin function could affect tumor formation. In vivo, squamous cell carcinoma can have normal, reduced or increased levels of integrin.[5,29] This indicates that there is no general role for integrins in tumor formation, but that the importance of integrins rather depends on the specific type of tumor cell.

A clear correlation between the expression level of specific integrins and tumor growth was demonstrated for α6β4 integrin in human keratinocytes transformed with oncogenic Ras and IkBα, an inhibitor of the NFkB pathway.[12] Antibodies against α6β4 or its extracellular matrix ligand laminin 5, prevented tumor formation by these cells in mice. Moreover, keratinocytes lacking either β4 integrin or laminin 5 could not be made tumorigenic by infection with oncogenic Ras and IkBα. Only after reexpression of β4 integrin or laminin 5, squamous cell carcinomas were formed.

Suprabasal expression of α6β4, which is found in human squamous cell carcinoma with poor prognosis,[45,49] reduces TGFβ induced growth inhibition of basal keratinocytes in vitro.[35] This effect is dependent on cadherin-mediated cell-cell contacts. In transgenic mice, suprabasal α6β4 increased the susceptibility towards chemical carcinogenesis, independent on the β4 cytoplasmic domain.[35]

Recently it was suggested that an activating mutation in the extracellular domain of β1 integrin suppresses differentiation of the squamous cell carcinoma line SCC4.[16] Since a low degree of differentiation is indicative of a poor prognosis, β1 integrins could contribute to the disease severity. However, cell proliferation was not affected by the mutation. Transgenic mice expressing such a mutated β1 integrin in skin should be made to test the hypothesis in vivo.

Strong evidence supports a role of integrins in tumor invasion. Particularly α6β4 integrin seems to be important. In squamous carcinoma cells, EGF stimulates the mobilization of α6β4 from the hemidesmosomes at the basal side to the lamellipodia at the leading edge.[39] This loss of hemidesmosomal attachment facilitates cell movement. Furthermore, α6β4 integrin indirectly associates with the actin cytoskeleton and activates, conceivably in collaboration with growth factor receptors, PI3-K and Rho GTPases favoring cell protrusion and contraction.[19,38,42] Finally, α6β4 clusters at filopodial extensions of carcinoma cells could mediate laminin attachment at the leading edge.

αvβ6 integrin is expressed de novo in wound healing, but also in oral squamous cell carcinoma. In vitro, overexpression of αvβ6 in oral squamous cell carcinoma lines promoted cell migration and invasion, suggesting that αvβ6 contributes to metastasis formation oral squamous cell carcinoma.[46]

β1 integrin mediates keratinocyte adhesion and can promote cell protrusion and contraction by activation of Rho GTPases.[7] It is crucial for the migration of keratinocytes in vitro and during wound healing in mice. In vivo, expression of β1 integrin in a tumorigenic β1 integrin deficient fibroblastoid cell line or in a T cell lymphoma line increased the metastatic potential in vivo.[9,44] Although a similar role for β1 integrins in skin tumor invasion is highly likely, no studies have been carried out with keratinocytes.

Concluding Remarks

Integrins are important for skin development and maintenance. In addition, they play a crucial role during wound healing and tumor formation. The main task for the future is to delineate the mechanistic pathways through which integrins mediate their effects using both in vivo and in vitro models. The improved understanding of integrin function might then enable us to specifically promote beneficial functions such as the migration of reepithelializing keratinocytes and to inhibit unwanted effects like the invasion of tumor cells.

References

1. Alonso L, Fuchs E. Stem cells in the skin: Waste not, Wnt not. Genes Dev 2003; 17:1189-1200.
2. Alonso L, Fuchs E. Stem cells of the skin epithelium. Proc Natl Acad Sci USA 2003; 100(Suppl.1):11830-11835.
3. Ashton GH, Sorelli P, Mellerio JE et al. α6β4 integrin abnormalities in junctional epidermolysis bullosa with pyloric atresia. Br J Dermatol 2001; 144:408-414.
4. Aumailley M, El Khal A, Knoss N et al. Laminin 5 processing and its integration into the ECM. Matrix Biol 2003; 22:49-54.
5. Bagutti C, Speight PM, Watt FM. Comparison of integrin, cadherin, and catenin expression in squamous cell carcinomas of the oral cavity. J Pathol 1998; 186:8-16.
6. Bezerra JA, Carrick TL, Degen JL et al. Biological effects of targeted inactivation of hepatocyte growth factor-like protein in mice. J Clin Invest 1998; 101:1175-1183.
7. Brakebusch C, Bouvard D, Stanchi F et al. Integrins in invasive growth. J Clin Invest 2002; 109:999-1006.
8. Brakebusch C, Grose R, Quondamatteo F et al. Skin and hair follicle integrity is crucially dependent on β1 integrin expression on keratinocytes. EMBO J 2000; 19:3990-4003.
9. Brakebusch C, Wennerberg K, Krell HW et al. β1 integrin promotes but is not essential for metastasis of ras-myc transformed fibroblasts. Oncogene 1999; 18:3852-3861.
10. Braun KM, Niemann C, Jensen UB et al. Manipulation of stem cell proliferation and lineage commitment: Visualisation of label-retaining cells in wholemounts of mouse epidermis. Development 2003; 130:5241-5255.
11. Chen J, Diacovo TG, Grenache DG et al. The α2 integrin subunit-deficient mouse: A multifaceted phenotype including defects of branching morphogenesis and hemostasis. Am J Pathol 2002; 161:337-344.
12. Dajee M, Lazarov M, Zhang JY et al. NF-kB blockade and oncogenic Ras trigger invasive human epidermal neoplasia. Nature 2003; 421:639-643.
13. DiPersio CM, Hodivala-Dilke KM, Jaenisch R et al. α3β1 Integrin is required for normal development of the epidermal basement membrane. J Cell Biol 1997; 137:729-742.
14. DiPersio CM, van der Neut R, Georges-Labouesse E et al. α3β1 and α6β4 integrin receptors for laminin-5 are not essential for epidermal morphogenesis and homeostasis during skin development. J Cell Sci 2000; 113:3051-62.
15. Dowling J, Yu QC, Fuchs E. β4 integrin is required for hemidesmosome formation, cell adhesion and cell survival. J Cell Biol 1996; 134:559-557.
16. Evans RD, Perkins VC, Henry A et al. A tumor-associated beta 1 integrin mutation that abrogates epithelial differentiation control. J Cell Biol 2003; 160:589-596.
17. Frisch SM, Screaton RA. Anoikis mechanisms. Curr Opin Cell Biol 2002; 13:555-562.
18. Gagnoux-Palacios L, Allegra M, Spirito F et al. The short arm of the laminin γ2 chain plays a pivotal role in the incorporation of laminin 5 into the extracellular matrix and in cell adhesion. J Cell Biol 2001; 153:835-850.
19. Gambaletta D, Marchetti A, Benedetti L et al. Cooperative signaling between α6β4 integrin and ErbB-2 receptor is required to promote phosphatidylinositol 3-kinase-dependent invasion. J Biol Chem 2000; 275:10604-10610.
20. Gandarillas A. Epidermal differentiation, apoptosis, and senescence: Common pathways? Exp Gerontol 2000; 35:53-62.
21 Grose R, Hutter C, Bloch W et al. A crucial role of β1 integrins for keratinocyte migration in vitro and during cutaneous wound repair. Development 2002; 129:2303-2315.
22. Hahn WC, Counter CM, Lundberg AS et al. Creation of human tumour cells with defined genetic elements. Nature 1999; 400:464-468.
23. Hahn WC, Weinberg RA. Rules for making human tumor cells. N Engl J Med 2002; 347:1593-1603.
24. Huang XZ, Wu JF, Cass D et al. Inactivation of the integrin β6 subunit gene reveals a role of epithelial integrins in regulating inflammation in the lung and skin. J Cell Biol 1996; 133:921-928.

25. Huang XZ, Wu JF, Ferrando R et al. Fatal bilateral chylothorax in mice lacking the integrin α9β1. Mol Cell Biol 2000a; 20:5208-5215.
26. Huang X, Griffiths M, Wu J et al. Normal development, wound healing, and adenovirus susceptibility in β5-deficient mice. Mol Cell Biol 2000b; 20:755-759.
27. Huang X, Wu J, Spong S et al. The integrin alphavbeta6 is critical for keratinocyte migration on both its known ligand, fibronectin, and on vitronectin. J Cell Sci 1998; 111:2189-2195.
28. Jensen UB, Lowell S, Watt FM. The spatial relationship between stem cells and their progeny in the basal layer of human epidermis: A new view based on whole-mount labelling and lineage analysis. Development 1999; 126:2409-2418.
29. Jones J, Watt FM, Speight PM. Changes in the expression of αv integrins in oral squamous cell carcinomas. J Oral Pathol Med 1997; 26:63-68.
30. Kreidberg JA, Donovan MJ, Goldstein SL et al. α3β1 integrin has a crucial role in kidney and lung organogenesis. Development 1996; 122:3537-3547.
31. Manohar A, Shome SG, Lamar J et al. Integrin promotes keratinocyte cell survival through activation of a MEK/ERK signaling pathway. J Cell Sci 2004.
32. Millar SE. Molecular mechanisms regulating hair follicle development. J Invest Dermatol 2002; 118:216-225.
33. Nguyen BP, Ren XD, Schwartz MA et al. Ligation of integrin α3β1 by laminin 5 at the wound edge activates Rho-dependent adhesion of leading keratinocytes on collagen. J Biol Chem 2001; 276:43860-43870.
34. Oshima H, Rochat A, Kedzia C et al. Morphogenesis and renewal of hair follicles from adult multipotent stem cells. Cell 2001; 104:233-234.
35. Owens DM, Romero MR, Gardner C et al. Suprabasal α6β4 integrin expression in epidermis results in enhanced tumourigenesis and disruption of TGFβ signalling. J Cell Sci 2003; 116:3783-3791.
36. Panteleyev AA, Jahoda CA, Christiano AM. Hair follicle predetermination. J Cell Sci 2001; 114:3419-3431.
37. Pulkkinen L, Uitto J. Hemidesmosomal variants of epidermolysis bullosa. Mutations in the α6β4 integrin and the 180-kD bullous pemphigoid antigen/type XVII collagen genes. Exp Dermatol 1998; 7:46-64.
38. Rabinovitz I, Mercurio AM. The integrin α6β4 functions in carcinoma cell migration on laminin-1 by mediating the formation and stabilization of actin-containing motility structures. J Cell Biol 1997; 139:1873-1884.
39. Rabinovitz I, Toker A, Mercurio AM. Protein kinase C-dependent mobilization of the α6β4 integrin from hemidesmosomes and its association with actin-rich cell protrusions drive the chemotactic migration of carcinoma cells. J Cell Biol 1999; 146:1147-1160.
40. Raghavan S, Bauer C, Mundschau G et al. Conditional ablation of β1 integrin in skin. Severe defects in epidermal proliferation, basement membrane formation, and hair follicle invagination. J Cell Biol 2000; 150:1149-1160.
41. Santoro MM, Gaudino G, Marchisio PC. The MSP receptor regulates α6β4 and α3β1 integrins via 14-3-3 proteins in keratinocyte migration. Dev Cell 2003; 5:257-271.
42. Shaw LM, Rabinovitz I, Wang HH et al. Activation of phosphoinositide 3-OH kinase by the α6β4 integrin promotes carcinoma invasion. Cell 1997; 91:949-960.
43. Stenn KS, Paus R. Controls of hair follicle cycling. Physiol Rev 2001; 81:449-494.
44. Stroeken PJ, van Rijthoven EA, van der Valk MA et al. Targeted disruption of the β1 integrin gene in a lymphoma cell line greatly reduces metastatic capacity. Cancer Res 1998; 58:1569-1577.
45. Tennenbaum T, Weiner AK, Belanger AJ et al. The suprabasal expression of α6β4 integrin is associated with a high risk for malignant progression in mouse skin carcinogenesis. Cancer Res 1993; 53:4803-4810.
46. Thomas GJ, Lewis MP, Whawell SA et al. Expression of the αvβ6 integrin promotes migration and invasion in squamous carcinoma cells. J Invest Dermatol 2001; 117:67-73.
47. Tumbar T, Guasch G, Greco V et al. Defining the epithelial stem cell niche in skin. Science 2004; 303:359-363.
48. van der Neut R, Krimpenfort P, Calafat J et al. Epithelial detachment due to absence of hemidesmosomes in integrin β4 null mice. Nat Genet 1996; 13:366-369.
49. van Waes C, Kozarsky KF, Warren AB et al. The A9 antigen associated with aggressive human squamous carcinoma is structurally and functionally similar to the newly defined integrin α6β4. Cancer Res 1991; 51:2395-2402.
50. Watt FM. Role of integrins in regulating epidermal adhesion, growth and differentiation. EMBO J 2002; 21:3919-3926.
51. Zhu AJ, Haase I, Watt FM. Signaling via β1 integrins and mitogen-activated protein kinase determines human epidermal stem cell fate in vitro. Proc Natl Acad Sci USA 1999; 96:6728-6733.

CHAPTER 8

Integrins in Skeletal Cell Function and Development

Michael A. Horton* and Miep H. Helfrich

Abstract

Genes for 18 α and 8 β integrin subunits have been identified in vertebrates, their protein products associating into 24 different heterodimeric membrane receptors. Skeletal cells—osteoclasts, osteoblasts and chondrocytes—exhibit characteristic integrin profiles and for some there is good evidence for a functional role. This is best exemplified by the osteoclast, the 'bone' resorbing cell of the skeleton which expresses high levels of an RGD peptide-sensitive integrin, the αvβ3 vitronectin receptor, that is intimately involved in recognition and adhesion to mineralised matrix and its degradation; this property has been utilised in the development of novel drugs that inhibit αvβ3 function for use in osteoporosis and arthritis. This article will review the expression profiles of integrins in the major cell types of bone and cartilage, evidence for integrin function therein, and their role in skeletal development where known.

Introduction

Connective tissue cells in general, and bone and cartilage cells in particular, are surrounded by an abundance of extra-cellular matrix. Chondroblasts and osteoblasts are responsible for the synthesis of the majority of the organic components of skeletal matrix, which subsequently calcifies, whereas osteoclasts degrade the matrix and dissolve mineral components (Fig. 1). The function of bone and cartilage cells reflects the matrix in which they are embedded and, conversely, the composition of the matrix is highly dependent upon the cellular function of skeletal cells. Cell-matrix and cell-cell interactions are mediated by adhesion molecules and in recent years much information has become available regarding the expression and role of integrin adhesion receptors in the skeleton. Adhesion molecules not only play a part in the function of fully differentiated cells in the skeleton, but they are also increasingly implicated in their developmental pathways. Skeletal biology is well reviewed in[1] and cell adhesion in the bone and cartilage in.[2-4] In this chapter, we will review the expression and function of integrins in bone and cartilage cells and discuss results from gene deletion studies.

Scope of the Chapter

This review has been restricted in scope to describing the integrin phenotype and function of the different cell types in the skeleton and the impact upon the skeleton of integrin gene deletion. Due to space limitations, we have decided not to extend our discussion, whilst it could be considered informative, to the role of associated matrix protein ligands for the

*Corresponding Author: Michael A. Horton—Bone and Mineral Centre, Department of Medicine, Rayne Building, 5 University St, London WC1E 6JJ, U.K.
Email: m.horton@ucl.ac.uk

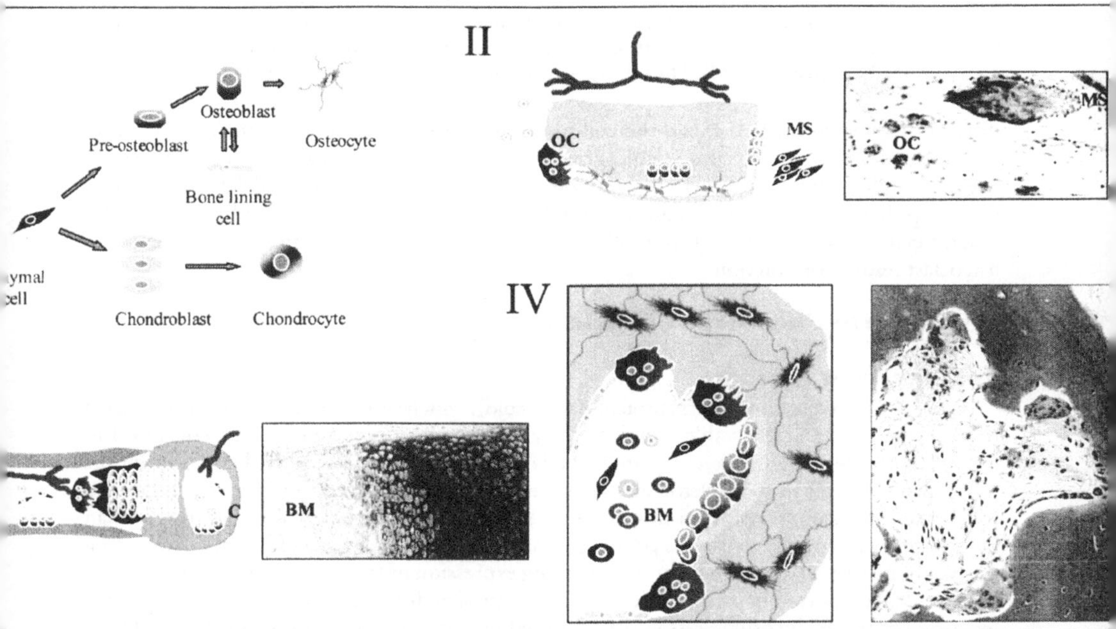

d chondroblasts are derived from a common precursor, the mesenchymal stem cell. Osteoblasts differentiate to a committed pr
izing osteoblast. When bone formation switches off, osteoblasts can revert to a bone lining cell, a state in which they do not sy
nd bone lining cells can rapidly be activated to become osteoblasts again. During bone formation a small number of osteoblasts bec
matrix. These cells do not die, but live on as osteocytes, connected with each other and with osteoblast on the bone surface. Chon
wth. When they start to generate cartilage matrix they are called chondrocytes. II) During embryogenesis two different types of
of the skull, pelvis and scapula intramembranous ossification occurs. Here, mesenchymal cells (MS) condense to form a fibrous pla
m from the mesenchymal stem cells and osteoclasts (OC) form from precursors delivered via the blood supply. Together they
) Long bones, such as femur, humerus and tibia, are formed by endochondral ossification. This means that first a cartilaginous mo
ells form an unmineralised "bone anlage", which is subsequently mineralised by hypertrophic chondrocytes (HC). Figure legen

Figure 1. Continued. This mineralized cartilage attracts blood vessels and they bring hemopoietic precursors and mesenchymal stem cells into the primitive bone marrow cavity (BM). Osteoclasts form and begin to resorb the mineralised cartilage, which acts as a template for osteoblasts, formed from the mesenchymal stem cells, that start to lay down bone matrix. With time a secondary ossification center forms above the growth plate. Articular cartilage (C) remains around the ends of the bone for smooth movement in joints.
IV: Mature bone is constantly "remodelled" (= renewed) to repair microfractures and also to release calcium during normal homeostasis. To illustrate the different processes, a piece of bone in which pathological remodelling is taking place is shown. Under normal conditions it would be rare to encounter three osteoclasts in one section of adult bone. The image on the right illustrates the histological appearance of the bone and on the left the different cell types are shown in their relative position to each other. Bone formation follows bone resorption and inactive surfaces (i.e., where neither formation nor resorption takes place) are covered by bone lining cells. Osteoclasts and osteoblast are constantly differentiating from precursors present in the bone marrow (BM). The cell types are depicted as in the lineage diagram in I.

different expressed integrins, cytoskeletal molecules and the numerous signaling pathways downstream of integrin receptors, unless crucial to the discussion. Readers can easily identify relevant publications on the web: for example, the fibronectin gene when deleted produces a phenotype like deletion of α5;[5] many clinical drug targets for osteoporosis modify intracellular F-actin cytoskeleton[6] and inhibition of c-src function leads to osteopetrosis due to defective osteoclast resorptive function.[7]

The Integrin Phenotype of Skeletal Cells

The Osteoblast Lineage

Osteoblasts secrete bone matrix proteins (as osteoid), which subsequently become mineralized to form the mechanically stable skeleton. They line bone surfaces and become entrapped in osteoid as it is deposited, when they are named osteocytes—in this capacity they act as sensors of mechanical forces in the skeleton and hence regulate bone turnover, though a comprehensive role for cell adhesion receptors in this process is still lacking.[8] Osteoblasts from normal human bone have been reported to express β1 and β5 integrins[9-11] but there are controversies regarding the α subunits associated with β1 and concerning expression of αvβ3.[4] β2, β4 and β6 integrins have not been reported in osteoblastic cells. There are also differences reported on the integrin repertoire expressed by osteoblasts in situ, cultured human osteoblasts, or osteoblastic cell lines and these are most likely due to the heterogeneity of the cell population under study and the different culture conditions employed. Moreover, information is beginning to accrue regarding the effect of osteoblast activation status (i.e., synthetic osteoblast versus bone lining cell, see (Fig. 1), anatomical site (e.g., endochondral versus intramembranous bone, (see Fig. 1) or disease on adhesion molecule expression,[12,13] though no definitive studies specifically on integrin expression have been published. For example, it has been clearly demonstrated that, in vitro, the substrate on which cells are cultured directly influences the pattern of integrin expression by osteoblasts.[14,15] Generally, cultured osteoblasts express a wider integrin repertoire than osteoblasts in situ and in particular increased expression of the collagen binding integrin α2β1, the fibronectin receptor α5β1 and the vitronectin receptor αvβ3 is seen.[4] As yet it is unclear whether increased expression of these integrins occurs in bone pathology, since osteoblast integrin expression in bone disease is relatively unexplored. Integrin expression in osteocytes has not been studied extensively, but there is some data from immuno-cytochemical analysis of human bone sections and from functional studies in the chick. In the chick, where isolation procedures for osteocytes have been developed, osteocytes bound to a range of matrix proteins in vitro in a β1- and partially RGD- dependent way[16] and expression of β1 integrin has been confirmed in mammalian osteocytes.[10,17] However, there are few definitive reports where integrin expression has been followed throughout osteoblastic differentiation to osteocytes within the same species. Difficulties in interpretation of immuno-cytochemical staining in bone sections, in particular for cells embedded within matrix, have been reported by many authors and this may well have contributed to the current controversies in integrin phenotype. From the

Table 1. Functional integrin receptor-ligand interactions in the skeleton

Receptor	Cell Type	Ligands Bound*	Possible Functions
αvβ3	osteoclast	osteopontin, bone sialoprotein, fibronectin, vitronectin, fibrinogen, denatured collagen	matrix adhesion, polarization, cessation of resorption
α1β1	chondrocyte	basement membrane collagen more than fibrillar collagens	matrix adhesion, fracture repair, osteoarthritis
α2β1	osteoclast	native collagen	matrix adhesion
	osteoblast	native collagen	matrix adhesion, differentiation
	chondrocyte	type II collagen	matrix adhesion
α5β1	osteoblast	fibronectin (RGD)	osteoblast differentiation
	chondrocyte	fibronectin (RGD)	cartilage breakdown, osteoarthritis
α11β1	chondrocyte	native fibril forming collagens	embryonic development of cartilage

*The range of ligands shown to be bound in in vitro adhesion assays; there is no definitive information on the natural ligands in bone for these molecules

limited reports on isolated cells at different stages of differentiation, it has become clear that the fibronectin-receptor α5β1 is the most abundantly expressed integrin throughout osteoblastic differentiation and that, maybe surprisingly, the collagen-receptor α2β1 is expressed at much lower levels, in particular in more differentiated cells.[18]

Functional studies (see Table 1) on osteoblast integrins have included adhesion assays to a variety of extra-cellular matrix molecules and inhibition by RGD-containing peptides. In keeping with their extensive repertoire of integrins, osteoblasts have been found to adhere to osteopontin, bone sialoprotein, vitronectin and fibronectin in an RGD-dependent way, whereas binding to type I collagen and thrombospondin was less inhibited by RGD peptides.[9,19,20] In organ cultures of mineralizing fetal rat parietal bone, RGD peptides decreased bone formation accompanied by a decrease in α2 and β1 expression and disruption of the organization in the osteoblast layer.[21] Functional studies on osteoblast integrins include reports on their role in osteoblast differentiation. First Moursi et al[22,23] demonstrated convincingly that osteoblast-fibronectin interaction was a critical event in the differentiation of fetal rat osteoblasts in an in vitro model and involved interaction between α5β1 and the central binding domain of fibronectin. These results confirmed earlier work suggesting a role for α5β1 in differentiation of the human osteoblastic cell line MG63.[24-26] Later, Ganta et al[27] found that ascorbic acid deficiency, which leads to under-hydroxylation of type I collagen, resulted in down-regulation of α2β1 in osteoblasts and dysregulation of differentiation and mineralization in cultures of rat calvaria. Jikko et al[28] showed that α1 and α2 integrins are mainly important in early stages of osteoblast differentiation to mediate differentiation signals, such as from BMP-2 and Mizuno et al[29] demonstrated a role for α2β1and collagen in osteoblast differentiation from early progenitors. These findings correlate well with the reduced levels of expression for collagen binding integrins in more differentiated cells, as mentioned above.[18] Even though mature osteoblasts adhere to collagen, it is likely that this is complemented by a nonintegrin dependent mechanism. Thus, in addition to providing an anchor for osteoblasts to the bone matrix, integrin-ligand interactions play a role in the differentiation of osteoblasts and via yet undefined signaling pathways in the expression of differentiation-associated genes.

Chondrocytes

Chondroblasts secrete the specialized extra-cellular matrix of cartilage and mature into chondrocytes that become entrapped therein (for a comprensive discussion of this complex tissue type see 1). As for cells of the osteoblast lineage, the integrin phenotype of chondrocytes is complex, with inconsistency between publications[30-34] (for a recent review see ref. 35). A synthesis of the literature shows that human chondrocytes express the β1 integrins α1, α2, α3, α5 and α6, but not α4. β2, β4 and β6 are absent; analysis of β7-9 and CD11 has not been reported. Some studies have shown high expression of αv integrin and, as with osteoblasts, this is mainly as αvβ5 and not the αvβ3 dimer; however, a subpopulation of superficial articular chondrocytes has been found to express αvβ3.[33] These complexities could well relate in part to variation in sampling site, use of fetal versus adult material, species differences, culture artifacts, or influences of disease on phenotypes. Indeed the first possibility is born out by the study of Salter et al[32] where the distribution of integrin is clearly different by site (human articular, epiphyseal and growth plate chondrocytes were studied) and changes have been reported in cultured chondrocytes.[36,37]

Two relatively newly discovered integrins, α10[38,39] and α11,[40,41] have been shown to be collagen receptors that are expressed in cartilage. α10 has a cellular distribution that differs from α1 and α2 and is the dominant integrin during embryonic development.[39] α10 preferentially binds basement membrane collagens, as α1, whereas α11 resembles α2 showing specificity for fibril forming collagens.[42]

The functional role of cell adhesion molecules in cartilage is beginning to be elucidated (see Table 1). Examples include roles in cartilage differentiation during fetal development; response to mechanical forces (for example in articular cartilage, menisci); maintenance of tissue architecture and integrity including by matrix synthesis and assembly and cell adhesion; and regulation of chondrocyte gene expression. Additionally, there is likely to be a role of cell adhesion molecules in responses in cartilage to injury and disease.[43-45] For example, up-regulation of β2 and α4 subunit expression in osteoarthritic versus normal and cartilage were reported by Ostergaard et al.[44] Changes in distribution of both integrin and matrix proteins[28] in different zones of cartilage suggest a role in chondrocyte differentiation from mesenchymal precursors[46,47] and/or interaction with matrix. Evidence is also accruing for a specialised function for integrins in responses of chondrocytes to mechanical stresses[48-50] (reviewed in ref. 51) that is mediated by α5β1 integrin in osteoarthritis cartilage.[52]

Extensive studies have been performed to address the role of α5β1 in chondrocyte interaction with fibronectin.[30,31,53-58] Function blocking antibodies and RGD peptides have been shown to inhibit cell adhesion to fibronectin and its fragments, thus modifying chondrocyte behaviour and cartilage function. Likewise, chondrocyte recognition of collagen (including types I, II and VI collagen) has been studied in vitro[30,31,53,56,59-61] and shown to be mediated via β1 integrins.

There is also increasing evidence for a connection between chondrocyte adhesion to extra-cellular matrix proteins, especially fibronectin, chondrocyte-synovial cell interaction,[62] chondrocyte cell signaling[63] and survival,[45,58,60] regulation of integrin expression and function by cytokines such as IL1, TGFβ and IGF1, and the release of matrix metalloproteinases[64] and hence cartilage breakdown.[55,61,65,66] The key role for downstream signaling following integrin interaction with matrix ligands is underscored by the phenotype of integrin-linked kinase (ILK) knockout mice.[67] ILK acts as a pleiotropic adapter interfacing beta integrin cytoplasmic domains into the PKB/Akt signaling pathway; mice that lack ILK have chondrodysplasia and isolated chondrocytes have matrix adhesion defects. Such events are likely to be involved in the pathogenesis of the cartilage destruction seen in osteoarthritis and rheumatoid arthritis.

Osteoclasts

Osteoclasts are bone cells derived from the monocyte-macrophage cell lineage under the influence of a membrane-bound osteoclast differentiation factor of the TNFα family, Receptor

Activator of NFκB Ligand (RANKL).[1] They are specialized to recognize, adhere to and degrade the proteinaceous and mineral phases of skeletal tissues[1] and utilize integrin cell adhesion receptors in these processes (Table 1); (reviewed in refs. 2,3 but first recognized by Beckstead et al, 68). The principle integrin of the osteoclast is the αvβ3 vitronectin receptor[2,3,69] which is present at high copy number in osteoclasts. αvβ3 binds promiscuously to several noncollagenous bone protein ligands in an RGD-dependent manner.[2,3] Other αv integrins, particularly αvβ5, are not conspicuously expressed by mature osteoclasts.[69] Recognition of native collagen is mediated by another integrin, α2β1.[70] Other integrins have not been found consistently in osteoclasts in extensive analyses of several species (reviewed in refs. 2,3).

Inhibition of osteoclast αvβ3 (reviewed in refs. 2,3) and α2β1 integrins[70] leads to reduced bone resorption in vitro and in a range of in vivo animal models. These features have encouraged the pharmaceutical industry to develop nonpeptidic RGD analogue drugs for use as inhibitors of osteoclast function in treating skeletal diseases, such as osteoporosis or bone cancer,[1,71] and are now in clinical trial.[72-75] These developments have emphasised the need to establish the impact of chronic inhibition or lack of αvβ3 on osteoclast function, but, despite this, the integrin phenotype of osteoclasts from animals treated long-term with αvβ3 antagonist drugs has not been reported. The question, thus, still remains unanswered as to whether a compensatory response will occur in the osteoclast to counteract the effects of chronic down-regulation or inhibition of αvβ3–ligand interactions, such as might occur via an up-regulation of an existing or 'new' integrin dimer (see discussion below in section on β3 null osteoclast function).

Detailed localization studies have tried to answer the question whether integrins are present in, and possibly responsible for maintenance of, the tight sealing zone, the specialized F-actin rich membrane domain seen in actively resorbing osteoclasts, but the data so far is conflicting. Although some studies have excluded presence of αvβ3, reviewed in refs. 2,76,77 and α2β1[78] in the sealing zone, others have reported the presence of αv and/or β3 subunits at this site.[79-81] It is clear, however that none of the integrin receptors are enriched in the clear zone of resorbing osteoclasts and that they are in fact most abundant on the opposite, basolateral membrane, where they may act as true receptors, coupled to signal transduction pathways.[78,82,83] There is also high expression of integrins on the ruffled border membrane in resorbing osteoclasts,[78,82] suggesting that cell-matrix interactions at that site may be involved in the progression, or cessation, of resorption.

The role of the vitronectin receptor αvβ3 in osteoclast biology has been extensively studied and gene knockout studies where β3 integrin has been deleted in the mouse[84,85] have underscored the central role of αvβ3 integrin in osteoclast biology (vide infra). It has become clear that mammalian osteoclasts can adhere to a wide range of extracellular matrix proteins known to be expressed in bone and bone marrow using this receptor (Table 1) and that an RGD peptide sequence forms the recognition site for the receptor in all these ligands. Which protein constitutes the natural ligand of osteoclasts in bone remains to be determined. Colocalization studies have suggested that osteopontin is a candidate,[86] since it was found to be enriched underneath the clear zones of resorbing osteoclasts. However, since osteoclasts actively synthesize osteopontin[87] this finding should be interpreted with some reservation. Osteopontin also has been suggested to have an important role in osteoclast motility, an effect mediated through the αvβ3 receptor.[88] RGD-containing peptides and snake venoms (echistatin and kistrin) have been shown to inhibit osteoclast polarization, activity and formation in vitro[81,89-91] and they, as with vitronectin receptor antibodies, also block bone resorption in various rodent models in vivo,[92-95] reviewed in ref. 2. In vitro, occupation of the vitronectin receptor by antibodies or RGD peptides causes osteoclasts to retract and detach from matrix, similar to the shape changes observed after administration of the potent anti-resorptive peptide hormone calcitonin. In mammalian osteoclasts this effect is preceded by a rise in intracellular calcium,[96,97] localized predominantly to the nucleus,[96] linked to the extensive downstream signaling pathways evoked by ligand engaging with integrin receptors (reviewed in refs. 98,99). The impressive effects of RGD peptide analogues developed by the pharmaceutical industry[75] on bone resorption (both in vitro assays and more recently in vivo animal models) (reviewed in ref. 2) has led to their introduction into clinical trails for bone diseases associated with excessive bone resorption such as osteoporosis.

The function of α2β1 in mammalian osteoclasts is predominantly as a receptor for type I collagen. In contrast to other cell types, adhesion of osteoclast to collagen appears to be RGD-dependent,[78,81] although this could possibly be explained as a dominant-negative effect of RGD occupation of the abundant αvβ3 receptors on osteoclasts. Antibodies to α2 and β1 integrin inhibit resorption by human osteoclasts in vitro, but not to the same extent as anti-vitronectin receptor antibodies. Avian osteoclasts express abundant β1 integrin, but do not adhere to collagen.[100] Avian osteoclasts probably use β1 integrin, predominantly in association with α5, as a fibronectin-binding receptor.

The role of αvβ1 on osteoclasts has not been explored in functional assays since no receptor complex-specific antibodies are available at present. This receptor is far less abundant than αvβ3 and α2β1 in osteoclasts.[69] By analogy with other cell types it is likely that αvβ1 functions as a receptor for collagen or fibronectin in osteoclasts.

Integrin Knockouts and Skeletal Cell Function

The best evidence for a role for integrins in skeletal development comes from gene knockout studies in mice,[101,102] complemented by studies of human diseases such as Glanzmann thrombasthenia (GT) that carry mutations in integrin genes.

αv and β3 Mouse Knockouts

Gene deletion of the key osteoclast integrin, αv, has been performed in mice.[103] The deletion of αv leads to a complex phenotype: 80% embryonic lethality is observed with the surviving pups dying in the perinatal period of cerebral and other vascular malformations. The skeleton appeared grossly normal and no obvious bone or cartilage cellular defects were observed, though cell function was not specifically addressed. This model only points to a redundant role for αv in skeletal modeling during development and cannot be used to study remodelling processes that might produce a skeletal phenotype on ageing.

The skeletal phenotype of β3 integrin knockout mice has recently been reported.[84,85] The skeleton at birth was relatively unaffected by the lack of β3 integrin, but mild osteosclerosis developed by adulthood. Some information on the function of mature β3 null osteoclasts has come from these murine studies—there is altered osteoclast adhesion to matrix and a reduction in bone resorption observed in vitro.[84,85] However, the fact that normal multinucleated osteoclasts form in the absence of β3 excluded an essential role of this integrin in osteoclast development and precursor fusion.

Mice from both deletions have shown normal skeletal development and/or patterning, contrary to the prediction that bone modeling would have been significantly affected. This suggests that other integrins may compensate in the process of bone recognition by osteoclasts. Indeed, this appears to be true as, in an equivalent study in humans using peripheral blood cultures, patients with GT carrying an integrin β3 null mutation[104] have upregulated α2β1 collagen-binding receptor that enables bone resorption to proceed, albeit to a reduced amount (vide infra). The main abnormality seen in β3 null osteoclasts is in the appearance of the ruffled border, which has wide and blunt folds in the knockout, rather than the numerous narrow membrane protrusions penetrating deep into the bone matrix, which are characteristic of normal osteoclasts. These deficiencies are seen both in vivo and in vitro and indicate that β3 plays an important role in normal ruffled border formation, possibly through a specific adhesive interaction helping the membrane to penetrate into the bone, which cannot be compensated for by other integrins. Inadequate ruffled border formation is known to affect bone resorption and is a well known phenomenon in osteopetrosis.[105]

Table 2. Alterations in integrin expression and osteoclast function in Glanzmann thrombastenia

a. *Expression of $\alpha_v\beta_3$ and $\alpha_2\beta_1$ integrins in β_3 null Glanzmann thrombasthenia**

	Integrin expression level	
	$\alpha_v\beta_3$	α_2
% of control value	4	243

b. *Bone resorption is reduced in β_3 null Glanzmann thrombasthenia**

	Pit number (per unit area)	Pit depth
% decrease from control	44	59

*Compared with data from normal control cultures. Resorption by osteoclasts was assessed by culturing peripheral blood derived mononuclear cells with permissive growth factors on slices of bovine cortical bone for up to 21 days and assessing the number of resorption pits formed and their depth by confocal microscopy in reflection mode. Data adapted from ref. 104.

Osteoclast Function in Glanzmann Thrombasthenia: A Human β3 Loss of Function Mutation

We have recently exploited the ability to generate osteoclasts from human peripheral blood mononuclear cells, under the influence of RANK,[106] to study genotype-phenotype relationships in human 'experiments of nature', the equivalent of mouse knockouts. GT is an autosomal recessive bleeding disorder defined by defective or quantitatively abnormal platelet αIIbβ3 receptors for fibrinogen; diagnosis is based upon the clinical syndrome of mucocutaneous bleeding in association with platelet aggregation abnormalities in response to physiological stimuli. Earlier biochemical and molecular genetic characterization of the αIIb and β3 integrin subunits and their genes in patients with GT has shed considerable light on the aetiology of this genetic disorder. One particular form is the extreme β3 null phenotype observed in the Iraqi-Jewish population in Israel.[107-109] Patients from this genetic isolate[110,111] have no platelet αIIbβ3 due to the complete absence of the integrin β3 protein caused by an 11 bp deletion in exon 12 in the β3 (CD61/gpIIIa) gene, the mutation producing a frame-shift leading to premature termination of translation amino-terminal to the trans-membrane domain of the β3 protein. This form of GT has been used by our group to derive and analyse osteoclasts, lacking the integrin β3 protein, by in vitro culture;[104] results have been compared with those reported for murine β3 -/- osteoclasts.[84,85] Deficiency of β3 leads to complete absence of the osteoclast integrin, αvβ3, that shares its β chain with αIIbβ3 of platelets (Fig. 2), (Table 2a). Other αv integrins, for example αvβ5 that is normally at low levels in mature osteoclasts,[69] are not up-regulated to compensate for lack of αvβ3, despite a similar but not identical range of ligand recognition. In contrast, the integrin α2β1 collagen receptor, normally expressed at an intermediate level in osteoclasts, is significantly up-regulated (Fig. 2), (Table 2a). β3 null osteoclasts polarised in vitro on bone substrate as with control cells and osteoclastic bone substrate resorption proceeded at approximately 50% of control levels, assessed by pit number and depth (Table 2b). The αv and β3 mouse knockout models aimed to address the same question, though, due to the absence of good reagents, integrin expression studies are more difficult in mice. Control levels of β5 integrin were seen on Northern analysis of bone marrow macrophages[84] suggesting, as in our studies, that αvβ5 is not up-regulated to compensate for the absence of β3. Bone resorption is compromised in β3 knockout mice,[84,85] as with human GT osteoclasts studied in vitro. There is, though, one difference evident between the data from the human and mouse models. In the mouse system there was a more marked effect upon cell adhesion and spreading. It could be that the role of α2β1 is less prominent in rodents, though our previous studies of rat osteoclasts would suggest otherwise;[70] or this effect could relate to experimental differences.

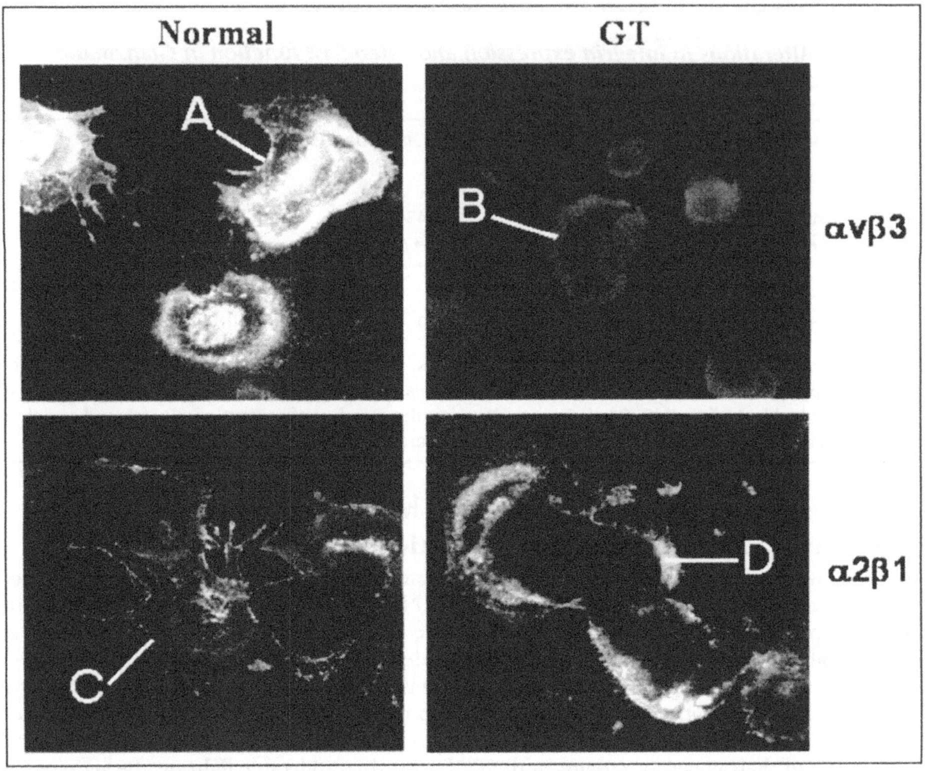

Figure 2. *αvβ3 and α2β1 integrin expression by human osteoclasts generated in cultures of normal and β3 null Glanzmann thrombasthenia peripheral blood mononuclear cells.* Control osteoclasts express high levels of αvβ3 (membrane fluoresescence marked, A), form F-actin rings ('light' ring-like structures seen in A) and resorb bone (not shown). In contrast, osteoclasts from β3 null Glanzmann thrombasthenia patients failed to express αvβ3 (no membrane staining seen associated with actin rings in B; (see Table 1a for quantification), though they make F-actin rings (B) and resorb bone sub-optimally (quantified in Table 2b). Control osteoclasts (C) express moderate amounts of α2 integrin (similar findings for β1 but not shown) which is increased in amount in β3 null Glanzmann thrombasthenia samples (D; see Table 2a) for quantification). Through-focus confocal images of osteoclasts stained with monoclonal antibody 23C6 to human αvβ3 complex (A) and α2 (C, D); F-actin 'rings' were demonstrated by staining with phalloidin-TRITC (A-D). Data adapted from ref. 104.

Interestingly, osteoclast numbers are increased in vivo in the bones of β3 knockout mice. This effect is due to the marked hypocalcemia in the animals, leading to increased levels of parathyroid hormone and through this, increased levels of M-CSF in the bone microenvironment.[112] It is not known if a similar increase in osteoclast number is found in the β3 null GT due to the impossibility of performing skeletal biopsies in the face of a severe hemorrhagic diathesis in humans. Increased osteoclast numbers in absence of β3 goes against earlier in vivo studies in oophorectomised rats treated with a small molecule αvβ3 integrin antagonist drug where osteoclast numbers were reduced in chronically in treated animals.[95] These latter data might point to fundamental differences between the dependence upon αvβ3 integrins during osteoclast development in the absence of αvβ3 and chronic exposure to 'antagonist' drugs that might have complex effects on cell function other than inhibition of cell adhesion. For example, direct effects of such drugs might induce signal transduction events and downstream cellular responses, unrelated to inhibition of cell adhesion (reviewed by ref. 98).

The Impact of Chronic αvβ3 Integrin 'Deficiency' on Skeletal Integrity: Implications of Integrin Antagonist Therapies

The establishment of the role of αvβ3 vitronectin receptor in osteoclast physiology (vide supra) led to its identification by the pharmaceutical industry as a potential drug development target as an osteoclast functional inhibitor for osteoporosis.[72-74] This has now been realised and at least one drug has remained in the clinic for this indication.[75]

What are the implications for skeletal function and calcium metabolism in β3 null GT patients—that is, do GT patients develop an age-dependent osteosclerosis as in mice lacking the β3 integrin? A preliminary study of bone mineral density in five Iraqi-Jewish women with GT showed values for spine and femur lying within the normal range.[111] More recently, there has been a single case report of the association of GT with osteopetrosis[113]—whether this was a chance association or the result of a single pathogenetic mechanism awaits further study. Thus, these data support the concept that a lifelong, genetic lack of αvβ3 integrin may not be especially deleterious to the skeleton and, in support, bone disease is not a generally reported feature of GT. It would still be of considerable interest to investigate bone and calcium metabolism in this group of β3 null GT patients in more detail as it could provide significant information upon the effects of chronic treatment with αvβ3 integrin antagonists in humans.

Does the information provided by our studies impact upon the clinical development of integrin antagonist drugs? If chronic inhibition of αvβ3 resulted in a significant compensatory up-regulation of other osteoclast adhesion receptors in vivo, such as of αvβ5 expressed by osteoclast precursors or the α2β1 collagen-binding integrin of mature osteoclasts, then only a transient therapeutic benefit in bone might be expected. The lack of αvβ5 up-regulation, if also seen in other tissues, might result in the inhibitory effect being preserved in clinical situations where αvβ5 plays a more prominent role, for example in vascular restenosis.[72,73] In contrast, the observed up-regulation of α2β1, an integrin that would not be expected to be inhibited by an RGD analogue drug, may result in significant loss of therapeutic response in target tissues such as bone. Thus, if data in mouse and human β3 gene deletion translates to αvβ3 antagonist drugs, then we can suggest that chronic exposure to αvβ3–selective antagonist drugs is unlikely to lead to total ablation of bone resorption. If similar compensatory mechanisms operate in normal osteoclasts at times of skeletal maturity, rather than during development as occurs in GT or knockout mice, then this response could curtail clinical efficacy. However, such a controlled inhibitory effect in the skeleton actually could be advantageous, since complete inhibition of bone turnover is deleterious as it leads to accumulation of microfactures in bone matrix. Early work with chronic dosing in ovariectomised rodent models, though, predicts that these issues may not be a major consideration in the clinical development of this new class of bone-active drugs.

The β1 Deletion Has an Embryonic Lethal Phenotype

Deletion of the β1 gene in mice leads to early embryonic lethality[114,115] before skeletal elements are formed and hence is uninformative. α5 knockout mice have a similar phenotype.[116] The finding of a role for β1-fibronectin interaction in osteoblast development[22,23] and studies on β1 integrins in osteoclast function (Horton, unpublished data) suggest that it would be informative to examining bone cell function in tissue-specific conditional β1 knockout mice; these have yet to be created with the appropriate targeting characteristics, though the Faessler group have created haemopoietic lineage-specific knockouts[117] suggesting that it should be possible to generate osteoclasts from β1 null myeloid precursors. An alternative strategy has been used by the Damsky group; they expressed a dominant negative form of β1 under the control of an osteoblast 'specific' osteocalcin promoter;[118] these mice had defects in osteoblast proliferation and matrix synthesis leading to low bone mass.

α1β1, α2β1 and Other Collagen-Binding Integrins of Osteoclasts

Despite extensive expression in mesenchymal cells; α1 knockout mice show normal skeletal development[119] following fracture, though, there is reduced callus formation compatible with an effect on mesenchymal precursors of chondrocytes and osteoblasts. Zemyo et al[120] have reported on the distribution of α1 in chondrocytes and have found that knockout mice have a propensity to develop significant age-related cartilage degeneration, similar to osteoarthritis, when compared with age-matched, wild-type litter mates. This implies that α1 might be a potential drug target for osteoarthritis, an area of unmet clinical need.

Of the four collagen-binding integrins found in vertebrates,[42] α2β1 is known to be expressed by osteoclasts at levels somewhat less that αvβ3. α1β1 is not expressed by osteoclasts and α10β1 and α11β1 have not been examined though they are known to be found in skeletal tissues. The biological function of α2β1 in osteoclasts has been studied but its role in osteoblasts is less well established. However, α2β1 integrin is involved in 3-D matrix remodeling in collagen gels including osteoblasts, compatible with a function in bone modelling.[29]

Recently, two groups have described the phenotype of α2 knockout mice.[121,122] As in the β3 null mouse, the skeleton is grossly normal at birth, though a detailed analysis of changes on skeletal maturation or of osteoclast or osteoblast function have not yet been reported. Given our results with β3 null GT patients, such studies should be undertaken to further elucidate the role of α2 integrin in bone and, by inference, its interplay with αvβ3.

Two new integrins, α10 and α11, collagen receptors have been identified recently and both are expressed in skeletal tissues, especially cartilage.[42] Little is known of function other than by implication—no knockout mice have been reported in the literature.

Other Integrin Knockouts and the Skeleton

Faessler has reviewed the impact of integrin gene deletion in a number of tissues including the skeleton;[123] of the other integrins that have been deleted in mice, two are informative. Amongst other features, the α4 knockout mouse[124] exhibits defective closure of cranial sutures, though whether this is due to a local osteoblast, or other cell type, defect has not been studied. Combined deletion of α3 and α6[125] results in severe skeletal abnormalities involving both the axial and peripheral skeleton that is not seen in single gene deletion animals; this phenotype is reminiscent of that seen on deletion of laminin α5 gene.

Osteoclasts develop from 'monocytic' precursor cells that express CD11/CD18 (LFA and β2 integrins).[106] Interestingly, given the severe myeloid functional defects seen in knockout mice and patients with leukocyte adhesion deficiency (LAD I), no bone (osteoclast) functional or development defect has been observed in vivo. There has, though, been a single report[126] showing an in vitro defect in osteoclast generation from LFA-1 deficient precursors at the level of interaction with the osteoblast counter-receptor, ICAM-1, for LFA. This is also true for mice deficient in the nonintegrin cell adhesion receptor, selectin, and combination knockouts where multiple adhesion proteins are deleted and major alterations in inflammatory responses are seen; the reason for the lack of a bone phenotype is not apparent.

Summary

There is extensive data on integrin expression in skeletal cells and increasing information on possible functions at the cellular level. The role of integrins in skeletal development, in contrast, is less well understood. Gene deletion studies have provided some indication as to redundant processes in cell-matrix and intercellular interactions that may be crucial for cell function in maturity, but little on whether integrins play a role in patterning and bone modeling in embryogenesis.

Acknowledgements

MAH is in receipt of a programme grant from The Wellcome Trust and MHH is supported by the Arthritis Research Campaign.

References

1. In: Bilezikian JP, Raisz LG, Rodan GA, eds. Principles of Bone Biology. 2nd ed. USA, San Diego: Academic Press, 2001.
2. Helfrich MH, Horton MA. Integrins and adhesion molecules. In: Seibel MJ, Robins SP, Bilezikian JP, eds. Dynamics of Bone and Cartilage Metabolism. USA, San Diego: Academic Press, 1999:111-125.
3. Horton MA, Nesbitt SA, Bennett JH et al. Integrins and other cell surface attachment molecules of bone cells. In: Bilezikian JP, Raisz LG, Rodan GA, eds. Principles of Bone Biology. 2nd ed. USA USA, San Diego: Academic Press, 2002:265-286.
4. Bennett JH, Moffatt S, Horton M. Cell adhesion molecules in human osteoblasts: Structure and function. Histol Histopathol 2001; 16:603-611.
5. George EL, Georges-Labouesse EN, Patel-King RS et al. Defects in mesoderm, neural tube and vascular development in mouse embryos lacking fibronectin. Development 1993; 119:1079–1109.
6. Rogers MJ. New insights into the molecular mechanisms of action of biophosphonates. Curr Pharma Design 2003; 9:2643-2658.
7. Soriano P, Montgomery C, Geske R et al. Targeted disruption of the c-src proto-oncogene leads to osteopetrosis in mice. Cell 1991; 64:693-702.
8. Charras GT, Horton MA. Single cell mechanotransduction and its modulation analyzed by atomic force microscope indentation. Biophys J 2002; 82:2970-2981.
9. Grzesik WJ, Robey PG. Bone matrix RGD glycoproteins: Immunolocalization and interaction with human primary osteoblastic bone cells in vitro. J Bone Miner Res 1994; 9:487-496.
10. Hughes DE, Salter DM, Dedhar S et al. Integrin expression in human bone. J Bone Miner Res 1993; 8:527-533.
11. Clover J, Dodds RA, Gowen M. Integrin subunit expression by human osteoblasts and osteoclasts in situ and in culture. J Cell Sci 1992; 103:267-271.
12. Everts V, Delaisse JM, Korper W et al. The bone lining cell: Its role in cleaning the Howship's lacunae and initiating bone formation. J Bone Min Res 2002; 17:77-90.
13. Tanaka Y, Maruo A, Fijii K et al. Intercelluar adhesion molecule 1 discriminates functionally different populations of human osteoblasts: Characteristic involvement of cell cylcle regulators. J Bone Min Res 2000; 15:1912-1923.
14. Sinha RK, Tuan RS. Regulation of human osteoblast integrin expression by orthopedic implant materials. Bone 1996; 18:451-457.
15. Gronowicz G, McCarthy MB. Response of human osteoblasts to implant materials: Integrin-mediated adhesion. J Orthop Res 1996; 14:878-887.
16. Aarden EM, Nijweide PJ, van der Plas A et al. Adhesive properties of isolated chick osteocytes in vitro. Bone 1996; 18:305-313.
17. Gohel AR, Hand A-R, Gronowicz GA. Immunogold localization of β1-integrin in bone: Effect of glucocorticoids and insulin-like growth factor I on integrins and osteocyte formation. J Histochem Cytochem 1995; 43:1085-1096.
18. Helfrich M, Van 't Hof R, MacPhee J. Expression and localization of integrins, NOS and caveolin-1 in osteoblasts and osteocytes. Calcif Tiss 2004; 74:S64-65.
19. Puleo DA, Bizios R. RGDS tetrapeptide binds to osteoblasts and inhibits fibronectin-mediated adhesion. Bone 1991; 12:271-276.
20. Majeska RJ, Port M, Einhorn TA. Attachment to extracellular matrix molecules by cells differing in the expression of osteoblastic traits. J Bone Miner Res 1993; 8:277-289.
21. Gronowicz GA, DeRome ME. Synthetic peptide containing Arg-Gly-Asp inhibits bone formation and resorption in a mineralizing organ culture system of fetal rat parietal bones. J Bone Miner Res 1994; 9:193-201.
22. Moursi AM, Globus RK, Damsky CH. Interactions between integrin receptors and fibronectin are required for calvarial osteoblast differentiation in vitro. J Cell Sci 1997; 110:2187-2196.
23. Damsky CH. Extracellular matrix-integrin interactions in osteoblast function and tissue remodeling. Bone 1999; 25:95-96.
24. Dedhar S, Mitchell MD, Pierschbacher MD. The osteoblast-like differentiated phenotype of a variant of MG-63 osteosarcoma cell line correlated with altered adhesive properties. Connect Tissue Res 1989; 20:49-61.
25. Dedhar S. Signal transduction via the β1 integrins is a required intermediate in interleukin-1 beta induction of alkaline phosphatase activity in human osteosarcoma cells. Exp Cell Res 1989; 183:207-214.
26. Dedhar S, Argraves WS, Suzuki S et al. Human osteosarcoma cells resistant to detachment by an Arg-Gly-Asp-containing peptide overproduce the fibronectin receptor. J Cell Biol 1987; 105:1175-1182.

27. Ganta DR, McCarthy MB, Gronowicz GA. Ascorbic acid alters collagen integrins in bone culture. Endocrinology 1997; 138:3606-3612.
28. Jikko A, Harris SE, Chen D et al. Collagen integrin receptors regulate early osteoblast differentiation induced by BMP-2. J Bone Miner Res 1999; 14:1075-1083.
29. Mizuno M, Fujisawa R, Kuboki Y. Type I collagen-induced osteoblastic differentiation of bone-marrow cells mediated by collagen-alpha2beta1 integrin interaction. J Cell Physiol 2000; 184:207-213.
30. Durr J, Goodman S, Potocnik A et al. Localization of β1-integrins in human cartilage and their role in chondrocyte adhesion to collagen and fibronectin. Exp Cell Res 1993; 207:235-244.
31. Enomoto M, Leboy PS, Menko AS et al. Beta 1 integrins mediate chondrocyte interaction with type I collagen, type II collagen, and fibronectin. Exp Cell Res 1993; 205:276-285.
32. Salter DM, Godolphin JL, Gourlay MS. Chondrocyte heterogeneity: Immunohistologically defined variation of integrin expression at different sites in human fetal knees. J Histochem Cytochem 1995; 43:447-457.
33. Woods VL, Schreck PJ, Gesink DS et al. Integrin expression by human articular chondrocytes. Arthritis Rheum 1994; 37:537-544.
34. Salter DM, Hughes DE, Simpson R et al. Integrin expression by human articular chondrocytes. Brit J Rheumatol 1992; 31:231-234.
35. Loeser RF. Chondrocyte integrin expression and function. Biorheology 2000; 37:109-116.
36. Loeser RF, Carlson CS, McGee MP. Expression of beta 1 integrins by cultured articular chondrocytes and in osteoarthritic cartilage. Exp Cell Res 1995; 217:248-257.
37. Shakibaei M. Integrin expression on epiphyseal mouse chondrocytes in monolayer culture. Histol Histopathol 1995; 10:339-349.
38. Camper L, Hellman U, Lundgren-Akerlund E. Isolation, cloning, and sequence analysis of the integrin subunit alpha10, a beta1-associated collagen binding integrin expressed on chondrocytes. J Biol Chem 1998; 273:20383-20389.
39. Camper L, Holmvall K, Wangnerud C et al. Distribution of the collagen-binding integrin alpha10beta1 during mouse development. Cell Tissue Res 2001; 306:107-116.
40. Lehnert K, Ni J, Leung E et al. Cloning, sequence analysis, and chromosomal localization of the novel human integrin alpha11 subunit (ITGA11). Genomics 1999; 60:179-187.
41. Tiger CF, Fougerousse F, Grundstrom G et al. Alpha11beta1 integrin is a receptor for interstitial collagens involved in cell migration and collagen reorganization on mesenchymal nonmuscle cells. Dev Biol 2001; 237:116-129.
42. Wright DJ, Puranen S, Johnson MS et al. The collagen receptor subfamily of the integrins. Int J Biochem Cell Biol 2004; 36:1405-1410.
43. Lapadula G, Iannone F, Zuccaro C et al. Integrin expression on chondrocytes: Correlations with the degree of cartilage damage in human osteoarthritis. Clin Exp Rheumatol 1997; 15:247-254.
44. Ostergaard K, Salter DM, Petersen J et al. Expression of alpha and beta subunits of the integrin superfamily in articular cartilage from macroscopically normal and osteoarthritic human femoral heads. Ann Rheum Dis 1998; 57:303-308.
45. Pulai JI, Del Carlo M, Loeser RF. The alpha5beta1 integrin provides matrix survival signals for normal and osteoarthritic human articular chondrocytes in vitro. Arthritis Rheum 2002; 46:1528-1535.
46. Hirsch MS, Svoboda KK. Beta 1 integrin antibodies inhibit chondrocyte terminal differentiation in whole sterna. Ann NY Acad Sci 1996; 785:267-270.
47. Tavella S, Bellesse G, Castagnola P et al. Regulated expression of fibronectin, laminin and related integrin receptors during the early chondrocyte differentiation. J Cell Sci 1997; 110:2261-2270.
48. Salter DM, Robb JE, Wright MO. Electrophysiological responses of human bone cells to mechanical stimulation: Evidence for specific integrin function in mechanotransduction. J Bone Miner Res 1997; 12:1133-1141.
49. Millward-Sadler SJ, Wright MO, Lee H et al. Integrin-regulated secretion of interleukin 4: A novel pathway of mechanotransduction in human articular chondrocytes. J Cell Biol 1999; 145:183-189.
50. Lucchinetti E, Bhargava MM, Torzilli PA. The effect of mechanical load on integrin subunits alpha5 and beta1 in chondrocytes from mature and immature cartilage explants. Cell Tissue Res 2004; 315:385-391.
51. Millward-Sadler SJ, Salter DM. Integrin-dependent signal cascades in chondrocyte mechanotransduction. Ann Biomed Eng 2004; 32:435-446.
52. Salter DM, Millward-Sadler SJ, Nuki G et al. Differential responses of chondrocytes from normal and osteoarthritic human articular cartilage to mechanical stimulation. Biorheology 2002; 39:97-108.
53. Shimizu M, Minakuchi K, Kaji S et al. Chondrocyte migration to fibronectin, type I collagen, and type II collagen. Cell Structure & Function 1997; 22:309-315.

54. Homandberg GA, Hui F. Arg-Gly-Asp-Ser peptide analogs suppress cartilage chondrolytic activities of integrin-binding and nonbinding fibronectin fragments. Arch Biochem Biophys 1994; 310:40-48.
55. Xie D, Homandberg GA. Fibronectin fragments bind to and penetrate cartilage tissue resulting in proteinase expression and cartilage damage. Biochem Biophys Acta 1993; 1182:189-196.
56. Enomoto-Iwamoto M, Iwamoto M, Nakashima K et al. Involvement of α5β1 integrin in matrix interactions and proliferation of chondrocytes. J Bone Miner Res 1997; 12:1124-1132.
57. Loeser RF. Integrin-mediated attachment of articular chondrocytes to extracellular matrix proteins. Arthritis Rheum 1993; 36:1103-1110.
58. Hirsch MS, Lunsford LE, Trinkaus-Randall V et al. Chondrocyte survival and differentiation in situ are integrin mediated. Dev Dyn 1997; 210:249-263.
59. Holmvall K, Camper L, Johansson S et al. Chondrocyte and chondrosarcoma cell integrins with affinity for collagen type II and their response to mechanical stress. Exp Cell Res 1995; 221:496-503.
60. Cao L, Lee V, Adams ME et al. Beta-Integrin-collagen interaction reduces chondrocyte apoptosis. Matrix Biol 1999; 18:343-355.
61. Loeser RF. Growth factor regulation of chondrocyte integrins. Differential effects of insulin-like growth factor 1 and transforming growth factor beta on α1b1 integrin expression and chondrocyte adhesion to type VI collagen. Arthritis Rheum 1997; 40:270-276.
62. Ramachandrula A, Tiku K, Tiku ML. Tripeptide RGD-dependent adhesion of articular chondrocytes to synovial fibroblasts. J Cell Sci 1992; 101:859-871.
63. Loeser RF. Integrins and cell signaling in chondrocytes. Biorheology 2002; 39:119-124.
64. Arner EC, Tortorella MD. Signal transduction through chondrocyte integrin receptors induces matrix metalloproteinase synthesis and synergizes with interleukin-1. Arthritis Rheum 1995; 38:1304-1314.
65. Yonezawa I, Kato K, Yagita H et al. VLA-5-mediated interaction with fibronectin induces cytokine production by human chondrocytes. Biochem Biophys Res Commun 1996; 219:261-265.
66. Loeser RF. Modulation of integrin-mediated attachment of chondrocytes to extracellular matrix proteins by cations, retinoic acid, and transforming growth factor beta. Exp Cell Res 1994; 211:17-23.
67. Grashoff C, Aszodi A, Sakai T et al. Integrin-linked kinase regulates chondrocyte shape and proliferation. EMBO Rep 2000; 4:432-438.
68. Beckstead JH, Stenberg PE, McEver RP et al. Immunohistochemical localization of membrane and alpha-granule proteins in human megakaryocytes: Application to plastic embedded bone marrow biopsy specimens. Blood 1986; 67:285-293.
69. Nesbitt S, Nesbit A, Helfrich M et al. Biochemical characterisation of human osteoclast integrins. Osteoclasts express αvβ3, α2β1 and αvβ1 integrins. J Biol Chem 1993; 268:16737-16745.
70. Helfrich MH, Nesbitt SA, Lakkakorpi PT et al. Beta 1 integrins and osteoclast function: Involvement in collagen recognition and bone resorption. Bone 1996; 19:317-328.
71. In: Marcus R, Feldman D, Kelsey J, eds. Osteoporosis. USA, San Diego: Academic Press, 1996.
72. Hartman GD, Duggan ME. αvβ3 integrin antagonists as inhibitors of bone resorption. Exp Opin Invest Drugs 2000; 9:1281-1291.
73. Miller WH, Keenan R, Willette RN et al. Identification and in vivo efficacy of small molecule antagonists of integrin αvβ3 (the 'vitronectin receptor'). Drug Discov Today 2000; 5:397-408.
74. Horton MA. Integrin antagonists as inhibitors of bone resorption: Implications for treatment. Proc Nutrit Soc 2001; 60:275-281.
75. Perkins JJ, Duong LT, Fernandez-Metzler C et al. Nonpeptide alpha(v)beta(3) antagonists: Identification of potent, chain-shortened RGD mimetics that incorporate a central pyrrolidinone constraint. Bioorg Med Chem Lett 2003; 13:4285-4288.
76. Väänänen HK, Horton M. The osteoclast clear zone is a specialized cell-extracellular matrix adhesion structure. J Cell Sci 1995; 108:2729-2732.
77. Stenbeck G, Horton MA. A new specialized cell-matrix interaction in actively resorbing osteoclasts. J Cell Sci 2000; 113:1577-1587.
78. Helfrich MH, Nesbitt SA, Lakkakorpi PT et al. Beta 1 integrins and osteoclast function: Involvement in collagen recognition and bone resorption. Bone 1996; 19:317-328.
79. Nakamura I, Gailit J, Sasaki T. Osteoclast integrin αvβ3 is present in the clear zone and contributes to cellular polarization. Cell Tissue Res 1996; 286:507-515.
80. Hultenby K, Reinholt FP, Heinegård D. Distribution of integrin subunits on rat metaphyseal osteoclasts and osteoblasts. Eur J Cell Biol 1993; 62:86-93.
81. Nakamura I, Takahashi N, Sasaki T et al. Chemical and physical properties of the extracellular matrix are required for the actin ring formation in osteoclasts. J Bone Miner Res 1996; 11:1873-1879.

82. Lakkakorpi PT, Helfrich MH, Horton MA et al. Spatial organization of microfilaments and vitronectin receptor, αvβ3, in osteoclasts. A study using confocal laser scanning microscopy. J Cell Sci 1993; 104:663-670.
83. Lakkakorpi PT, Horton MA, Helfrich MH et al. Vitronectin receptor has a role in bone resorption but does not mediate tight sealing zone attachment of osteoclasts to the bone surface. J Cell Biol 1991; 115:1179-1186.
84. McHugh KP, Hodivala-Dilke K, Zheng MH et al. Mice lacking β3 integrins are osteosclerotic because of dysfunctional osteoclasts. J Clin Invest 2000; 105:433-440.
85. Feng X, Novack DV, Faccio R et al. A Glanzmann's mutation in β3 integrin specifically impairs osteoclast function. J Clin Invest 2001; 107:1137-1144.
86. Reinholt FP, Hultenby K, Oldberg Å et al. Osteopontin—a possible anchor of osteoclasts to bone. Proc Natl Acad Sci USA 1990; 87:4473-4475.
87. Dodds RA, Connor JR, James IE et al. Human osteoclasts, not osteoblasts, deposit osteopontin onto resorption surfaces: An in vitro and ex vivo study of remodeling bone. J Bone Miner Res 1995; 10:1666-1680.
88. Chellaiah MA, Kizer N, Biswas R et al. Osteopontin deficiency produces osteoclast dysfunction due to reduced CD44 surface expression. Molecular Biol Cell 2003; 14:173-189.
89. Horton MA, Taylor ML, Arnett TR et al. Arg-Gly-Asp (RGD) peptides and the anti-vitronectin receptor antibody 23C6 inhibit dentine resorption and cell spreading by osteoclasts. Exp Cell Res 1991; 195:368-375.
90. van der Pluijm G, Mouthaan H, Baas C et al. Integrins and osteoclastic resorption in three bone organ cultures: Differential sensitivity to synthetic Arg-Gly-Asp peptides during osteoclast formation. J Bone Miner Res 1994; 9:1021-1028.
91. Sato M, Sardana MK, Grasser WA et al. Echistatin is a potent inhibitor of bone resorption in culture. J Cell Biol 1990; 111:1713-1723.
92. Fisher JE, Caulfield MP, Sato M et al. Inhibition of osteoclastic bone resorption in vivo by echistatin, an "arginyl-glycyl-aspartyl" (RGD)-containing protein. Endocrinology 1993; 132:1411-1413.
93. King KL, D'Anza JJ, Bodary S et al. Effects of kistrin on bone resorption in vitro and serum calcium in vivo. J Bone Miner Res 1994; 9:381-387.
94. Crippes BA, Engleman VW, Settle SL et al. Antibody to β3 integrin inhibits osteoclast-mediated bone resorption in the thyroparathyroidectomized rat. Endocrinology 1996; 137:918-924.
95. Engleman VW, Nickols GA, Ross FP et al. A peptidomimetic antagonist of the αvβ3 integrin inhibits bone resorption in vitro and prevents osteoporosis in vivo. J Clin Invest 1997; 99:2284-2292.
96. Shankar G, Davison I, Helfrich MH et al. Integrin receptor-mediated mobilisation of intranuclear calcium in rat osteoclasts. J Cell Sci 1993; 105:61-68.
97. Zimolo Z, Wesolowski G, Tanaka H et al. Soluble αvβ3-integrin ligands raise $[Ca^{2+}]_i$ in rat osteoclasts and mouse-derived osteoclast-like cells. Am J Physiol 1994; 266:C376-C381.
98. Duong LT, Lakkakorpi P, Nakamura I et al. Integrins and signaling in osteoclast function. Matrix Biol 2000; 19:97-105.
99. Xiong WC, Feng X. PYK2 and FAK in osteoclasts. Front Biosci 2003; 8:d1219-d1226
100. Ross FP, Chappel J, Alvarez JI et al. Interactions between the bone matrix proteins osteopontin and bone sialoprotein and the osteoclast integrin αvβ3 potentiate bone resorption. J Biol Chem 1993; 268:9901-9907.
101. Sheppard D. In vivo lessons of integrins: Lessons from mull mutations in mice. Matrix Biol 2000; 19:203-209.
102. McCauley LK. Transgenic models of metabolic bone disease. Curr Opin Rheum 2001; 13:316-325.
103. Bader BL, Rayburn H, Crowley D et al. Extensive vasculogenesis, angiogenesis, and organogenesis precede lethality in mice lacking all alpha v integrins. Cell 1998; 95:507-519.
104. Horton MA, Massey HM, Rosenberg N et al. Upregulation of osteoclast alpha2beta1 integrin compensates for lack of alphavbeta3 vitronectin receptor in Iraqi-Jewish-type Glanzmann thrombasthenia. Br J Haematol 2003; 122:950-957.
105. Stenbeck G. Formation and function of the ruffled border in osteoclasts. Sem Cell Develop Biol 2002; 13:285-292.
106. Lader CS, Scopes J, Horton MA et al. Generation of human osteoclasts in stromal cell-free and stromal cell-rich cultures: Differences in osteoclast CD11c/CD18 integrin expression. Br J Haematol 2001; 112:430-437.
107. Newman PJ, Seligsohn U, Lyman S et al. The molecular genetic basis of Glanzmann's thrombasthenia in the Iraqi-Jewish and Arab populations in Israel. Proc Natl Acad Sci (USA) 1991; 88:3160-3164.

108. Coller BS, Seligsohn U, Zivelin A et al. Immunologic and biochemical characterization of homozygous and heterozygous Glanzmann thrombasthenia in the Iraqi-Jewish and Arab populations of Israel: Comparison of techniques for carrier detection. Br J Haematol 1986; 62:723-735.
109. Rosenberg N, Dardik R, Rosenthal E et al. Mutations in the αIIB and β3 genes that cause Glanzmann thrombasthenia can be distinguished by a simple procedure using transformed B-lymphocytes. Thromb Haemost 1998; 79:244-248.
110. Seligsohn U, Rososhansky S. A Glanzmann's thrombasthenia cluster among Iraqi Jews in Israel. Thromb Haemost 1984; 52:230-231.
111. Coller BS, Seligsohn U, Peretz H et al. Glanzmann thrombasthenia: New insights from an historical perspective. Semin Hematol 1994; 31:301-311.
112. Faccio R, Takeshita S, Zallone A et al. c-Fms and the αv β3 integrin collaborate during osteoclast differentiation. J Clin Invest 2003; 111:749-758.
113. Yarali N, Fisgin T, Duru F et al. Osteopetrosis and Glanzmann's thrombasthenia in a child. Ann Hematol 2003; 82:254-256.
114. Fässler R, Meyer M. Consequences of lack of β_1 integrin gene expression in mice. Genes Dev 1995; 9:1896–1908.
115. Stephens LE, Sutherland AE, Klimanskaya IV et al. Deletion of β_1 integrins in mice results in inner cell mass failure and peri-implantation lethality. Genes Dev 1995; 9:1883–1895.
116. Yang JT, Rayburn H, Hynes RO. Embryonic mesodermal defects in α5 integrin-deficient mice. Development 1993; 119:1093–1105.
117. Nieswandt B, Brakebusch C, Bergmeier W et al. Glycoprotein VI but not alpha2beta1 integrin is essential for platelet interaction with collagen. EMBO J 2001; 20:2120-2130.
118. Zimmerman D, Jin F, Leboy P et al. Impaired bone formation in transgenic mice resulting from altered integrin function in osteoblasts. Dev Biol 2000; 220:2-15.
119. Ekholm E, Hankenson KD, Uusitalo H et al. Diminished callus size and cartilage synthesis in alpha 1 beta 1 integrin-deficient mice during bone fracture healing. Am J Pathol 2002; 160:1779-1785.
120. Zemmyo M, Meharra EJ, Kuhn K et al. Accelerated, aging-dependent development of osteoarthritis in alpha1 integrin-deficient mice. Arthritis Rheum 2003; 48:2873-2880.
121. Holtkotter O, Nieswandt B, Smyth N et al. Integrin alpha 2-deficient mice develop normally, are fertile, but display partially defective platelet interaction with collagen. J Biol Chem 2002; 277:10789-10794.
122. Chen J, Diacovo TG, Grenache DG et al. The alpha(2) integrin subunit-deficient mouse: A multifaceted phenotype including defects of branching morphogenesis and hemostasis. Am J Pathol 2002; 161:337-344.
123. Bouvard D, Brakebusch C, Gustafsson E et al. Functional consequences of integrin gene mutations in mice. Circ Res 2001; 89:211-23.
124. Yang JT, Rayburn H, Hynes RO. Cell adhesion events mediated by α4 integrins are essential in placental and cardiac development. Development 1995; 121:549–560.
125. De Arcangelis A, Mark M, Kreidberg J et al. Synergistic activities of α_3 and α_6 integrins are required during apical ectodermal ridge formation and organogenesis in the mouse. Development 1999; 126:3957–3968.
126. Tani-Ishii N, Penninger JM, Matsumoto G et al. The role of LFA-1 in osteoclast development induced by cocultures of mouse bone marrow cells and MC3T3-G2/PA6 cells. J Periodontal Res 2002; 37:184-191.

Chapter 9

Insights into Integrin Function in Skeletal Muscle

Lisa K. Rogers and Ulrike Mayer*

Introduction

Skeletal muscle accounts for 40% of the human body and is the major consumer of body fuels. Contraction within muscle fibers by ATP consumption generates the force required for our body movement, but the coordinated transmission of this force through the muscle cell membrane to the surrounding basement membrane and through the tendon and into the bone is as important as the contraction itself. Cell adhesion molecules are thought to provide the link between the muscle cell interior and the environment, both as a means of transferring signals from the outside into the cell as well as maintaining the structural integrity of skeletal muscle upon repeated contraction. But this also leaves the muscle susceptible to inherited or spontaneously occurring mutations in genes of transmembrane spanning complexes. Integrins and the dystrophin-glycoprotein complex are known to be crucial in this connection and most of what is known about their function in skeletal muscle development and in the adult has been learned in recent years through the targeted inactivation of the genes in embryonic stem cells.

An Overview of Skeletal Muscle Development, Structure and Function

The first sign of commitment to the myogenic lineage occurs, when precursor cells migrate from part of the somite, the dermomyotome, to their destination where they differentiate into skeletal muscle. It has become evident that myogenic determination is dependant on the coordinated expression of members of the helix-loop-helix myogenic regulatory factor (MRF) family, followed by complex transcription of muscle specific proteins.[1,2] Once committed to the muscle cell lineage, myoblasts migrate to the region of muscle formation and start to proliferate. Some align with each other, withdraw from the cell cycle and fuse to form multinucleated primary myotubes. During secondary myogenesis, subsequent migrating myoblasts then fuse to the primary scaffold. The secondary myotubes continue to grow and elongate by cell fusion before eventually separating and maturing into specialized adult muscle fibers (Fig. 1A).

Over the course of this process, the structural proteins and enzymes are synthesized and assembled into the muscle compartments or secreted to contribute to the specialized basement membrane surrounding the muscle cell plasma membrane, the so-called sarcolemma. Muscle cells become innervated at the neuromuscular junction (NMJ) intercalate with the tendon to form myotendinous junction (MTJ) and specialize into slow/oxidative (type I) and fast/glycolytic

*Corresponding Author: Ulrike Mayer—Wellcome Trust Centre for Cell-Matrix Research, Faculty of Life Sciences, University of Manchester, Michael Smith Building, Oxford Road, Manchester M13 9PT, U.K. Email: Ulrike.Mayer@manchester.ac.uk

Integrins and Development, edited by Erik H.J. Danen. ©2006 Landes Bioscience.

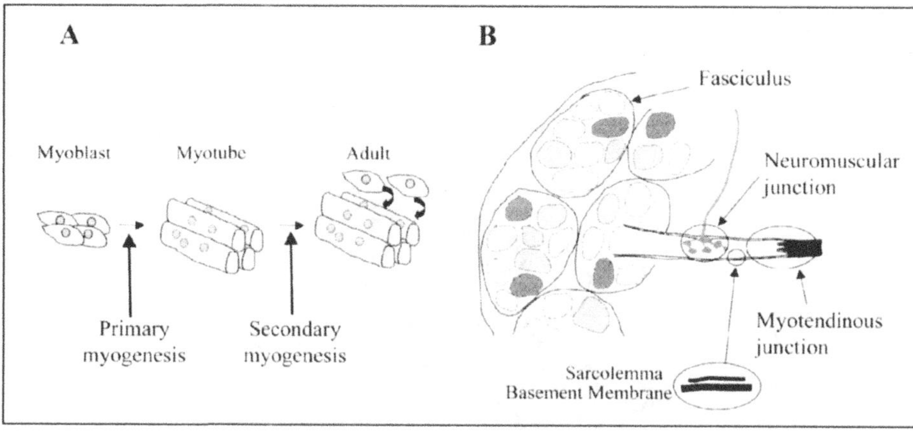

Figure 1. Formation of a mature muscle. A) Proliferating myoblasts migrate to the site of muscle formation. Withdrawal from the cell cycle and fusion to multinucleated primary myotubes marks primary myogenesis. Fusion of subsequent migrating myoblasts to this primary structure is termed secondary myogenesis. B) In the mature muscle fiber the basement membrane surrounds its sarcolemma. The fiber has formed intercalations with the tendon at the myotendinous junction. Each muscle fiber is innervated once at the neuromuscular junction. Individual muscle fibers are then bundled alongside others to form a fasciculus, a bundle of fasciculi forms an adult muscle.

(type II) fibers. By these processes, skeletal muscle cells become multinucleated and cylindrical in shape and are termed muscle fibers (Fig. 1B).

Transmembrane Receptors in Skeletal Muscle

Two major transmembrane receptors are known to be present in skeletal muscle, namely the dystrophin-glycoprotein complex (DGC)[3] and integrins.[4] In addition, recent data suggest a distinct role for the members of the syndecan family in skeletal muscle maintenance and regeneration.[5,6] The DGC is a multi-component complex consisting of α- and β-dystroglycan, dystrophin and the sarcoglycan family of proteins.[7] Dystrophin binds both to the actin cytoskeleton and to β-dystroglycan, which then establishes the link to the extracellular matrix (ECM) via α-dystroglycan binding to laminin, a major component of basement membranes.[8] The functional importance of the DGC in skeletal muscle is demonstrated by mutation of the human dystrophin gene as seen in Duchenne (DMD) and Becker muscular dystrophies (BMD), severe muscle wasting diseases.[9,10] Genes encoding members of the associated sarcoglycan family are mutated in some limb-girdle muscular dystrophies (LMGD).[11] Like the DGC, integrin receptors are capable of forming extracellular/cytoskeletal connections through adaptor molecules linking their cytoplasmic tails to actin filaments[12] (Fig. 2).

Integrins are noncovalently associated heterodimeric cell surface receptors mediating both cell-cell and cell-matrix interactions. Currently, there are 18 α and 8 β chains identified that form at least 24 different combinations,[13] but the possible combinations are further increased by the existence of different intra- and extracellular splice variants in several chains. The integrins differ from each other in the ligands they bind, how their binding activity is regulated, the cytoplasmic components and associated complexes they interact with and the types of signalling pathways they mediate.[14] The integrin family probably comprises the most functionally diverse family of cell adhesion molecules, but their composition and functional role in skeletal muscle has remained ambiguous for a long time, largely because most information has been gathered from a variety of species at a time when reagents were limited.

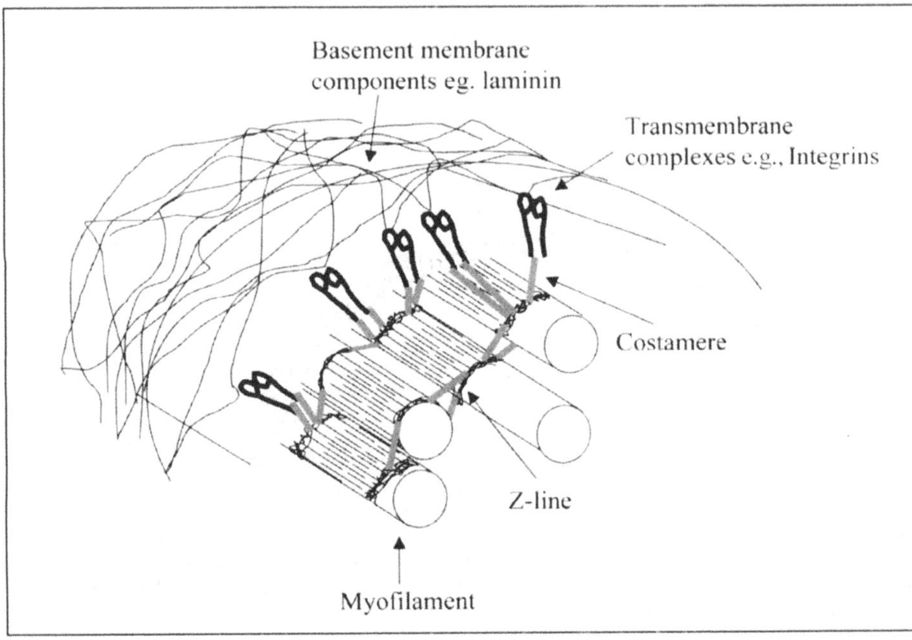

Figure 2. Organisation of the contractile unit and mediation of force transmission across the sarcolemma. Myosin slides across actin filaments in a process requiring ATP hydrolysis to produce muscle contraction. Actin is anchored at the Z-line, the point at which contacts are made to other myofilaments and the sarcolemma by costameric proteins such as vinculin, α-actinin and talin. At the sarcolemma, a link is formed to the intracellular domains of transmembrane complexes such as integrin α7β1 and the DGC. These contact laminin in the basement membrane via their extracellular domains, where the collagen fibers of the tendon are also anchored. The transmission of force from the contractile unit, across the sarcolemma to the tendon and eventually the bone is therefore a complex, multi-unit link.

Integrins in Skeletal Muscle

The Integrin β1 Subunit in Vitro and in Vivo Muscle Development

The formation of muscle fibers in vivo is a complex process that can be mimicked in in vitro culture systems and easily be analyzed by the alignment and fusion of mononucleated myoblasts into multinucleated myotubes. As the formation of myotubes involves both cell-cell and cell-ECM interactions, it seemed logical that integrins could control key stages in myogenesis. Indeed, pioneering experiments of Menko and Boettiger[15] demonstrated that integrin β1 function blocking antibodies inhibited myoblast fusion, suggesting an essential role for members of the β1 integrin family in the differentiation process. In contrast, in vivo studies of wild-type and β1 integrin null composite muscle in chimeric mice derived from β1 integrin-deficient embryonic stem (ES) cells showed normal muscle development. β1 integrin-deficient myoblasts and ES cells also differentiated normally in vitro, though the latter showed a delayed and medium-dependent fusion behavior.[16] The discrepancy between in vitro and in vivo studies has recently been solved.[17] Cre-mediated skeletal muscle-specific ablation of β1 integrins in mice resulted in perinatal death, due to impaired muscle fiber differentiation. The data led to the conclusion that β1 integrins are regulating a protein complex, necessary to reorganize the plasma membrane during cell fusion.[17] Interestingly, wild type cells rescued myotube formation, when cocultured with β1 integrin null cells, while their conditioned medium failed to do so,

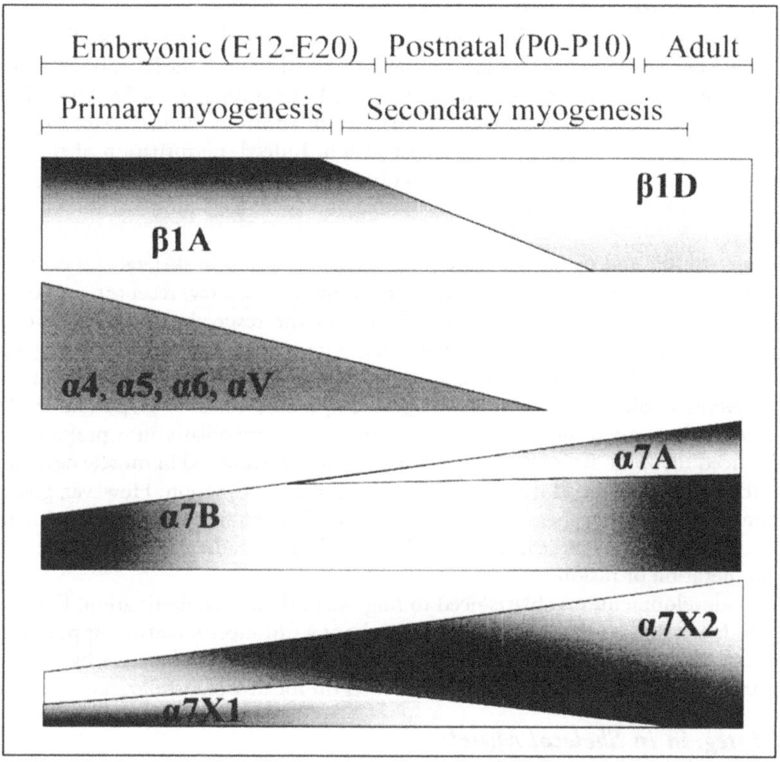

Figure 3. Integrin subunit expression patterns in skeletal muscle development. The expression of integrin subunits and their splice variants is developmentally regulated. Integrin β1A is ubiquitously expressed during primary myogenesis and replaced by the β1D subunit in secondary myogenesis. Integrins α4, α5, α6 and αV are all highly expressed in early stages of myogenesis but are down-regulated from secondary myogenesis and are no longer seen in adult muscle. Integrin α7 is expressed from the early stages of myogenesis but is up-regulated at secondary myogenesis. At this point the expression profile of its extra- and intra-cellular splice variants also changes. In early stages the intracellular α7B variant is exclusively expressed and although it remains the predominant form in adult muscle, the α7A variant is induced at secondary myogenesis. Both X1 and X2 extra-cellular variants are expressed in early stages of myogenesis, though in secondary myogenesis the expression of the X1 variant decreases and disappears in the adult.

suggesting that the defect is intrinsic to β1 integrin function and not to soluble factors secreted in an integrin β1-dependant manner.

In the mouse, two intracellular splice variants exist for the integrin β1 subunit (β1A and D) and these are also the most abundant forms in human.[18] The two variants are not functionally equal and their expression is controlled in a spatio-temporal manner. While the β1A variant is ubiquitously present, β1D is restricted to skeletal and cardiac muscle[19,20] and replaces β1A shortly before birth (Fig. 3), at a time when muscle mass is mainly gained through secondary myogenesis and myoblast withdrawal from the cell cycle.[21,22] In line with its onset of expression, β1D has been shown to interfere with cell migration and proliferation[23,24] and replacement of β1A by β1D in the murine genome resulted in reduced muscle mass due to impaired primary myogenesis,[25] further supporting the notion that members of the β1 integrin family play a critical role in muscle development.

Integrin α Subunits in Muscle Development

The β1 integrins comprise the largest family of cell adhesion receptors of ECM proteins and there are twelve different α subunits known to associate with the β1 subunit. The α4, α5, α6, α7, αv subunits have been convincingly verified in mammalian skeletal muscle, either in development and/or the adult. Integrin α4β1 can in addition mediate cell-cell contact and thus became a prime candidate for myoblast fusion. Indeed, perturbation of the interaction with its counter receptor VCAM-1 on neighboring cells inhibited myoblast alignment and fusion,[26] suggesting that the heterophilic interaction is its predominant function in muscle development.

The α5, α6, α7 and αv subunits are predominantly found in skeletal and cardiac muscle. Interestingly, all are either laminin (α6, α7) or fibronectin (α5, αv) receptors and their level of expression in skeletal muscle coincides with that of the respective ligand. The α6 and α5 subunits are prominent in proliferating myoblasts, down-regulated during secondary myogenesis and have disappeared from maturing muscle by postnatal day 10 in the mouse.[27,29] Overexpression studies in quail muscle suggested opposing roles for α5β1 and α6β1. While α5β1 inhibited terminal differentiation and maintained myoblasts in a proliferative phase, α6β1 induced myoblast fusion.[30] As both integrins are coexpressed in muscle development, at first glance it appears a paradox that they are so contrary in function. However, given the fact that during secondary myogenesis terminal differentiation and myoblast proliferation take place concurrently, it is conceivable that subtle differences and/or gradients in their expression favour either proliferation or fusion.

During development, myoblasts need to migrate to their final destination. Early studies in the 1980s demonstrated that only a laminin and not a fibronectin matrix supported myoblast migration.[31] Later studies demonstrated that myoblasts adhere to laminin by integrin α7β1 but not α6β1[32] and suggested α7β1 as a key integrin for myoblast migration.

α7β1 Integrin in Skeletal Muscle

Most knowledge about a specific integrin in skeletal muscle has been gathered for integrin α7β1, as it is the only, if not exclusive member of the β1 integrin family in the adult skeletal muscle and its absence leads to a muscular dystrophy both in human and mouse.[33,34]

The integrin α7 subunit is predominantly expressed in skeletal and cardiac muscle, though it is found in various other cell types.[4] In the mouse it is first detected in the myotome at embryonic day (E) 10[35] and is strongly upregulated after muscle fiber formation in secondary myogenesis.[36,37] In the adult, integrin α7 is associated with β1D and enriched at MTJs and NMJs, but is also present along the sarcolemma.[35,38,40]

The integrin α7 subunit is alternatively spliced in both the cytoplasmic (α7A and α7B) and the extracellular domain (α7X1 and α7X2),[36,41,42] resulting in four possible variants in human and mouse. In skeletal muscle development the splice forms are differentially regulated. The extracellular X1 and X2 domains are present in equal amounts in myoblasts, while X2 is exclusively present in the adult.[42] The B form is the first intracellular variant expressed and remains predominant throughout development and in the adult, while α7A is only induced during terminal differentiation and weakly present in mature muscle, where it localizes at MTJs.[37]

The complex regulation of alternative splicing in skeletal muscle development suggested structurally and functionally distinct roles for each integrin α7β1 variant for ligand-binding and induction of signalling cascades. Indeed, transfection of the integrin α7 subunit in nonmuscle cells increased cell motility on laminin substrates. Surprisingly however, the cytoplasmic α7A and α7B forms conferred motility to a similar extent and to date no functional differences between them have been reported.[43,45] Conversely, the extracellular domain is functionally distinct dependent on whether X1 or X2 is present. X2 favors binding to laminin-1 whereas X1 preferably interacts with laminin-8 and only adheres to laminin-1 after activation.[46,47] However the X1 and X2 variants bind with equal affinity to laminin-2.[47] In analogy to the

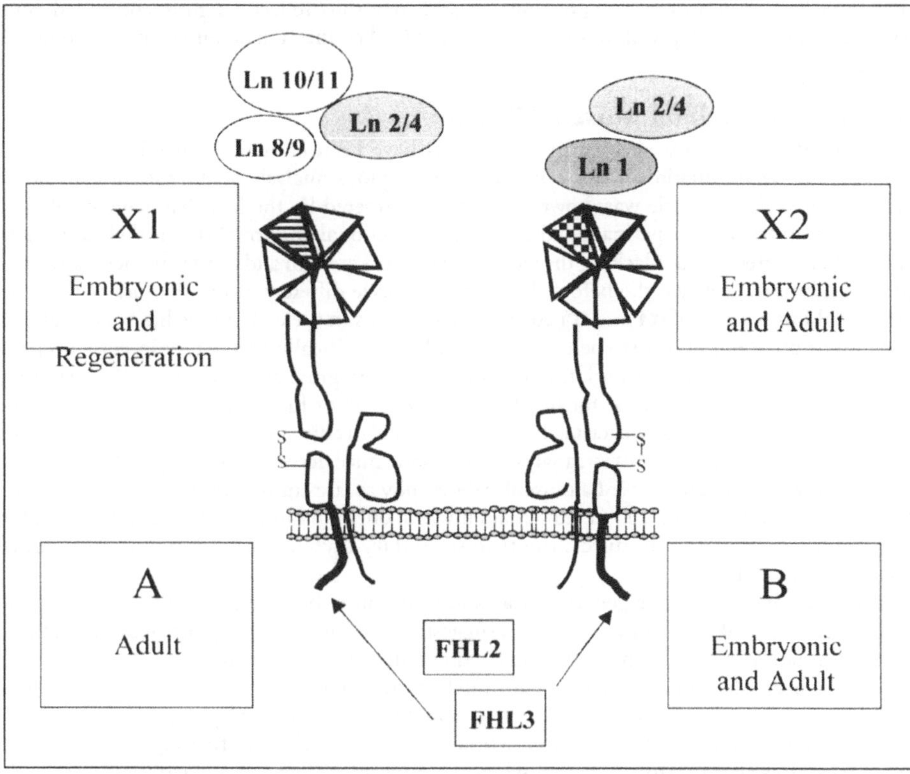

Figure 4. Intra- and extracellular interaction specificities of integrin α7 splice variants. Integrin α7 is alternatively spliced in its extra- and intracellular domains. The X1 and X2 variants differ in a region located between blades 3 and 4 of the β-propeller domain and differences in binding to laminin isoforms exist. Although both bind with equal affinity to laminin-2, the X2 variant preferably interacts with laminin-1. The X1 variant preferentially binds to laminin-8/9 and 10. Binding of FHL2 and FHL3 to the integrin α7A and B cytoplasmic domains has been shown.

3-dimentional structure determined for αvβ3,[48] the mutual exclusive insertion of X1 or X2 results in alternative interfaces between blade 3 and 4 of the seven-bladed propeller extracellular domain of the α7 subunit and the differences in ligand specificity make this region likely to be directly involved in ligand recognition (Fig. 4).

Basement membranes are a thin layer of specialized ECM proteins surrounding, amongst many other cells, skeletal muscle fibers. They are mainly comprised of members of the laminin family, collagen type IV, the heparan sulphate proteoglycan perlecan, nidogen.[49] Laminin has the most significant effect on cell behavior, including the regulation of adhesion and migration and changes to gene expression patterns and differentiation state by the specific induction of signalling cascades.[50,51] Laminin is a heterotrimer, composed of genetically distinct α, β and γ chains. To date five α, three β and three γ chains have been identified, which can assemble into at least 14 different laminin isomers.[52] Laminin 2, -4, -8, -9, -10 and -11 containing the α2, α4, and α5 in conjunction with the γ1 and β1 or β2 chains, respectively, are those identified in the skeletal muscle basement membrane during development, but in the adult the laminin α4 and α5 chains are confined to NMJs and laminin-2 and -4, containing the α2 chain become the predominant laminin isoform.[53,54] Interestingly, the switch in the extracellular domain of the α7 subunit that occurs either in development or after muscle injury[55] coincides with a change in laminin isoforms composition[56,57] and may thus reflect a tightly controlled gene regulatory system in the specialization of the muscle-ECM connection in muscle cell development and function.

Integrins in Skeletal Muscle Disease

In recent years almost all integrin α subunits have been eliminated in mice by targeted homologous recombination in ES cells. An unambiguous analysis of the functional role of integrins in skeletal muscle was, however, in part hampered by the fact that genetic ablation resulted in embryonic or perinatal lethality.[58] No functional abnormalities in skeletal muscle have been reported in mice lacking the α6 or αv subunits at birth and in the absences of the α5 and α4 subunits, embryos die before the onset of muscle differentiation. Because of this, the functional importance of α4 and α5 containing integrins in skeletal muscle has been analysed in chimeric mice and in in vitro behavior of myoblasts specifically selected for integrin α chain deficiency. Growth and differentiation did not reveal any gross abnormalities[59,60] suggesting that neither integrin is required for myoblast proliferation or fusion in either primary or secondary myogenesis. Similarly, skeletal muscle was normal in chimeric mice derived from integrin α4-deficient ES cells.[59] In contrast, analysis of chimeric mice derived from integrin α5-deficient ES cells showed characteristics of a muscular dystrophy at late embryonic stages.[60] The reasons for the degenerative changes observed remain unclear, but could imply a structural function for integrin α5β1 to maintain muscle fibers in secondary myogenesis rather than to support myoblast proliferation.

Muscular dystrophies are genetic diseases resulting in either congenital or late onset progressive muscle weakness and wasting. It is suggested that inherited or spontaneous mutations in approximately 100 different genes could potentially cause muscular dystrophy. So far ~40 genes have been identified, which code for nuclear membrane proteins, cytoplasmic enzymes and proteases, transmembrane and associated proteins as well as ECM proteins.[7,61,62]

The importance of integrins to muscle function and integrity was first highlighted by the progressive muscular dystrophy caused when the integrin α7 gene was inactivated in mice.[33] Dystrophic symptoms, variable fiber size and centrally located nuclei were, however, only evident in the deeply located soleus muscle of the hindlimbs and the diaphragm. The most striking phenotype of integrin α7-deficient muscle was seen at the MTJs. The interdigitations of the muscle fiber with the tendon were lost and the cytoskeleton retracted from the sarcolemma.[33] Subsequently, mutations in the integrin α7 gene in human patients with previously uncharacterised muscular dystrophies were reported[34] and the overall data suggest that the progression of the disease is likely similar between mouse and human.

None of the gene targeting experiments for the individual members of the integrin β1 family has so far reflected the drastic effect seen upon ablation of all members by the tissue-specific inactivation of the β1 subunit. Interestingly, all α subunits have been shown to be most highly expressed at the onset of secondary myogenesis, while their levels drop at postnatal stages and become, with the exception of α7β1, undetectable in the adult (Fig. 3). Together, this suggests that two or more integrins have similar functions in muscle development of which the absence of one can be tolerated. Alternatively, another integrin can functionally compensate. Combinations of double or triple deletions of integrin α subunits in the murine genome are therefore necessary to resolve this question finally.

Integrin α7β1 in Muscular Dystrophies

The pathology of the MTJ is specific to integrin α7 deficient muscle and is unseen in DMD patients or in the corresponding *mdx* mouse model[63] in which all muscles are effected by the absence of dystrophin.[64] This led to the conclusion that the DGC and integrin α7β1 are independent transmembrane protein complexes fulfilling distinct functions at the lateral site of the muscle fiber and in the junctional area of the MTJs, respectively. Interestingly, both protein complexes need to be absent in mice to lead to muscular dystrophy with similar severity to human DMD patients lacking only dystrophin,[65] though an understanding for the discrepancy is still missing. A simple explanation might be a different muscle load in human and mouse, or that the expression profile of the DGC and integrin α7β1 differs slightly between the species.

Consistent with it being the main integrin in skeletal muscle, the level of α7β1 integrin changes in the various muscular dystrophies, dependent on whether mutations affect its ligand or components of the DGC. In congenital muscular dystrophies, in which the laminin α2 chain is either lacking or greatly diminished, a decreased amount of the α7 subunit has been detected at the sarcolemmal membrane.[66,68] Yet, it is still controversial whether its regulation is controlled at the transcriptional level or whether the absence of its ligand leads to reduced protein stability.[66,67] Conversely, laminin is unaltered in the basement membrane upon mutations in components of the DGC.[69] However, in DMD patients and the corresponding *mdx* mouse model the level of integrin α7 is elevated.[66,68] In addition, in dystroglycan-depleted chimeric mice increased amounts of integrin α7 has been observed[70] and therefore supported the hypothesis that integrin α7β1 compensates for the absence of the DGC.

As integrin α7 is naturally elevated in response to the absence of dystrophin, it seems reasonable to consider integrin α7β1 as a potential therapeutic target for DMD patients. Indeed, several lines of evidence corroborated that integrin α7β1 enhances muscle function in dystrophin deficiency. Transgenic overexpression of the integrin α7X2B variant in mice, lacking dystrophin and its orthologue utrophin[71,72] markedly increased life span and mobility, although the number of fibers with centrally located nuclei and thus the dystrophic phenotype, was not significantly reduced.[73,74] Also, it has been demonstrated that overexpression of the disintegrin and metalloproteinase ADAM12 in mice increased the level of integrin α7β1 at the sarcolemmal membrane by a post-transcriptional mechanism, which has yet to be discovered.[75] Accordingly, its overexpression on the *mdx* background ameliorated the dystrophic phenotype.[76] The current model predicts that integrin α7β1 prevents structural defects upon continuous muscle contraction that are thought to further lead to membrane instability and necrosis.[77] More work is required to further our understanding of the mechanism underlying its beneficial effects in dystrophin-deficient muscle and, ideally, to identify transcription factors that are involved in regulating integrin α7β1 expression.

Intracellular Integrin Binding Proteins in Skeletal Muscle

Integrins provide a means to physically connect the actin cytoskeleton to the extracellular matrix. By this, a mechanical stabilization of the cell and integrin-mediated signalling in response to their environment are achieved (Fig. 2). While integrins directly interact with their ligands on the outside of the cell, integrin-binding adaptor proteins are necessary to make the connection to the cytoskeleton. The majority of adaptor proteins so far identified bind to the cytoplasmic tail of the β1 subunit. α-actinin, filamin and talin are thought to provide the major structural link to the cytoskeleton, while others such as focal adhesion kinase (FAK) and integrin-linked kinase (ILK) are involved in integrin-mediated signalling. Several proteins have also been shown to interact with the cytoplasmic domain of integrin α subunits.[12,78] Although the number of integrin cytoplasmic domain-binding proteins identified has steadily increased, little is known at present as to which of these are essential for muscle development or muscle maintenance in the adult.

By the end of last century evidence was provided that the integrin β1 cytoplasmic tail is crucial in determining cell fate in myoblast culture. Sastry and colleagues[79] demonstrated that the β1A cytoplasmic tail autonomously inhibited myoblast differentiation. Furthermore, their data indicated that phosphorylation of FAK and paxillin, both of which bind to the β1 cytoplasmic domain[12] inhibited myoblast differentiation. Consistent with these results, the integrin β1D cytoplasmic tail interfered with myoblast proliferation and failed to activate mitogen-activated protein kinase.[24] Together, these data suggested the involvement of specific integrin cytoplasmic tail interacting proteins and their downstream signalling cascades as key to muscle cell differentiation.

Since then several proteins have been identified that interact with either the β1 or α chain cytoplasmic tails in skeletal muscle. So far their functional importance has mainly been addressed in C2C12 muscle cell differentiation, while genetic information has been gathered from studies in flies and worms.[80] Deficiency of talin in Drosophila resulted in the same severe wing blistering phenotype at muscle termini, where muscle and tendon meet, as in integrin mutants. Yet, the initial alteration differs in that cell/ECM contact is lost in mutants lacking integrins, while upon talin deficiency the actin filaments fail to attach to the cell membrane.[81] Deletion of ILK in Drosophila resulted in a similar, though weaker phenotype than the talin mutants.[82]

ILK is described as an intracellular multidomain protein, which interacts with its C-terminal serine/threonine protein kinase domain to the integrin β1 cytoplasmic tail.[83] Several lines of evidence suggest ILK as a mediator of integrin-mediated signal transduction as part of a multiprotein complex composed of ILK, PINCH and parvin.[84] Functional studies in myoblasts resulted in controversial results. Overexpression of ILK in C2C12 impaired fusion into multinucleated myotubes, while deletion mutants lacking either the kinase or the PINCH-binding domain did not interfere with terminal differentiation.[85] Conversely, overexpression of ILK in a rat myoblast cell line enhanced fusion, while a dominant-negative mutation in its kinase domain inhibited fusion.[86] It is not known why the two studies resulted in conflicting results. However, both studies imply that ILK and its kinase domain are crucial for myoblast differentiation. A null mutation of ILK in mice led to an early embryonic phenotype[87] and a skeletal muscle-specific ablation of the ILK gene is therefore needed to further clarify ILK's function in skeletal muscle.

Tensin is broadly expressed and also found in skeletal muscle at the sarcolemma and at MTJs and NMJs. It has two actin-binding domains (ABD) at its N-terminal end and interacts with its C-terminal phosphotyrosine-binding domain to the NPXY motif common to all integrin β cytoplasmic tails. With its src homology 2 domain (SH2) tensin is capable of recruiting tyrosine-phosporylated proteins and could thus organize a multiprotein complex, which links integrins to the actin cytoskeleton and into which a variety of signalling molecules can be integrated.[88] Mice with a targeted deletion of tensin are viable but die from kidney failure.[89] In addition, an increased number of centrally located nuclei in muscle fibers and impaired skeletal muscle regeneration was reported,[90] suggesting a role for tensin in skeletal muscle maintenance and reorganization. Similarly to the mouse, tensin null-mutations in Drosophila are viable and develop mechanically induced wing blisters at muscle attachment sites. Interestingly, the typical localization of tensin at muscle ends is lost both in integrin, talin and ILK mutant flies.[91] Together, these data led to the conclusion that tensin stabilizes the integrin link to the cytoskeleton.

Melusin and MIBP (muscle integrin binding protein) were both identified in yeast two-hybrid screens to interact with fragments encoding either β1A and β1D specific amino acids or the conserved, membrane proximal part of the integrin β1 cytoplasmic domain, respectively. MIBP is exclusively present in myoblasts, while melusin is only detected upon myoblast fusion both in vivo and in vitro and is still present in the adult.[92,93] In addition, it was demonstrated that the interaction of integrin β1 with MIBP is modulated by the integrin α7 but not the α5 subunit.[94] As both proteins are expressed in a mutual exclusive manner in striated muscle, it made them prime candidates in regulating integrin function and signalling in development and the adult. Yet, inactivation of melusin in mice resulted in viable and fertile mice with no obvious skeletal muscle pathology[95] and at present, there is no in vivo analysis for MIBP available. Colocalization with integrin α7β1 in skeletal muscle and interaction with α7 and β1 cytoplasmic tails has also been reported for FHL2 and FHL3, members of the LIM-only family of proteins, comprised of four and one-half LIM (FHL) domains[96] (Fig. 4). Yet, their overexpression had no adverse effects in C2C12 cells and the knockout of FHL2 in mice resulted only in a mild cardiac phenotype.[97]

In addition, several other intracellular proteins have been identified, which may directly or indirectly be associated with integrin cytoplasmic domains and potentially be linked to muscle diseases. For example, Cypher, a PDZ and LIM domain containing protein, is exclusively expressed in striated muscle both in development and in the adult.[98] It binds to α-actinin and associates with the cytoskeleton. A null-mutation in mice resulted in perinatal lethality due to a severe form of congenital myopathy.[99] Of particular importance are the filamins, known as actin-crosslinking proteins. There are three genetically distinct filamin proteins, filamin 1-3, which share high sequence homology and a similar domain organization: an ABD at the N-terminus, followed by 24 repetitive segments of ~100 amino acids, which are interrupted by two flexible hinge regions between repeats 15/16 and 23/24.[100,101] More than 30 ligands have been reported to interact with filamins and mutations in filamin genes have been associated with human genetic diseases.[102] Filamin 1 is ubiquitously distributed but does not bind to any integrin cytoplasmic tail, while filamin 3 interacts with the cytoplasmic tail of integrin β1A. It is expressed in skeletal muscle development but absent in the adult.[103,104] Binding to β1A was enhanced and additional binding to the β1D cytoplasmic domain was detected, when 40 amino acids were deleted between the repetitive repeats 19/20. In the genome this area is encoded by one exon, which can be alternatively spliced in skeletal muscle development.[105] In contrast filamin 2 is skeletal muscle specific and the only filamin isoform expressed in the adult.[103,104] Accordingly, filamin 2 has been proposed as a candidate muscular disease gene and its chromosomal localization maps to a region on chromosome 7, which has been linked to a subtype of LMGD.[106] So far, γ- and δ-sarcoglycan have been identified to interact with filamin 2 at the sarcolemmal membrane,[107] but there are no reports, which describe an interaction with integrin cytoplasmic domains. Interestingly, the integrin β1-binding repetitive repeat region 19/20 differs in filamin 2 from the other isoforms in that it contains an additional insertion of 80 amino acids and makes it thus likely that this region confers specific functions to filamin 2 in skeletal muscle. Indeed, it was shown that the extended 19/20 repeat region of filamin 2 specifically interacts with myotilin, an α-actinin-binding protein.[104] Although no direct binding to integrin cytoplasmic tails has been proven yet, indirect evidence supports a potential interaction. In the absence of γ- and δ-sarcoglycan in DMD and LMGD patients or the corresponding mouse models, increased amounts of filamin 2 are deposited at the sarcolemmal membrane[107] suggesting that filamin 2 binds to membrane bound proteins other than the sarcoglycans. A logical candidate is α7β1 integrin and this work supports the idea that filamin 2 also interacts with integrins in normal conditions.

Conclusions

Formation of skeletal muscle is a complex process, which requires a dynamic adaptation of the cell to its environment when myoblasts migrate and proliferate, a dramatic cytoskeleton rearrangement during myoblast fusion and a more static and stable built cell, which can withstand the forces generated during continuous muscle contraction in the adult. Integrins are key organizers in establishing and coordinating the communication between the ECM and the cell interior, the myofibrils, composed of actin and myosin filaments. Our knowledge about integrin function in skeletal muscle has steadily increased since integrins were first identified. The challenge now is to integrate individual and/or combined integrin function and the cytoplasmic domain-binding protein complexes that link the sarcolemma to the actin cytoskeleton. This will further our understanding of integrin-mediated signalling in muscle development and of mechanisms to maintain muscle architecture in the adult. With this knowledge it will then be possible to further exploit integrins as potential targets for gene therapy of muscular dystrophies in the future.

Acknowledgement

The authors would like to thank lab members for their input. Work in UMs laboratory is supported by the Wellcome Trust (#060549). LKR is supported by a MRC studentship.

References

1. Buckingham M, Bajard L, Chang T et al. The formation of skeletal muscle: From somite to limb. J Anat 2003; 202:59-68.
2. Francis-West PH, Antoni L, Anakwe K. Regulation of myogenic differentiation in the developing limb bud. J Anat 2003; 202:69-81.
3. Ibraghimov-Beskrovnaya O, Ervasti JM, Leveille CJ et al. Primary structure of dystrophin-associated glycoproteins linking dystrophin to the extracellular matrix. Nature 1992; 355:696-702.
4. Mayer U. Integrins: Redundant or important players in skeletal muscle? J Biol Chem 2003; 278:14587-14590.
5. Cornelison DD, Filla MS, Stanley HM et al. Syndecan-3 and syndecan-4 specifically mark skeletal muscle satellite cells and are implicated in satellite cell maintenance and muscle regeneration. Dev Biol 2001; 239:79-94.
6. Cornelison DD, Wilcox-Adelman SA, Goetinck PF et al. Essential and separable roles for Syndecan-3 and Syndecan-4 in skeletal muscle development and regeneration. Genes Dev 2004; 18:2231-2236.
7. Durbeej M, Campbell KP. Muscular dystrophies involving the dystrophin-glycoprotein complex: An overview of current mouse models. Curr Opin Genet Dev 2002; 12:349-361.
8. Ervasti JM, Campbell KP. A role for the dystrophin-glycoprotein complex as a transmembrane linker between laminin and actin. J Cell Biol 1993; 122:809-823.
9. Cohn RD, Campbell KP. Molecular basis of muscular dystrophies. Muscle Nerve 2000; 23:1456-1471.
10. Blake DJ, Weir A, Newey SE et al. Function and genetics of dystrophin and dystrophin-related proteins in muscle. Physiol Rev 2002; 82:291-329.
11. Bushby KM. The limb-girdle muscular dystrophies-multiple genes, multiple mechanisms. Hum Mol Genet 1999; 8:1875-1882.
12. Liu S, Calderwood DA, Ginsberg MH. Integrin cytoplasmic domain-binding proteins. J Cell Sci 2000; 113(Pt 20):3563-3571.
13. van der Flier A, Sonnenberg A. Function and interactions of integrins. Cell Tissue Res 2001; 305:285-298.
14. Hynes RO. Integrins: Bidirectional, allosteric signaling machines. Cell 2002; 110:673-687.
15. Menko AS, Boettiger D. Occupation of the extracellular matrix receptor, integrin, is a control point for myogenic differentiation. Cell 1987; 51:51-57.
16. Hirsch E, Lohikangas L, Gullberg D et al. Mouse myoblasts can fuse and form a normal sarcomere in the absence of beta1 integrin expression. J Cell Sci 1998; 111:2397-2409.
17. Schwander M, Leu M, Stumm M et al. Beta1 integrins regulate myoblast fusion and sarcomere assembly. Dev Cell 2003; 4:673-685.
18. Baudoin C, van der Flier, Borradori L et al. Genomic organization of the mouse beta 1 gene: Conservation of the beta 1D but not of the beta 1B and beta 1C integrin splice variants. Cell Adhes Commun 1996; 4:1-11.
19. van der Flier A, Kuikman I, Baudoin C et al. A novel beta 1 integrin isoform produced by alternative splicing: Unique expression in cardiac and skeletal muscle. FEBS Lett 1995; 369:340-344.
20. Zhidkova NI, Belkin AM, Mayne R. Novel isoform of beta 1 integrin expressed in skeletal and cardiac muscle. Biochem Biophys Res Commun 1995; 214:279-285.
21. Brancaccio M, Cabodi S, Belkin AM et al. Differential onset of expression of alpha 7 and beta 1D integrins during mouse heart and skeletal muscle development. Cell Adhes Commun 1998; 5:193-205.
22. van der Flier A, Gaspar AC, Thorsteinsdottir S et al. Spatial and temporal expression of the beta1D integrin during mouse development. Dev Dyn 1997; 210:472-486.
23. Gimond C, Baudoin C, Sonnenberg A. Defects in adhesion and migration, but not in proliferation and differentiation, of embryonic stem cells upon replacement of integrin subunit beta1A by beta1D. Differentiation 2000; 66:93-105.
24. Belkin AM, Retta SF. beta1D integrin inhibits cell cycle progression in normal myoblasts and fibroblasts. J Biol Chem 1998; 273:15234-15240.
25. Cachaco AS, Chuva de Sousa Lopes SM, Kuikman I et al. Knock-in of integrin beta 1D affects primary but not secondary myogenesis in mice. Development 2003; 130:1659-1671.
26. Rosen GD, Sanes JR, LaChance R et al. Roles for the integrin VLA-4 and its counter receptor VCAM-1 in myogenesis. Cell 1992; 69:1107-1119.
27. Blaschuk KL, Holland PC. The regulation of alpha 5 beta 1 integrin expression in human muscle cells. Dev Biol 1994; 164:475-483.
28. McDonald KA, Lakonishok M, Horwitz AF. Alpha v and alpha 3 integrin subunits are associated with myofibrils during myofibrillogenesis. J Cell Sci 1995; 108:2573-2581.

29. Nawrotzki R, Willem M, Miosge N et al. Defective integrin switch and matrix composition at alpha 7-deficient myotendinous junctions precede the onset of muscular dystrophy in mice. Hum Mol Genet 2003; 12:483-495.
30. Sastry SK, Lakonishok M, Thomas DA et al. Integrin alpha subunit ratios, cytoplasmic domains, and growth factor synergy regulate muscle proliferation and differentiation. J Cell Biol 1996; 133:169-184.
31. Öcalan M, Goodman SL, Kuhl U et al. Laminin alters cell shape and stimulates motility and proliferation of murine skeletal myoblasts. Dev Biol 1988; 125:158-167.
32. von der Mark H, Dürr J, Sonnenberg A et al. Skeletal myoblasts utilize a novel beta 1-series integrin and not alpha 6 beta 1 for binding to the E8 and T8 fragments of laminin. J Biol Chem 1991; 266:23593-23601.
33. Mayer U, Saher G, Fässler R et al. Absence of integrin alpha 7 causes a novel form of muscular dystrophy. Nat Genet 1997; 17:318-323.
34. Hayashi YK, Chou FL, Engvall E et al. Mutations in the integrin alpha7 gene cause congenital myopathy. Nat Genet 1998; 19:94-97.
35. Velling T, Collo G, Sorokin L et al. Distinct alpha 7A beta 1 and alpha 7B beta 1 integrin expression patterns during mouse development: Alpha 7A is restricted to skeletal muscle but alpha 7B is expressed in striated muscle, vasculature, and nervous system. Dev Dyn 1996; 207:355-371.
36. Song WK, Wang W, Sato H et al. Expression of alpha 7 integrin cytoplasmic domains during skeletal muscle development: Alternate forms, conformational change, and homologies with serine/threonine kinases and tyrosine phosphatases. J Cell Sci 1993; 106:1139-1152.
37. Yao CC, Ziober BL, Sutherland AE et al. Laminins promote the locomotion of skeletal myoblasts via the alpha 7 integrin receptor. J Cell Sci 1996; 109:3139-3150.
38. Bao ZZ, Lakonishok M, Kaufman S et al. Alpha 7 beta 1 integrin is a component of the myotendinous junction on skeletal muscle. J Cell Sci 1993; 106:579-589.
39. Martin PT, Kaufman SJ, Kramer RH et al. Synaptic integrins in developing, adult, and mutant muscle: Selective association of alpha1, alpha7A, and alpha7B integrins with the neuromuscular junction. Dev Biol 1996; 174:125-139.
40. Belkin AM, Zhidkova NI, Balzac F et al. Beta 1D integrin displaces the beta 1A isoform in striated muscles: Localization at junctional structures and signaling potential in nonmuscle cells. J Cell Biol 1996; 132:211-226.
41. Collo G, Starr L, Quaranta V. A new isoform of the laminin receptor integrin alpha 7 beta 1 is developmentally regulated in skeletal muscle. J Biol Chem 1993; 268:19019-19024.
42. Ziober BL, Vu MP, Waleh N et al. Alternative extracellular and cytoplasmic domains of the integrin alpha 7 subunit are differentially expressed during development. J Biol Chem 1993; 268:26773-26783.
43. Echtermeyer F, Schöber S, Pöschl E et al. Specific induction of cell motility on laminin by alpha 7 integrin. J Biol Chem 1996; 271:2071-2075.
44. Schöber S, Mielenz D, Echtermeyer F et al. The role of extracellular and cytoplasmic splice domains of alpha7-integrin in cell adhesion and migration on laminins. Exp Cell Res 2000; 255:303-313.
45. Yao CC, Ziober BL, Squillace RM et al. Alpha7 integrin mediates cell adhesion and migration on specific laminin isoforms. J Biol Chem 1996; 271:25598-25603.
46. Ziober BL, Chen Y, Kramer RH. The laminin-binding activity of the alpha 7 integrin receptor is defined by developmentally regulated splicing in the extracellular domain. Mol Biol Cell 1997; 8:1723-1734.
47. von der Mark H, Williams I, Wendler O et al. Alternative splice variants of alpha 7 beta 1 integrin selectively recognize different laminin isoforms. J Biol Chem 2002; 277:6012-6016.
48. Xiong JP, Stehle T, Diefenbach B et al. Crystal structure of the extracellular segment of integrin alpha Vbeta3. Science 2001; 294:339-345.
49. Timpl R, Brown JC. Supramolecular assembly of basement membranes. Bioessays 1996; 18:123-132.
50. Miner JH, Yurchenco PD. Laminin functions in tissue morphogenesis. Ann Rev Cell Dev Biol 2004; 20:255-284.
51. Yurchenco PD, Amenta PS, Patton BL. Basement membrane assembly, stability and activities observed through a developmental lens. Matrix Biol 2004; 22:521-538.
52. Colognato H, Yurchenco PD. Form and function: The laminin family of heterotrimers. Dev Dyn 2000; 218:213-234.
53. Gullberg D, Tiger CF, Velling T. Laminins during muscle development and in muscular dystrophies. Cell Mol Life Sci 1999; 56:442-460.
54. Sanes JR. The basement membrane/basal lamina of skeletal muscle. J Biol Chem 2003; 278:12601-12604.

55. Käariainen M, Nissinen L, Kaufman S et al. Expression of alpha7beta1 integrin splicing variants during skeletal muscle regeneration. Am J Pathol 2002; 161:1023-1031.
56. Ringelmann B, Röder C, Hallmann R et al. Expression of laminin alpha1, alpha2, alpha4, and alpha5 chains, fibronectin, and tenascin-C in skeletal muscle of dystrophic 129ReJ dy/dy mice. Exp Cell Res 1999; 246:165-182.
57. Sorokin LM, Maley MA, Moch H et al. Laminin alpha4 and integrin alpha6 are upregulated in regenerating dy/dy skeletal muscle: Comparative expression of laminin and integrin isoforms in muscles regenerating after crush injury. Exp Cell Res 2000; 256:500-514.
58. Sheppard D. In vivo functions of integrins: Lessons from null mutations in mice. Matrix Biol 2000; 19:203-209.
59. Yang JT, Rando TA, Mohler WA et al. Genetic analysis of alpha 4 integrin functions in the development of mouse skeletal muscle. J Cell Biol 1996; 135:829-835.
60. Taverna D, Disatnik MH, Rayburn H et al. Dystrophic muscle in mice chimeric for expression of alpha5 integrin. J Cell Biol 1998; 143:849-859.
61. Rando TA. The dystrophin-glycoprotein complex, cellular signaling, and the regulation of cell survival in the muscular dystrophies. Muscle Nerve 2001; 24:1575-1594.
62. Burton EA, Davies KE. Muscular dystrophy—reason for optimism? Cell 2002; 108:5-8.
63. Miosge N, Klenczar C, Herken R et al. Organization of the myotendinous junction is dependent on the presence of alpha7beta1 integrin. Lab Invest 1999; 79:1591-1599.
64. Bulfield G, Siller WG, Wight PA et al. X chromosome-linked muscular dystrophy (mdx) in the mouse. Proc Natl Acad Sci USA 1984; 81:1189-1192.
65. Allikian MJ, Hack AA, Mewborn S et al. Genetic compensation for sarcoglycan loss by integrin alpha7beta1 in muscle. J Cell Sci 2004; 117:3821-3830.
66. Vachon PH, Xu H, Liu L et al. Integrins (alpha7beta1) in muscle function and survival. Disrupted expression in merosin-deficient congenital muscular dystrophy. J Clin Invest 1997; 100:1870-1881.
67. Hodges BL, Hayashi YK, Nonaka I et al. Altered expression of the alpha7beta1 integrin in human and murine muscular dystrophies. J Cell Sci 1997; 110:2873-2881.
68. Cohn RD, Mayer U, Saher G et al. Secondary reduction of alpha7B integrin in laminin alpha2 deficient congenital muscular dystrophy supports an additional transmembrane link in skeletal muscle. J Neurol Sci 1999; 163:140-152.
69. Allamand V, Campbell KP. Animal models for muscular dystrophy: Valuable tools for the development of therapies. Hum Mol Genet 2000; 9:2459-2467.
70. Cote PD, Moukhles H, Carbonetto S. Dystroglycan is not required for localization of dystrophin, syntrophin, and neuronal nitric-oxide synthase at the sarcolemma but regulates integrin alpha 7B expression and caveolin-3 distribution. J Biol Chem 2002; 277:4672-4679.
71. Grady RM, Teng H, Nichol MC et al. Skeletal and cardiac myopathies in mice lacking utrophin and dystrophin: A model for Duchenne muscular dystrophy. Cell 1997; 90:729-738.
72. Deconinck AE, Rafael JA, Skinner JA et al. Utrophin-dystrophin-deficient mice as a model for Duchenne muscular dystrophy. Cell 1997; 90:717-727.
73. Burkin DJ, Wallace GQ, Milner DJ et al. Transgenic expression of {alpha}7{beta}1 integrin maintains muscle integrity, increases regenerative capacity, promotes hypertrophy, and reduces cardiomyopathy in dystrophic mice. Am J Pathol 2005; 166:253-263.
74. Burkin DJ, Wallace GQ, Nicol KJ et al. Enhanced expression of the alpha 7 beta 1 integrin reduces muscular dystrophy and restores viability in dystrophic mice. J Cell Biol 2001; 152:1207-1218.
75. Moghadaszadeh B, Albrechtsen R, Guo LT et al. Compensation for dystrophin-deficiency: ADAM12 overexpression in skeletal muscle results in increased alpha 7 integrin, utrophin and associated glycoproteins. Hum Mol Genet 2003; 12:2467-2479.
76. Kronqvist P, Kawaguchi N, Albrechtsen R et al. ADAM12 alleviates the skeletal muscle pathology in mdx dystrophic mice. Am J Pathol 2002; 161:1535-1540.
77. Engvall E, Wewer UM. The new frontier in muscular dystrophy research: Booster genes. FASEB J 2003; 17:1579-1584.
78. Geiger B, Bershadsky A, Pankov R et al. Transmembrane crosstalk between the extracellular matrix—cytoskeleton crosstalk. Nat Rev Mol Cell Biol 2001; 2:793-805.
79. Sastry SK, Lakonishok M, Wu S et al. Quantitative changes in integrin and focal adhesion signaling regulate myoblast cell cycle withdrawal. J Cell Biol 1999; 144:1295-1309.
80. Bökel C, Brown NH. Integrins in development: Moving on, responding to, and sticking to the extracellular matrix. Dev Cell 2002; 3:311-321.
81. Brown NH, Gregory SL, Rickoll WL et al. Talin is essential for integrin function in Drosophila. Dev Cell 2002; 3:569-579.
82. Zervas CG, Gregory SL, Brown NH. Drosophila integrin-linked kinase is required at sites of integrin adhesion to link the cytoskeleton to the plasma membrane. J Cell Biol 2001; 152:1007-1018.

83. Hannigan GE, Leung-Hagesteijn C, Fitz-Gibbon L et al. Regulation of cell adhesion and anchorage-dependent growth by a new beta 1-integrin-linked protein kinase. Nature 1996; 379:91-96.
84. Brakebusch C, Fässler R. The integrin-actin connection, an eternal love affair. EMBO J 2003; 22:2324-2333.
85. Huang Y, Li J, Zhang Y et al. The roles of integrin-linked kinase in the regulation of myogenic differentiation. J Cell Biol 2000; 150:861-872.
86. Miller MG, Naruszewicz I, Kumar AS et al. Integrin-linked kinase is a positive mediator of L6 myoblast differentiation. Biochem Biophys Res Commun 2003; 310:796-803.
87. Sakai T, Li S, Docheva D et al. Integrin-linked kinase (ILK) is required for polarizing the epiblast, cell adhesion, and controlling actin accumulation. Genes Dev 2003; 17:926-940.
88. Lo SH. Tensin. Int J Biochem Cell Biol 2004; 36:31-34.
89. Lo SH, Yu QC, Degenstein L et al. Progressive kidney degeneration in mice lacking tensin. J Cell Biol 1997; 136:1349-1361.
90. Ishii A, Lo SH. A role of tensin in skeletal-muscle regeneration. Biochem J 2001; 356:737-745.
91. Torgler CN, Narasimha M, Knox AL et al. Tensin stabilizes integrin adhesive contacts in Drosophila. Dev Cell 2004; 6:357-369.
92. Brancaccio M, Guazzone S, Menini N et al. Melusin is a new muscle-specific interactor for beta(1) integrin cytoplasmic domain. J Biol Chem 1999; 274:29282-29288.
93. Li J, Mayne R, Wu C. A novel muscle-specific beta 1 integrin binding protein (MIBP) that modulates myogenic differentiation. J Cell Biol 1999; 147:1391-1398.
94. Li J, Rao H, Burkin D et al. The muscle integrin binding protein (MIBP) interacts with alpha7beta1 integrin and regulates cell adhesion and laminin matrix deposition. Dev Biol 2003; 261:209-219.
95. Brancaccio M, Fratta L, Notte A et al. Melusin, a muscle-specific integrin beta1-interacting protein, is required to prevent cardiac failure in response to chronic pressure overload. Nat Med 2003; 9:68-75.
96. Samson T, Smyth N, Janetzky S et al. The LIM-only proteins FHL2 and FHL3 interact with alpha- and beta-subunits of the muscle alpha7beta1 integrin receptor. J Biol Chem 2004; 279:28641-28652.
97. Kong Y, Shelton JM, Rothermel B et al. Cardiac-specific LIM protein FHL2 modifies the hypertrophic response to beta-adrenergic stimulation. Circulation 2001; 103:2731-2738.
98. Zhou Q, Ruiz-Lozano P, Martone ME et al. Cypher, a striated muscle-restricted PDZ and LIM domain-containing protein, binds to alpha-actinin-2 and protein kinase C. J Biol Chem 1999; 274:19807-19813.
99. Zhou Q, Chu PH, Huang C et al. Ablation of Cypher, a PDZ-LIM domain Z-line protein, causes a severe form of congenital myopathy. J Cell Biol 2001; 155:605-612.
100. van der Flier A, Sonnenberg A. Structural and functional aspects of filamins. Biochim Biophys Acta 2001; 1538:99-117.
101. Stossel TP, Condeelis J, Cooley L et al. Filamins as integrators of cell mechanics and signalling. Nat Rev Mol Cell Biol 2001; 2:138-145.
102. Feng Y, Walsh CA. The many faces of filamin: A versatile molecular scaffold for cell motility and signalling. Nat Cell Biol 2004; 6:1034-1038.
103. van der Ven PF, Obermann WM, Lemke B et al. Characterization of muscle filamin isoforms suggests a possible role of gamma-filamin/ABP-L in sarcomeric Z-disc formation. Cell Motil Cytoskeleton 2000; 45:149-162.
104. van der Ven PF, Wiesner S, Salmikangas P et al. Indications for a novel muscular dystrophy pathway. gamma-filamin, the muscle-specific filamin isoform, interacts with myotilin. J Cell Biol 2000; 151:235-248.
105. van der Flier A, Kuikman I, Kramer D et al. Different splice variants of filamin-B affect myogenesis, subcellular distribution, and determine binding to integrin [beta] subunits. J Cell Biol 2002; 156:361-376.
106. Speer MC, Vance JM, Grubber JM et al. Identification of a new autosomal dominant limb-girdle muscular dystrophy locus on chromosome 7. Am J Hum Genet 1999; 64:556-562.
107. Thompson TG, Chan YM, Hack AA et al. Filamin 2 (FLN2): A muscle-specific sarcoglycan interacting protein. J Cell Biol 2000; 148:115-126.

Chapter 10

The Role of Integrins in Vasculogenesis and Angiogenesis

Andrew R. Reynolds, Kate A. Scott and Kairbaan M. Hodivala-Dilke*

This chapter aims to summarise the current literature regarding the role of integrins in blood vessel growth. Blood vessels arise via vasculogenesis and angiogenesis and we begin by giving a brief introduction to these processes. We then introduce the methods most commonly used for studying the function of integrins in this field, before reviewing the role of individual integrin subunits. In particular, we have focused on the role of $\alpha v\beta 3$ integrin, for which most literature is available. Finally, we suggest future directions for research in this field.

Introduction

How Does the Vascular System Develop and Grow?

The vascular system is one of the first organs to develop in the embryo. Cells of the mesoderm form blood islands, consisting of endothelial progenitors (angioblasts) and haematopoietic cells. These islands fuse and angioblasts differentiate into a primitive vascular network of endothelial tubes, a process known as 'vasculogenesis'[1] (see Fig. 1). These tubes are sheathed by a basement membrane consisting of extracellular matrix (ECM) molecules such as laminin, collagen, nidogen/entactin and various proteoglycans.[2]

The immature vascular network subsequently undergoes remodelling, a process mediated largely by the sprouting of new blood vessels from the existing vasculature, termed 'angiogenesis' (see Figs. 1 and 2). During angiogenesis, the vascular basement membrane is locally degraded, allowing proliferating and migrating endothelial cells to invade the surrounding stroma (see Fig. 1). These cells differentiate into endothelial tubes that eventually, develop into mature vessels which allow blood to flow (see Fig. 1 and 2). In the fully developed vascular system, vessels are differentiated into arteries, arterioles, capillaries, venules and veins (see Fig. 1D). In large vessels, the endothelial tube is completely sheathed by vascular smooth muscle cells (VSMCs), whilst smaller vessels possess only a sparse covering of pericyctes.[3-5]

The endothelial cells of the adult vascular system remain largely quiescent. However, angiogenesis occurs naturally in the adult during the menstrual cycle and in skeletal muscle growth or in response to injury e.g., during wound healing or following cardiac ischaemia. New blood vessel growth also occurs in diseases such as diabetic retinopathy, where aberrant vascularisation of the retina causes blindness, and in cancer, where tumour vascularisation supports tumour growth. Importantly, new blood vessel growth in the adult is considered to arise mainly through sprouting angiogenesis. Since angiogenesis contributes to numerous pathological

*Corresponding Author: Kairbaan M. Hodivala-Dilke—Cell Adhesion and Disease Group, Tumour Biology Laboratory, Cancer Research U.K. Clinical Centre, Barts & The London, Queen Mary's School of Medicine & Dentistry, John Vane Science Centre, Charterhouse Square, London EC1M 6BQ, U.K. Email: kairbaan.hodivala-dilke@cancer.org.uk

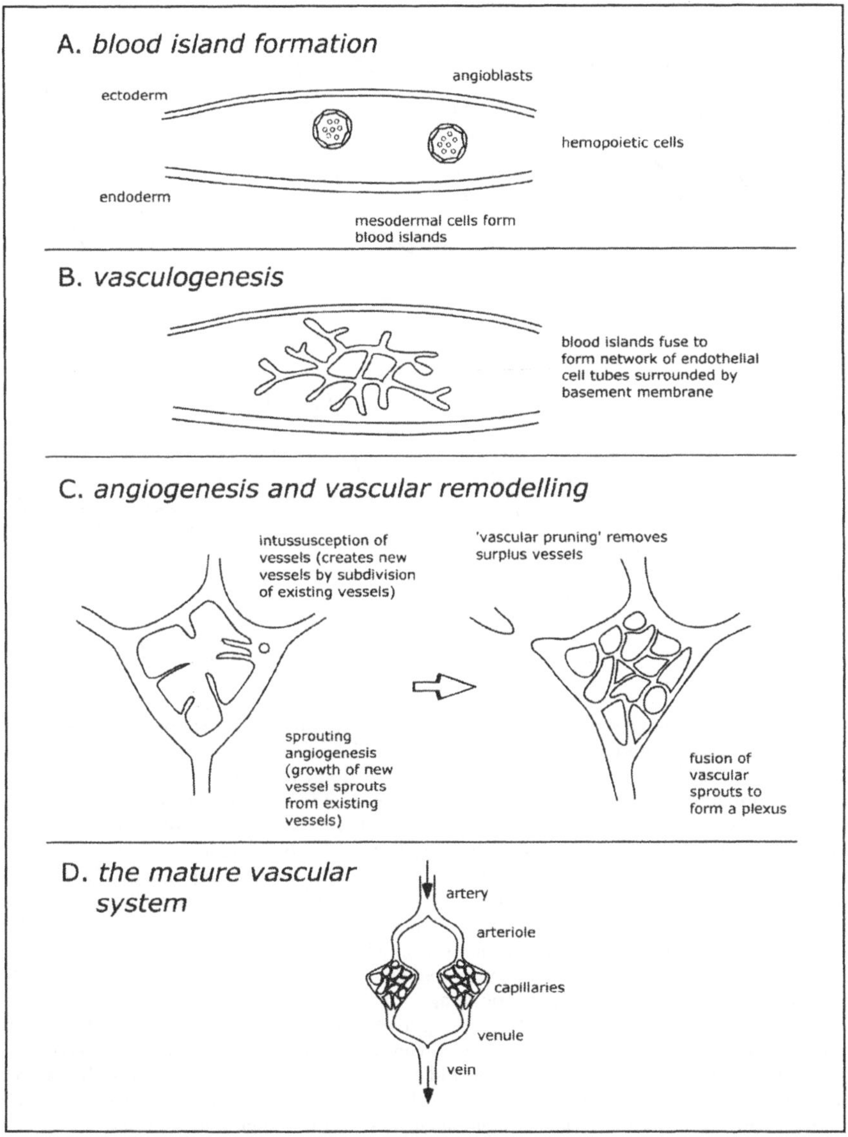

Figure 1. Summary of vasculogenesis and angiogenesis during development A) Blood island formation. Cells of the mesoderm differentiate into angioblasts and hemapoietic cells, which cluster to form blood islands with the angioblasts on the outside and the hemopoietic cells on the inside. B) Vasculogenesis. The blood islands fuse and the angioblasts differentiate into endothelial cells which form a network of tubes, supported by a basement membrane, thus forming a primitive vascular plexus. C) Angiogenesis and vascular remodelling. New blood vessels develop from the primitive vascular plexus via several processes, including 'interssusception' (the subdivision of existing blood vessels) and 'sprouting angiogenesis' (the growth of new blood vessels from existing blood vessels). Vessels produced by sprouting angiogenesis eventully fuse with other sprouts and mature into vessels which can allow blood to flow. Vessels can also regress, a process often described as 'vascular pruning'. D) The mature vascular system. In the fully developed vascular system, vessels are differentiated into arteries, arterioles, capillaries, venules and veins (arrows indicate direction of blood flow). In large vessels, the endothelial tube is completely sheathed by vascular smooth muscle cells (VSMCs), whilst smaller vessels possess only a sparse covering of pericyctes.

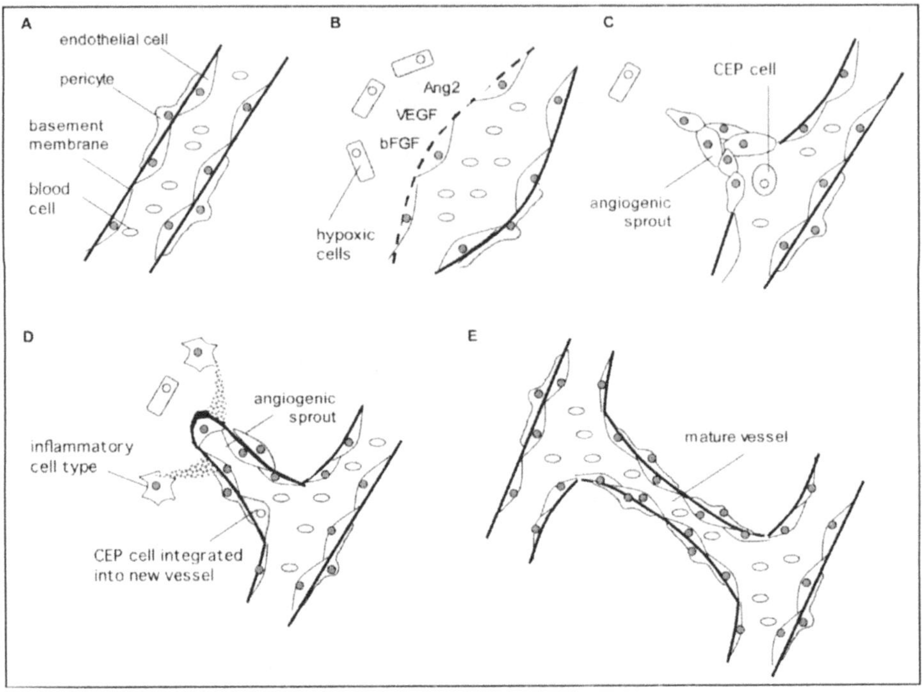

Figure 2. Sprouting angiogenesis. A) Mature blood vessels are composed of endothelial cells and vascular smooth muscle cells, also known as pericytes. The endothelial cells form a tube and the lumen of this tube acts as a conduit for blood. The abluminal face of the endothelial tube is associated with the vascular basement membrane and also forms contacts with the surrounding vascular smooth muscle cells. B) Angiogenesis is initiated by the release of pro-angiogenic growth factors, such as VEGF, bFGF and angiopoietin 2 from a neighbouring tissue or cancer cell. Typically, the release of these factors is driven by tissue hypoxia, but it may also occur due to oncogenic changes in tumour cells. An early response to VEGF, i.e., within minutes, manifests as vessel dilation and an increase in vessel permeability. C) In the longer term, typically hours to days, the pro-angiogenic growth factors induce localised proliferation and migration of endothelial cells and up-regulation of other cellular processes required for new vessel formation such as altered matrix remodelling activity and a change in adhesive properties. These activated endothelial cells invade through the basement membrane and into the surrounding stroma, to form an 'angiogenic sprout'. The release of VEGF may also lead to the recruitment of circulating endothelial progenitor (CEP) cells to the zone of active angiogenesis. D) The angiogenic sprout becomes organised into a vessel-like structure that is composed of an endothelial cell tube, basement membrane and associated pericytes. Whilst most of the new vessel is believed to be derived from endothelial cells that are local to the active zone of angiogenesis, CEP cells can also integrate into new vessels. A balance of pro-angiogenic and anti-angiogenic signals in favour of angiogenesis will ensure that the new vessel continues to invade into the stroma towards the source of the pro-angiogenic factor. During pathological angiogenesis it is believed that local inflammatory cells may also regulate the growth of the angiogenic sprout. E) Under normal circumstances, angiogenesis is coordinated so that the new blood vessel is eventually connected with the arterial system at one end and the venous system at the other. This allows blood to flow and ensures that the new vessel is competent to supply the tissue with oxygen and nutrients. Local down regulation of pro-angiogenic signals may then occur, the functional vessel becomes stable and the endothelial cells become relatively quiescent. In the case of tumour angiogenesis the formation of functional vessels can occur, but these vessels are typically chaotically organised, can be blind-ended and are often leaky and therefore may be compromised in their capability to provide adequate tissue perfusion.

conditions, the mechanism of vascular development and growth has received considerable attention.[3,4]

Vasculogenesis and angiogenesis are coordinated by several growth factors; e.g., the angiopoetin/Tie system, the Eph/Ephrin system, basic fibroblast growth factor (bFGF), transforming growth factor-β (TGFβ) and, most notably, vascular endothelial growth factor (VEGF).[4,6] VEGF binds to several cognate receptors on endothelial cells, including VEGF receptor 1 (VEGFR1) and VEGF receptor 2 (VEGFR2); genetic ablation of VEGF or either receptor inhibits vascular development in mice. In addition, VEGF is a key player in pathological angiogenesis and antagonists that inhibit VEGF-mediated angiogenesis are currently in clinical trials.[7,8] Finally, although embryonic and adult angiogenesis share many similarities, such as a requirement for angiogenic growth factors, there are also subtle differences. For example, the exact combination of growth factors and ECM molecules involved in angiogenesis may vary between the adult and embryo and may be further influenced by the type of disease involved. Also, in contrast to developmental vascular development, infiltrating inflammatory cells and circulating endothelial progenitor cells may play a role in adult angiogenesis.[3,4,9-11]

What Is the Role of Integrins in Vasculogenesis and Angiogenesis?

Integrins are important in vasculogenesis and angiogenesis, since they mediate interactions with the ECM that are essential for endothelial cell adhesion, migration, proliferation and survival. Furthermore, during embryonic and adult angiogenesis, integrin function may change as endothelial cells disengage from the basement membrane and encounter new ECM proteins in the stroma.[2,12] Although VSMCs, pericytes and inflammatory cells also play a role in blood vessel formation, and undoubtedly express integrins, this chapter will focus primarily on the role of endothelial cell integrins in vascular development and growth.

Vascular endothelial cells express a subset of integrins including: α1β1, α2β1,α3β1, α5β1, α6β1, α6β4, αvβ3 and αvβ5 which are known to bind distinct, yet overlapping, groups of ECM ligands.[13] Furthermore, recent research has revealed several new integrin ligands, which may be specific to angiogenesis, such as secreted growth factors and proteolytic fragments of matrix.[2,14,15] Histological analysis of tissues reveals that the expression of certain integrins correlates with new blood vessel formation. αvβ3, αvβ5 and α2β1 are barely detectable in quiescent vessels, but their expression is elevated greatly in new vessel sprouts[16-18] with αvβ3 expression enhanced particularly upon bFGF treatment.[19] The α6 and β4 subunits are expressed in capillary loops and have been observed in putative vessel sprouts[17] although in another study, α6β4 expression was not observed in new blood vessels during an ex vivo model of angiogenesis.[20] α1β1, α3β1 and α5β1 are expressed at low levels in quiescent vessels and α5β1 up-regulation has been reported during angiogenesis.[21] However, elucidating the functional role that these integrins play during angiogenesis is a more complex task.

One approach to study vasculogenesis is to delete the gene of interest and observe the morphological defects in vascular development that may occur within the embryo. Providing the integrin-null embryos develop and the mice are viable and fertile there are several in vivo and in vitro assays that can be used to study adult angiogenesis (See Tables 1 and 2 for further details). Angiogenesis assays can also be performed in the presence of integrin antagonists, which block integrin ligand binding function. Antagonists can be monoclonal antibodies directed against the extracellular domain of the integrin (e.g., LM609, an αvβ3-specific antagonist), soluble ligands (e.g., cyclic RGD-peptides) or small molecule inhibitors. In addition, information concerning the role of various ECM components can be obtained by altering the composition of the matrix within in vitro assays. Collectively, these assays have provided insight into how integrins regulate both vascular development in the embryo and adult angiogenesis in normal and pathological processes.

β1-Integrins

The β1-integrin subunit forms part of several integrin heterodimers known to be expressed on endothelial cells. β1-integrin-deficient embryos die early in gestation before developmental

Table 1. In vivo assays of angiogenesis

	Assay Description
Teratoma	Embryonic stem (ES) cells, which contain derivatives of all primitive germ layers, are injected subcutaneously into mice. The ES cells divide and differentiate into a heterogeneous tissue, known as a teratoma (normally takes 28-56 days). The teratoma is excised and the contribution of the ES cells to blood vessel formation can be determined.
Chick chorioallantoic membrane (CAM)	The test substance is implanted on the CAM of a 7-8 day old chick embryo either in a slow release pellet, absorbed onto gelatin sponge or air-dried onto a filter disc. The angiogenic effect can be measured by counting the number of blood vessels in a given area.
Retinopathy of prematurity (ROP)	Mice are incubated in a high oxygen environment between post-natal day 7 and 12 to induce central retinal avascularization. Mice are subsequently returned to normoxia, which induces stromal VEGF production. After 1-5 days the level of retinal neovascularisation can be measured.
Tumour growth	Tumour cells that have been cultured in vitro are injected subcutaneously into mice. When tumours have formed (typically after 10-100 days dependent on the cell line) the tumour is excised and blood vessel formation can be analysed.
Matrigel plug	Matrigel is impregnated with bFGF and/or VEGF and injected subcutaneously into mice. After angiogenic vessels have grown into the Matrigel pellet (typically 5-10 days), the pellet is excised and the number of angiogenic vessels can be counted.
Sponge assay	Similar to Matrigel plug assay, except that a piece of sponge soaked in bFGF and/or VEGF is implanted subcutaneously instead of Matrigel.

Summary of the various models commonly used to study angiogenesis in vivo.

angiogenesis can occur[22,23] so embryoid bodies and teratomas have been used to study the role of β1-integrin in angiogenesis. β1-null teratoma growth was much slower than teratomas derived from normal ES cells and the β1-null teratomas had less vessels to support tumour growth. In addition, the vessels in β1-deficient teratomas were all host derived as no β1-null endothelial cells were detected.[24] In vitro studies have revealed that although β1-null ES cells can differentiate into endothelial cells the formation of a complex vascular network is significantly delayed and of poor quality. Furthermore, β1-null embryoid bodies were resistant to VEGF-induced proliferation and branching.[24] These data suggest that β1-integrins are crucial in angiogenesis. However, endothelial cells express several integrins that contain the β1-integrin subunit. Genetic ablation of the individual α-integrin subunits has allowed investigators to further dissect the role of each specific integrin heterodimer in angiogenesis.

α5β1 Integrin

Genetic ablation of α5-integrin and subsequent loss of the fibronectin receptor, α5β1, results in a lethal phenotype where embryos die at day 10-11 of gestation due to incomplete formation of the yolk sac and embryonic vascular network.[25] These α5-integrin deficient mice share many similarities with fibronectin knockout mice[26] which also have mid-gestation vascular abnormalities. In both cases endothelial cells form from their angioblast progenitors, but the endothelial tubes in the α5-null embryo display abnormalities in vessel patterning, dilation and integrity.[25,27,28] However, vascular defects in the fibronectin-null embryos are far more severe than in the α5-null embryos. Loss of fibronectin results in either severe disruption of the

Table 2. In vitro and ex vivo angiogenesis models

	Assay Description
Embryoid bodies (EB)	To induce the formation of EB, ES cells are cultured in an environment to promote cell aggregation. After several days incubation, EBs are fixed and blood vessel formation can be measured by immunostaining.
Aortic ring	The aorta is isolated from adult mice, cut into rings and embedded in Matrigel or collagen. Rings are cultured (for up to 14 days) and the extent of angiogenesis is measured by quantification of the number and length of microvessel outgrowths.
In vitro tubulogenesis	A 3D angiogenesis system is prepared by seeding endothelial cells onto Matrigel or collagen in the presence of VEGF or bFGF. After several hours, the level of angiogenesis can be determined by quantifying the number and length of tubules that form.
Proliferation and migration	The ability of endothelial cells to proliferate or migrate in response to angiogenic stimuli in vitro is also often used as a simple measure of angiogenesis.

Summary of the various assays commonly used to study angiogenesis in vitro.

extraembryonic and intraembryonic vessels and heart or complete loss of heart development altogether.[26,29] In agreement with the α5-null embyro studies, vessel development defects were also evident in α5-null embryoid bodies[27] and in α5-null teratocarcinomas.[30] The role of α5β1 in angiogenesis has also been studied in vivo using specific inhibitors. Antagonists of α5β1 can inhibit angiogenesis in chick chorioallantoic membrane (CAM) assays as well as in murine tumour models.[21]

Recent reports show that in addition to binding its ligand fibronectin, α5β1 can bind to other molecules involved in angiogenesis. α5β1 on endothelial cells has been shown to bind directly to the pro-angiogenic factor angiopoietin-1 to enhance migration[31] as well as VEGFR1 to mediate adhesion and migration.[32] α5β1 can also interact with endostatin, a potent anti-angiogenic molecule, to inhibit endothelial cell migration.[33] Collectively, these data suggest that α5β1 appears to play a multi-functional role in angiogenesis although further studies are necessary.

α1β1 and α2β1 Integrin

There are several studies that have investigated the role of the collagen receptors, α1β1 and α2β1, in angiogenesis. Genetic ablation of α1-integrin or α2-integrin does not result in embryonic lethality and α1-null and α2-null mice are both viable and fertile.[34,35] Although these results imply that α1β1 and α2β1 are not required for blood vessel development there are several pieces of in vitro and in vivo evidence that suggest otherwise.

VEGF stimulation of endothelial cells has been shown to increase expression of α1β1 and α2β1 and antagonists of these integrins significantly inhibit angiogenesis.[36,37] Interestingly, a combined treatment with α1-blocking and α2-blocking antibodies markedly reduced VEGF-induced neovascularisation in vivo but had no effect on preexisting vasculature.[36] In addition, ligation of α1β2 and α2β1 to type I collagen appears to be a critical step in initiating cord formation of endothelial cells.[38] These cords closely imitate the solid precapillary cords of embryonic angiogenesis and these data suggest that α1β1 and α2β1 are important for angiogenesis. Furthermore, a novel inhibitor of α2β1, E7820, can successfully inhibit tumour angiogenesis in vivo.[39]

The picture becomes more complex when the roles of α1β1 and α2β1 have been studied in pathological angiogenesis using the integrin-null animals. Surprisingly, a deficiency of α2-integrin does not alter the rate of wound healing in mice.[34] This may be explained by the fact that in vivo wound healing involves several growth factors so inhibition of VEGF-mediated angiogenesis in α2-null mice may be masked during neovascularisation. An alternative approach to address whether α2β1 is specifically involved in VEGF-mediated angiogenesis in vivo would be to use the ROP or Matrigel plug assay. Based on the antagonist data mentioned above we would predict that angiogenesis would be reduced in α2-null mice. In contrast, α1β1 appears to play a role in pathological angiogenesis. Tumour growth assays in α1-null mice have revealed that loss of α1-integrin is associated with reduced vascularisation.[40] α1-integrin deficiency resulted in an increase in production of metalloproteinases MMP-7 and MMP-9. These metalloproteinases can cleave circulating plasminogen to generate angiostatin, a known inhibitor of angiogenesis. The authors conclude that the reduced tumour angiogenesis observed was probably due to elevated MMP production, since plasma from tumour-bearing α1-null mice inhibited endothelial cell growth in an MMP-dependent fashion.[40]

Thus, whilst α5β1 is essential for new blood vessel formation in the developing embryo, α1β1 and α2β1 seem to not be required for developmental angiogenesis. Although further studies are necessary the current data suggest that the lethal phenotype observed in β1-null mice may be mainly due to loss of α5β1. However, α1β1 and α2β1 appear to certainly be involved in pathological angiogenesis and but more data is needed to clarify their exact roles.

α3β1 and α6β4 Integrin

There are very few reports in the literature regarding the precise roles of the laminin receptors, α3β1 and α6β4, in angiogenesis. Genetic ablation of either α3-integrin or α6-integrin results in a lethal phenotype within hours of birth.[41,42] α3-null mice display a combination of defects including abnormal branching in the bronchi of the lungs, glomeruli of the kidneys and neurons,[42] as well as microblister formation in the skin.[43] Most of these defects are associated with aberrant basement membrane organisation. Developmental angiogenesis has not been formally examined apart from the observation that capillary loops associated with α3-null defective kidney glomeruli are dilated and reduced in number.[42] α6-null and β4-null mice also display severe epidermal blistering but to date there are no reports showing that these integrins play a role in developmental angiogenesis.[41,44-46] Mice deficient in both α3-integrin and α6-integrin have been generated and although they display severe skin blistering no blood vessel defects have been reported.[47]

More recently, laminin-8 has been detected in the basement membrane of embryonic blood vessels[48] and deletion of laminin-8 results in enhanced neovascularisation coupled with increased tumour growth and metastasis.[49] Cell adhesion to laminin-8 occurs via α3-integrin and α6-integrin, but in one study cell migration on laminin-8 was only inhibited by antibodies directed against α3-integrin.[50] These results suggest possible roles for laminin receptors in blood vessel formation. In another study, addition of laminin-1 stimulated tube formation by endothelial cells in vitro and the authors concluded that it may have been α3β1 ligation with laminin-1 that promoted tubulogenesis.[51] In contrast, spontaneous overexpression of α3β1 in a transformed cell line of human umbilical vein endothelial cells (ECV304) has been associated with retarded tube formation[52] although other researchers have shown enhanced tube formation with the same cell line.[53] Another piece of evidence to suggest that α3β1 may play an inhibitory role in angiogenesis is the finding that TIMP-2, a metalloproteinase inhibitor, can bind α3β1 directly to inhibit endothelial cell proliferation.[54] Thus, a high level of α3β1 on endothelial cells may prevent angiogenesis.

A role for α6β4 in pathological angiogenesis has been demonstrated using mice that express a mutant form of the integrin β4 subunit in place of the wild-type gene.[107] The mice carry a C-terminal deletion of the β4-integrin cytoplasmic tail which disables some of the functions of α6β4. The mutant mice exhibit normal developmental angiogenesis and are viable, yet they display a reduced angiogenic response to bFGF in the Matrigel plug assay and a reduced angiogenic response in the ROP assay. Tumours implanted in the mutant mice also grew smaller and were less angiogenic. The authors demonstrated that α6β4 signaling is required for the growth of β4-negative microvessels from β4-positive vessels without exerting a direct effect on endothelial cell proliferation or survival. α6β4 signalling was shown to induce endothelial cell migration and invasion by promoting nuclear translocation of phosphorylated ERK and NF-κB. These results suggest that α6β4 plays a role in the invasive phase of pathological angiogenesis. In summary, the roles for α3β1 and α6β4 in angiogenesis are incompletely understood and although initial results look interesting, further studies are required to determine their exact function.

αvβ3 and αvβ5 Integrin

The vast majority of literature concerning the involvement of integrins in angiogenesis has focused on the vitronectin receptors, αvβ3 and αvβ5. Expression of these integrins is up-regulated on angiogenic blood vessels and αvβ3/αvβ5 antagonists inhibit blood vessel growth in several angiogenesis assays including ROP assay, CAM assay and in murine tumour models.[16,55-58] These data identified αvβ3 and αvβ5 as prominent anti-angiogenic therapeutic targets and led to the natural assumption that αvβ3 and αvβ5 integrins are essential for angiogenesis.[59,60]

In contrast, β3-null, β5-null and β3/β5-doubly-deficient mice are all viable, fertile and exhibit normal vascular development.[61-63] In αv-integrin-null mice, which do not express αvβ1, αvβ3, αvβ5, αvβ6 or αvβ8, normal developmental vasculogenesis and angiogenesis is observed in all organs, except for the brain and intestines.[64,65] Interestingly, the phenotype of β8-integrin-null mice is very similar to the αv-integrin-null mice.[65,66] During development, β8-integrin expression is found mainly in cells which provide support for the vasculature, such as the parenchyma of the brain, but is not expressed by endothelial cells themselves.[66] These data suggest that the phenotype of the αv-integrin-deficient mice maybe associated with defects in the function of parenchymal cells in certain organs and that endothelial cell expression of αv-integrins may not be essential for angiogenesis.

In addition to exhibiting normal developmental angiogenesis, β3-integrin-deficient adult mice and β3/β5-integrin doubly-deficient adult mice display enhanced VEGF-induced angiogenesis and increased tumour growth.[63] These data have prompted much debate regarding the precise role of β3-integrin in angiogenesis.[4,14,67,68] The picture is complicated, since β3-integrin may promote or inhibit angiogenesis depending on the ligand to which it is bound and the type of signalling pathway that it activates.[14,68] It is important to understand the opposing functions of β3-integrin before we can understand why angiogenesis is enhanced in β3- and β3/β5-null mice.

Pro-Angiogenic Functions of αvβ3-Integrin

RGD Ligand Binding

It is well established that RGD-containing ligands, such as fibronectin, vitronectin and osteopontin can enhance endothelial cell migration, tube formation and angiogenesis via binding to αvβ3.[12,69] Interestingly, VEGF can increase the affinity of αvβ3 and α5β1 for fibronectin, via so-called inside-out signalling, thereby enhancing endothelial cell adhesion and migration.[70] Integrin ligation also provides a survival signal which protects cells from apoptosis by activating PI-3 kinase/Akt signalling and in the case of αvβ3, by the activation of NF-κB Furthermore, αvβ3 antagonists induce endothelial cell apoptosis, providing further evidence that αvβ3 ligation promotes endothelial cell survival.[67,71] There are several other molecules which contain RGD motifs that have also been shown to bind αvβ3 and enhance angiogenesis in various assays. These include a proteolytic fragment of the ECM molecule collagen type IV,[72] the coagulation factor, thrombin[73] and secreted factors, such as the developmental endothelial locus 1 protein (Del-1; 74,75), CYR61,[76-78] bone sialoprotein (BSP; 79) and angiopoetin 3

(ANGPTL3; 80). This ligand binding promiscuity may be advantageous to angiogenic blood vessels, since it might increase the likelihood that αvβ3 is ligated to a pro-angiogenic ligand and therefore ensure endothelial cell survival and migration.[69] Alternatively, each of these 'pro-angiogenic' αvβ3 ligands may elicit subtly different signals, which have yet to be fully elucidated. In any case, there is extensive evidence that αvβ3 integrin ligation can transmit pro-angiogenic signals by mediating endothelial cell migration and survival.

Synergism with Growth Factor Receptor Signalling

Growth factors control many aspects of endothelial cell behaviour including: adhesion, migration, proliferation, survival and vascular permeability. It is clear that cross talk between integrins and growth factor receptors plays a role in modulating these effects.[81-83] Cheresh and coworkers have used αvβ3 and αvβ5-selective antagonists in the CAM assay to show that αvβ3-integrin is required for bFGF-mediated angiogenesis, whilst αvβ5 is required for VEGF-mediated angiogenesis.[84] Further work has demonstrated that following VEGF stimulation Src associates with αvβ5 integrin, but not αvβ3 integrin, and this is required for activation of FAK in response to VEGF.[85] Also, αvβ3 is required to activate c-Raf and achieve sustained activation of ERK in response to bFGF, whereas αvβ5 is required to activate Ras and ERK in reponse to VEGF.[86,87] However, αvβ3 can also play a role in regulating VEGF signalling. αvβ3 binding to vitronectin or fibronectin can stimulate VEGFR2 phosphorylation and signalling and the direct physical interaction of αvβ3 and VEGFR2 has been proposed to mediate this effect.[70,88,89] It is important to consider that these studies are based upon the effect that αvβ3 and αvβ5 antagonists have on angiogenic signalling and it will be interesting to determine whether genetic deletion or RNAi-mediated suppression of these integrins has the same effect. In any case, this work has clearly demonstrated that αv integrins can regulate signalling processes downstream of growth factor signals.

Anti-Angiogenic Functions of αvβ3-Integrin

Tumstatin and Endostatin Binding

Proteolytic fragments of certain matrix proteins have antiangiogenic activity, including tumstatin, a fragment of type IV collagen, and endostatin, a fragment of type XVIII collagen.[2,90] Although endostatin can bind to both αvβ3 and α5β1, it inhibits angiogenesis by competing for RGD ligand binding to α5β1.[90] However, tumstatin binds to αvβ3 in a novel site outside the RGD binding region and induces a signalling cascade that causes endothelial cell apoptosis via the inhibition of protein synthesis.[91,92] Interestingly, although tumstatin can inhibit tumour angiogenesis in wild-type mice, tumour angiogenesis in β3-integrin-null mice is not blocked by tumstatin, highlighting the importance of β3-integrin in mediating the anti-angiogenic activity of tumstatin in vivo.[93] It is conceivable that other proteolytic fragments of matrix proteins, which can inhibit angiogenesis through αvβ3, will be discovered.

Thrombospondin Binding

The matricellular proteins thrombospondin 1 (TSP1) and thrombospondin 2 (TSP2) block angiogenesis via several mechanisms. The most characterised endothelial cell receptor for TSP1 is CD36, which is required for the antiangiogenic effects of TSP1 in vitro and in vivo. However, TSP1 and TSP2 can bind αvβ3 directly and can also bind CD47, a G-protein-coupled receptor found in association with αvβ3 and α2β1.[94,95] Importantly, TSP2-null mice display increased tumour growth due to enhanced angiogenesis.[96,97] The precise role of αvβ3 in the biological action of thrombospondins is unclear, but it is possible that the interaction with αvβ3 may help to 'present' TSP on the endothelial cell surface. Given that tumstatin is ineffective in β3-integrin-null mice, similar studies analysing the effects of thrombospondins in β3-integrin-null mice would be highly informative.

Integrin-Mediated Death

It has recently been shown that unligated αvβ3 can promote apoptosis by directly recruiting and activating caspase 8, a process which has been termed 'integrin-mediated death' (IMD). This implies that endothelial cells expressing αvβ3 are prone to cell death in the absence of an appropriate ligand.[15,67,98] Therefore, it is possible that αvβ3 can play an active role in inducing apoptosis, as well as providing pro-survival signals.

Trans-Dominant Inhibition of Other Integrins

Certain integrin heterodimers can inhibit the function of other distinct integrins, which has been termed 'transdominant inhibition.' For example, ligand binding to αllbβ3 suppresses the ligand binding capability of α5β1 and α2β1.[99] αvβ3 has been shown to suppress the function of α5β1 and vice-versa[100] and cross talk between αvβ3 and α5β1 has been suggested to affect angiogenesis.[101,102] However, there is currently no evidence for enhanced α5β1 function in β3-integrin-null mice.[63]

Inhibition of Growth Factor Receptor Signalling

Work from our lab has shown that β3-integrin expression in endothelial cells can suppress VEGFR2 expression[63] and that VEGFR2 phosphorylation and signalling is actually enhanced in β3-null endothelial cells.[103] These data suggest that β3-integrin can enhance or inhibit VEGF signalling via different mechanisms: vitronectin binding to β3-integrin enhances VEGF signalling,[89] but β3-integrin expression can limit the amount of VEGFR2 available for signalling.

Why Is Enhanced Angiogenesis Observed in β3-Integrin-Null Mice?

One possible explanation for the enhanced angiogenesis observed is that deletion of β3 integrin results in up regulation of other integrins. However, our laboratory has demonstrated that both the levels, and function, of other integrins are unaltered in β3-null endothelial cells suggesting that integrin compensation does not occur following β3-integrin deletion.[63] Interestingly, VEGF signalling via VEGFR2 is enhanced in β3-null endothelial cells and VEGFR2 inhibitors completely block the enhanced angiogenesis observed in β3-null mice.[63,103] Furthermore, the anti-angiogenic effects of tumstatin are compromised in β3-null mice[93] and β3-null endothelial cells may not possess the means to activate IMD.[67] Therefore, the absence of β3-integrin may give angiogenic blood vessels a key advantage, because they have an enhanced response to VEGF, cannot respond to tumstatin and are unable to activate IMD. This would then lead to enhanced angiogenesis in the β3-null mice and increased tumour growth.

Given that αvβ3 can play multiple roles in angiogenesis, another obvious question is, how do β3-antagonists function? One possibility is that the effect of β3-integrin antagonists is more complex than merely blocking adhesion. Several reports have shown that integrin antagonists can activate novel signalling pathways.[99,104,105] Most notably, RGD peptides have been shown to activate caspase 3 leading to apoptosis[106] and β3-antagonists can block VEGF receptor signalling.[89] It will therefore be important to understand the mechanisms via which integrin antagonists mediate their effects if integrin antagonists are to be used effectively as anti-angiogenic reagents in the treatment of disease.

Conclusions and Future Perspectives

We have presented the existing evidence describing the roles of different integrin subunits in vasculogenesis and angiogenesis. Although integrin gene ablation and integrin antagonist studies are largely in agreement with each other several pieces of contradictory data exist, which highlights our poor understanding of these molecules. The current use of integrin antagonists assumes that integrins merely function as static mediators of cell adhesion, migration and survival, but this underestimates the ability of integrins to regulate diverse cellular processes during angiogenesis depending on the type of ligand bound. There is a need to characterise the actions of all pro- and anti-angiogenic integrin ligands on endothelial cell behaviour both in vitro and in vivo.

Integrin-null mice have been useful in determining a requirement for various integrins in angiogenesis, but this approach is limiting since certain deletions result in lethality and the potential to analyse integrin function in endothelial cells alone is also compromised. The creation of blood vessel-specific integrin-null mice, using Cre/Lox strategies, will allow the exact contribution of each endothelial cell integrin in angiogenesis to be investigated. Mice could also be generated to express integrins with specific mutations, that alter distinct ligand binding and signal transduction pathways, in order to further define the precise mechanisms of integrin function in angiogenesis. It will also be informative to perform angiogenesis studies in mice which have been engineered to express mutated forms of the various integrin ligands.

Finally, there are very few reports concerning the roles of mural cell (VSMC and pericyte) integrins in angiogenesis. Moreover, there is increasing evidence that circulating endothelial progenitor cells may contribute to adult angiogenesis. Since these cells also contribute to angiogenesis the role of integrins in these cell types should be investigated, perhaps by using Cre/Lox strategies to delete specific integrins in these cell types.

References

1. Risau W, Flamme I. Vasculogenesis. Annu Rev Cell Dev Biol 1995; 11:73-91.
2. Kalluri R. Basement membranes: Structure, assembly and role in tumour angiogenesis. Nat Rev Cancer 2003; 3:422-33.
3. Carmeliet P. Mechanisms of angiogenesis and arteriogenesis. Nat Med 2000; 6:389-95.
4. Carmeliet P. Angiogenesis in health and disease. Nat Med 2003; 9:653-60.
5. Risau W. Mechanisms of angiogenesis. Nature 1997; 386:671-4.
6. Rossant J, Howard L. Signaling pathways in vascular development. Annu Rev Cell Dev Biol 2002; 18:541-73.
7. Matsumoto T, Claesson-Welsh L. VEGF receptor signal transduction. Sci STKE 2001; RE21.
8. Neufeld G, Cohen T, Gengrinovitch S et al. Vascular endothelial growth factor (VEGF) and its receptors. Faseb J 1999; 13:9-22.
9. Dor Y, Djonov V, Keshet E. Making vascular networks in the adult: Branching morphogenesis without a roadmap. Trends Cell Biol 2003; 13:131-6.
10. Ishida S, Usui T, Yamashiro K et al. VEGF164-mediated inflammation is required for pathological, but not physiological, ischemia-induced retinal neovascularization. J Exp Med 2003; 198:483-9.
11. Rabbany SY, Heissig B, Hattori K et al. Molecular pathways regulating mobilization of marrow-derived stem cells for tissue revascularization. Trends Mol Med 2003; 9:109-17.
12. Hood JD, Cheresh DA. Role of integrins in cell invasion and migration. Nat Rev Cancer 2002; 2:91-100.
13. Albelda SM, Daise M, Levine EM et al. Identification and characterization of cell-substratum adhesion receptors on cultured human endothelial cells. J Clin Invest 1989; 83:1992-2002.
14. Hodivala-Dilke KM, Reynolds AR, Reynolds LE. Integrins in angiogenesis: Multitalented molecules in a balancing act. Cell Tissue Res 2003; 314:131-44.
15. Stupack DG, Cheresh DA. ECM remodeling regulates angiogenesis: Endothelial integrins look for new ligands. Sci STKE 2002; E7.
16. Brooks PC, Montgomery AM, Rosenfeld M et al. Integrin alpha v beta 3 antagonists promote tumor regression by inducing apoptosis of angiogenic blood vessels. Cell 1994; 79:1157-64.
17. Enenstein J, Kramer RH. Confocal microscopic analysis of integrin expression on the microvasculature and its sprouts in the neonatal foreskin. J Invest Dermatol 1994; 103:381-6.
18. Max R, Gerritsen RR, Nooijen PT et al. Immunohistochemical analysis of integrin $\alpha v\beta 3$ expression on tumor-associated vessels of human carcinomas. Int J Cancer 1997; 71:320-4.
19. Sepp NT, Li LJ, Lee KH et al. Basic fibroblast growth factor increases expression of the alpha v beta 3 integrin complex on human microvascular endothelial cells. J Invest Dermatol 1994; 103:295-9.
20. Hiran TS, Mazurkiewicz JE, Kreienberg P et al. Endothelial expression of the alpha6beta4 integrin is negatively regulated during angiogenesis. J Cell Sci 2003; 116:3771-81.
21. Kim S, Bell K, Mousa SA et al. Regulation of angiogenesis in vivo by ligation of integrin alpha5beta1 with the central cell-binding domain of fibronectin. Am J Pathol 2000; 156:1345-62.

22. Fassler R, Pfaff M, Murphy J et al. Lack of beta 1 integrin gene in embryonic stem cells affects morphology, adhesion, and migration but not integration into the inner cell mass of blastocysts. J Cell Biol 1995; 128:979-88.
23. Stephens LE, Sutherland AE, Klimanskaya IV et al. Deletion of beta 1 integrins in mice results in inner cell mass failure and peri-implantation lethality. Genes Dev 1995; 9:1883-95.
24. Bloch W, Forsberg E, Lentini S et al. Beta 1 integrin is essential for teratoma growth and angiogenesis. J Cell Biol 1997; 139:265-78.
25. Yang JT, Rayburn H, Hynes RO. Embryonic mesodermal defects in alpha 5 integrin-deficient mice. Development 1993; 119:1093-105.
26. George EL, Georges-Labouesse EN, Patel-King RS et al. Defects in mesoderm, neural tube and vascular development in mouse embryos lacking fibronectin. Development 1993; 119:1079-91.
27. Francis SE, Goh KL, Hodivala-Dilke K et al. Central roles of alpha5beta1 integrin and fibronectin in vascular development in mouse embryos and embryoid bodies. Arterioscler Thromb Vasc Biol 2002; 22:927-33.
28. Goh KL, Yang JT, Hynes RO. Mesodermal defects and cranial neural crest apoptosis in alpha5 integrin-null embryos. Development 1997; 124:4309-19.
29. George EL, Baldwin HS, Hynes RO. Fibronectins are essential for heart and blood vessel morphogenesis but are dispensable for initial specification of precursor cells. Blood 1997; 90:3073-3081.
30. Taverna D, Hynes RO. Reduced blood vessel formation and tumor growth in alpha5-integrin-negative teratocarcinomas and embryoid bodies. Cancer Res 2001; 61:5255-61.
31. Carlson TR, Feng Y, Maisonpierre PC et al. Direct cell adhesion to the angiopoietins mediated by integrins. J Biol Chem 2001; 276:26516-25.
32. Orecchia A, Lacal PM, Schietroma C et al. Vascular endothelial growth factor receptor-1 is deposited in the extracellular matrix by endothelial cells and is a ligand for the alpha 5 beta 1 integrin. J Cell Sci 2003; 116:3479-89.
33. Rehn M, Veikkola T, Kukk-Valdre E et al. Interaction of endostatin with integrins implicated in angiogenesis. Proc Natl Acad Sci USA 2001; 98:1024-9.
34. Chen J, Diacovo TG, Grenache DG et al. The alpha(2) integrin subunit-deficient mouse: A multifaceted phenotype including defects of branching morphogenesis and hemostasis. Am J Pathol 2002; 161:337-44.
35. Gardner H, Kreidberg J, Koteliansky V et al. Deletion of integrin alpha 1 by homologous recombination permits normal murine development but gives rise to a specific deficit in cell adhesion. Dev Biol 1996; 175:301-13.
36. Senger DR, Claffey KP, Benes JE et al. Angiogenesis promoted by vascular endothelial growth factor: Regulation through alpha1beta1 and alpha2beta1 integrins. Proc Natl Acad Sci USA 1997; 94:13612-7.
37. Hong YK, Lange-Asschenfeldt B, Velasco P et al. VEGF-A promotes tissue repair-associated lymphatic vessel formation via VEGFR-2 and the alpha1beta1 and alpha2beta1 integrins. FASEB J 2004; 18:1111-3.
38. Whelan MC, Senger DR. Collagen I initiates endothelial cell morphogenesis by inducing actin polymerization through suppression of cyclic AMP and protein kinase A. J Biol Chem 2003; 278:327-34.
39. Funahashi Y, Sugi NH, Semba T et al. Sulfonamide derivative, E7820, is a unique angiogenesis inhibitor suppressing an expression of integrin alpha2 subunit on endothelium. Cancer Res 2002; 62:6116-23.
40. Pozzi A, Moberg PE, Miles LA et al. Elevated matrix metalloprotease and angiostatin levels in integrin alpha 1 knockout mice cause reduced tumor vascularization. Proc Natl Acad Sci USA 2000; 97:2202-7.
41. Georges-Labouesse E, Messaddeq N, Yehia G et al. Absence of integrin alpha 6 leads to epidermolysis bullosa and neonatal death in mice. Nat Genet 1996; 13:370-3.
42. Kreidberg JA, Donovan MJ, Goldstein SL et al. Alpha 3 beta 1 integrin has a crucial role in kidney and lung organogenesis. Development 1996; 122:3537-47.
43. DiPersio CM, Hodivala-Dilke KM, Jaenisch R et al. alpha3beta1 Integrin is required for normal development of the epidermal basement membrane. J Cell Biol 1997; 137:729-42.
44. Dowling J, Yu QC, Fuchs E. Beta4 integrin is required for hemidesmosome formation, cell adhesion and cell survival. J Cell Biol 1996; 134:559-72.
45. van der Flier A, Sonnenberg A. Function and interactions of integrins. Cell Tissue Res 2001; 305:285-98.
46. van der Neut R, Krimpenfort P, Calafat J et al. Epithelial detachment due to absence of hemidesmosomes in integrin beta 4 null mice. Nat Genet 1996; 13:366-9.

47. DiPersio CM, van der Neut R, Georges-Labouesse E et al. alpha3beta1 and alpha6beta4 integrin receptors for laminin-5 are not essential for epidermal morphogenesis and homeostasis during skin development. J Cell Sci 2000; 113:3051-62.
48. Spessotto P, Yin Z, Magro G et al. Laminin isoforms 8 and 10 are primary components of the subendothelial basement membrane promoting interaction with neoplastic lymphocytes. Cancer Res 2001; 61:339-47.
49. Zhou Z, Doi M, Wang J et al. Deletion of laminin-8 results in increased tumor neovascularization and metastasis in mice. Cancer Res 2004; 64:4059-63.
50. Fujiwara H, Kikkawa Y, Sanzen N et al. Purification and characterization of human laminin-8. Laminin-8 stimulates cell adhesion and migration through alpha3beta1 and alpha6beta1 integrins. J Biol Chem 2001; 276:17550-8.
51. Kanda S, Tomasini-Johansson B, Klint P et al. Signaling via fibroblast growth factor receptor-1 is dependent on extracellular matrix in capillary endothelial cell differentiation. Exp Cell Res 1999; 248:203-13.
52. Shin EY, Lee JY, Park MK et al. Overexpressed alpha3beta1 and constitutively activated extracellular signal-regulated kinase modulate the angiogenic properties of ECV304 cells. Mol Cells 1999; 9:138-45.
53. Hughes SE. Functional characterization of the spontaneously transformed human umbilical vein endothelial cell line ECV304: Use in an in vitro model of angiogenesis. Exp Cell Res 1996; 225:171-85.
54. Seo DW, Li H, Guedez L et al. TIMP-2 mediated inhibition of angiogenesis: An MMP-independent mechanism. Cell 2003; 114:171-80.
55. Brooks PC, Stromblad S, Klemke R et al. Antiintegrin $\alpha v\beta 3$ blocks human breast cancer growth and angiogenesis in human skin. J Clin Invest 1995; 96:1815-1822.
56. Friedlander M, Theesfeld CL, Sugita M et al. Involvement of integrins $\alpha v\beta 3$ and $\alpha v\beta 5$ in ocular neovascular diseases. Proc Natl Acad Sci USA 1996; 93:9764-9769.
57. Hammes H-P, Brownlee M, Jonczyk A et al. Subcutaneous injection of a cyclic peptide antagonist of vitronectin receptor-type integrins inhibits retinal neovascularization. Nature Medicine 1996; 2:529-533.
58. Kumar CC, Malkowski M, Yin ZZ et al. Inhibition of angiogenesis and tumor growth by SCH221153, a dual alpha(v)beta(3) and alpha(v)beta(5) integrin receptor antagonist. Cancer Research 2001; 61:2232-2238.
59. Eliceiri BP, Cheresh DA. The role of αv integrins during angiogenesis: Insights into potential mechanisms of action and clinical development. The Journal of Clinical Investigation 1999; 103:1227-1230.
60. Varner KA, Cheresh DA. Integrins and cancer. Current Opinion in Cell Biology 1996; 8:724-730.
61. Hodivala-Dilke KM, McHugh KP, Tsakiris DA et al. Beta3-integrin-deficient mice are a model for Glanzmann thrombasthenia showing placental defects and reduced survival. J Clin Invest 1999; 103:229-238.
62. Huang X, Griffiths M, Wu J et al. Normal development, wound healing, and adenovirus susceptibility in beta5-deficient mice. Mol Cell Biol 2000; 20:755-759.
63. Reynolds LE, Wyder L, Lively JC et al. Enhanced pathological angiogenesis in mice lacking beta3 integrin or beta3 and beta5 integrins. Nat Med 2002; 8:27-34.
64. Bader BL, Rayburn H, Crowley D et al. Extensive vasculogenesis, angiogenesis, and organogenesis precede lethality in mice lacking all αv integrins. Cell 1998; 95:507-519.
65. McCarty JH, Monahan-Earley RA, Brown LF et al. Defective associations between blood vessels and brain parenchyma lead to cerebral hemorrhage in mice lacking alphav integrins. Mol Cell Biol 2002; 22:7667-77.
66. Zhu J, Motejlek K, Wang D et al. $\beta 8$ integrins are required for vascular morphogenesis in mouse embryos. Develpoment 2002; 129:2891-2903.
67. Cheresh DA, Stupack DG. Integrin-mediated death: An explanation of the integrin- knockout phenotype? Nature Medicine 2002; 8:193-194.
68. Hynes RO. A re-evaluation of integrins as regulators of angiogenesis. Nature Medicine 2002; 8:918-921.
69. Ruegg C, Dormond O, Mariotti A. Endothelial cell integrins and COX-2: Mediators and therapeutic targets of tumor angiogenesis. Biochim Biophys Acta 2004; 4:51-67.
70. Byzova TV, Goldman CK, Pampori N et al. A mechanism for modulation of cellular responses to VEGF: Activation of the integrins. Mol Cell 2000; 6:851-860.
71. Stupack DG, Cheresh DA. Apoptotic cues from the extracellular matrix: Regulators of angiogenesis. Oncogene 2003; 22:9022-9.

72. Xu JS, Rodriguez D, Petitclerc E et al. Proteolytic exposure of a cryptic site within collagen type IV is required for angiogenesis and tumor growth in vivo (vol 154, pg 1069, 2001). J Cell Biol 2001; 155:859-859.
73. Byzova TV, Plow EF. Activation of $\alpha v\beta 3$ on vascular cells controls recognition of prothrombin. J Cell Biol 1998; 143:2081-2092.
74. Aoka Y, Johnson FL, Penta K et al. The embryonic angiogenic factor Del1 accelerates tumor growth by enhancing vascular formation. Microvascular Research 2002; 64:148-161.
75. Hidai C, Zupancic T, Penta K et al. Cloning and characterization of developmental endothelial locus-1: An embryonic endothelial cell protein that binds the alpha v beta 3 integrin receptor. Genes and Development 1998; 12:21-33.
76. Babic AM, Kireeva ML, Kolesnikova TV et al. CYR61, a product of a growth factor-inducible immediate early gene, promotes angiogenesis and tumor growth. Proceedings of the National Academy of Sciences of the United States of America 1998; 95:6355-6360.
77. Babic AM, Chen CC, Lau LF. Fisp12/mouse connective tissue growth factor mediates endothelial cell adhesion and migration through integrin $\alpha v\beta 3$ promotes endothelial cell survival, and induces angiogenesis in vivo. Molecular and Cellular Biology 1999; 19:2958-2966.
78. Leu SJ, Lam SCT, Lau LF. Pro-angiogenic activities of CYR61 (CCN1) mediated through integrins alpha(v)beta(3) and alpha(6)beta(1) in human umbilical vein endothelial cells. Journal of Biological Chemistry 2002; 277:46248-46255.
79. Bellahcene A, Bonjean K, Fohr B et al. Bone sialoprotein mediates human endothelial cell attachment and migration and promotes angiogenesis. Circulation Research 2000; 86:885-891.
80. Camenisch G, Pisabarro MT, Sherman D et al. ANGPTL3 stimulates endothelial cell adhesion and migration via integrin $\alpha v\beta 1$ and induces blood vessel formation in vivo. Journal of Biological Chemistry 2002; 277:17281-17290.
81. Schwartz MA, Ginsberg MH. Networks and crosstalk: Integrin signalling spreads. Nature Cell Biology 2002; 4:E65-E68.
82. Smyth SS, Patterson C. Tiny dancers: The integrin-growth factor nexus in angiogenic signaling. J Cell Biol 2002; 158:17-21.
83. Yamada KM, Even-Ram S. Integrin regulation of growth factor receptors. Nature Cell Biology 2002; 4:E75-E76.
84. Friedlander M, Brooks PC, Shaffer RW et al. Definition of two angiogenic pathways by distinct αv integrins. Science 1995; 270:1500-1502.
85. Eliceiri BP, Puente XS, Hood JD et al. Src-mediated coupling of focal adhesion kinase to integrin $\alpha v\beta 5$ in vascular endothelial growth factor signaling. J Cell Biol 2002; 157:149-60.
86. Eliceiri BP, Klemke R, Stromblad S et al. Integrin alphavbeta3 requirement for sustained mitogen-activated protein kinase activity during angiogenesis. J Cell Biol 1998; 140:1255-63.
87. Hood JD, Frausto R, Kiosses WB et al. Differential alphav integrin-mediated Ras-ERK signaling during two pathways of angiogenesis. J Cell Biol 2003; 162:933-43.
88. Schneller M, Vuori K, Ruoslahti E. alpha v beta 3 integrin associates with activated insulin and PDGF beta receptors and potentiates the biological activity of PDGF. EMBO Journal 1997; 16:5600-5607.
89. Soldi R, Mitola S, Strasly M et al. Role of $\alpha v\beta 3$ integrin in the activation of vascular endothelial growth factor receptor-2. EMBO 1999; 18:882-892.
90. Sudhakar A, Sugimoto H, Yang C et al. Human tumstatin and human endostatin exhibit distinct antiangiogenic activities mediated by alpha v beta 3 and alpha 5 beta 1 integrins. Proc Natl Acad Sci USA 2003; 100:4766-71.
91. Maeshima Y, Yerramalla UL, Dhanabal M et al. Extracellular matrix-derived peptide binds to alpha(v)beta(3) integrin and inhibits angiogenesis. J Biol Chem 2001; 276:31959-31968.
92. Maeshima Y, Sudhakar A, Lively JC et al. Tumstatin, an endothelial cell-specific inhibitor of protein synthesis. Science 2002; 295:140-143.
93. Hamano Y, Zeisberg M, Sugimoto H et al. Physiological levels of tumstatin, a fragment of collagen IV alpha3 chain, are generated by MMP-9 proteolysis and suppress angiogenesis via alphaV beta3 integrin. Cancer Cell 2003; 3:589-601.
94. Lawler J. Thrombospondin-1 as an endogenous inhibitor of angiogenesis and tumor growth. J Cell Mol Med 2002; 6:1-12.
95. Armstrong LC, Bornstein P. Thrombospondins 1 and 2 function as inhibitors of angiogenesis. Matrix Biol 2003; 22:63-71.
96. Kyriakides TR, Zhu YH, Smith LT et al. Mice that lack thrombospondin 2 display connective tissue abnormalities that are associated with disordered collagen fibrillogenesis, an increased vascular density, and a bleeding diathesis. J Cell Biol 1998; 140:419-430.

97. Lawler J, Sunday M, Thibert V et al. Thrombospondin-1 is required for normal murine pulmonary homeostasis and its absence causes pneumonia. Journal of Clinical Investigation 1998; 101:982-992.
98. Stupack DG, Puente XS, Boutsaboualoy S et al. Apoptosis of adherent cells by recruitment of caspase-8 to unligated integrins. J Cell Biol 2001; 155:459-470.
99. Díaz-Gonzalez F, Forsyth J, Steiner B et al. Trans-dominant inhibition of integrin function. Molecular Biology of the Cell 1996; 7:1939-1951.
100. Blystone SD, Graham IL, Lindberg FP et al. Integrin alpha v beta 3 differentially regulates adhesive and phagocytic functions of the fibronectin receptor alpha 5 beta 1. J Cell Biol 1994; 127:1129-37.
101. Kim S, Bakre M, Yin H et al. Inhibition of endothelial cell survival and angiogenesis by protein kinase A. Journal of Clinical Investigation 2002; 110:933-941.
102. Kim YM, Lee YM, Kim HS et al. TNF-related activation-induced cytokine (TRANCE) induces angiogenesis through the activation of Src and phospholipase C (PLC) in human endothelial cells. Journal of Biological Chemistry 2002; 277:6799-6805.
103. Reynolds AR, Reynolds LE, Nagel TE et al. Elevated Flk1 (VEGF receptor 2) signalling mediates enhanced angiogenesis in β3-integrin-deficient mice. Cancer Research 2004; 64:8643-8650.
104. Du X, Gu M, Weisel JW et al. Long range propagation of conformational changes in integrin alpha IIb beta 3. J Biol Chem 1993; 268:23087-92.
105. Legler DF, Wiedle G, Ross FP et al. Superactivation of integrin alpha v beta 3 by low antagonist concentrations. Journal of Cell Science 2001; 114:1545-1553.
106. Buckley CD, Pilling D, Henriquez NV et al. RGD peptides induce apoptosis by direct caspase-3 activation. Nature 1999; 397:534-539.
107. Nikolopoulos SN, Blaikie P, Yoshioka T et al. Integrin beta4 signaling promotes tumor angiogenesis. Cancer Cell 2004; 6(5):471-83.

Chapter 11

Integrins in Urogenital Development

Jordan A. Kreidberg*

Introduction

The urogenital system is a complex system composed of multiple organs, most notably the kidneys and gonads, as well as many other glandular components and connective tissues. This review will concentrate on the roles of integrins in the developing kidney, as integrins have been less well studied in the development of other components of the urogenital system, such as the gonads, prostate, bladder, ureters etc.

Development of the Urogenital System

The vertebrate urogenital system emerges from the urogenital ridge, an elongated structure along the posterior aspect of the abdominal cavity in the developing embryo. Both the gonads and kidneys develop from the urogenital ridge. The excretory apparatus develops in three stages along the urogenital ridge, from rostral to caudal. First, pronephric tubules emerge at the rostral end of the ridge, followed by mesonephric tubules along the middle of the ridge, and finally the metanephric kidney develops at the caudal end of the ridge, to give rise to the adult kidney. Although they become the adult excretory apparatus in lower vertebrates, the mesonephric and pronephric tubules are vestigial structures in mammals, and are lost prior to birth. Furthermore, there has been little analysis of integrin expression or function in these structures, and they will not be considered further, except to note that there is similarity between mesonephric tubules and single nephrons, such that integrin expression in mesonephric tubules may be similar to that of individual nephrons.

The gonads emerge from the coelem of the mid- urogenital ridge, adjacent to the mesonephric tubules, and eventually become distinct structures. In the male, cells closely associated with the mesonephric tubules migrate into the developing testes and may contribute to Leydig and peritubular myoid cells. No such migration is known to occur to the developing ovary in female embryos. Additionally, Primordial Germ Cells (PGCs) migrate to the urogenital ridge from the allantois and then into the developing gonad where they become spermatogonia or oocytes.

Development of the Metanephric Kidney

The metanephric kidney is the product of a series of inductive interactions between the metanephric mesenchyme and the epithelial ureteric bud.[1] Initially, the ureteric bud grows out from the Wolffian duct into the mesenchymal portion of the urogenital ridge. Simultaneous with bud outgrowth, the metanephric mesenchyme becomes histologically distinct from the surrounding mesenchyme. Upon contact of the ureteric bud with the metanephric mesenchyme, at E11.5, the mesenchyme is induced to condense along the surface of the bud (Fig. 1). After the initial condensation of induced cells around the bud, a portion of the condensed

*Jordan A. Kreidberg—Department of Medicine, Children's Hospital Boston; Department of Pediatrics, Harvard Medical School, Boston, Massachusetts, U.S.A.
Email: Jordan.Kreidberg@childrens.harvard.edu

Integrins and Development, edited by Erik H.J. Danen. ©2006 Landes Bioscience.

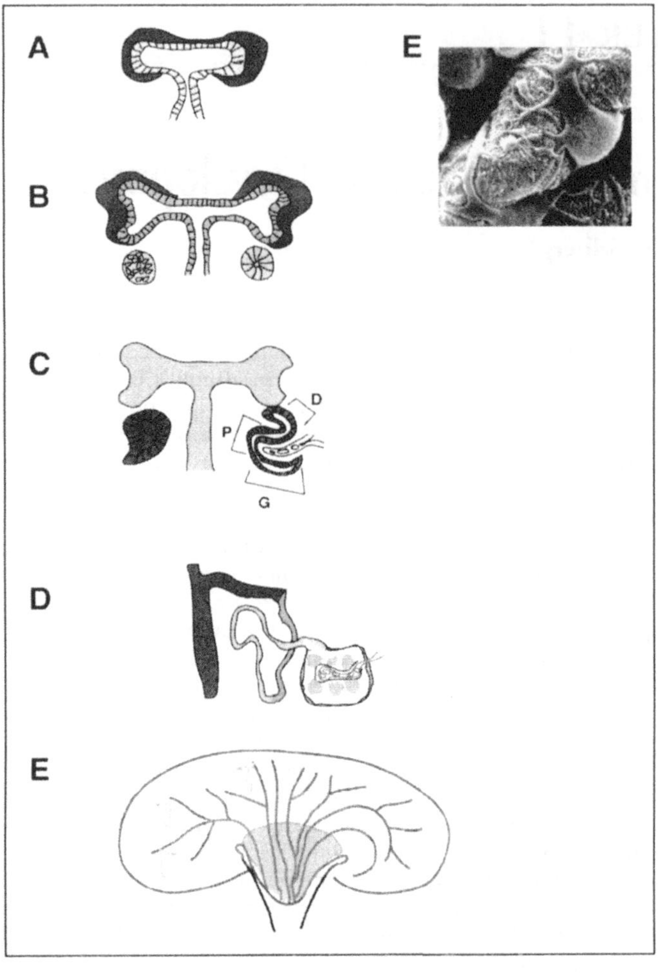

Figure 1. A schematic of kidney development. These are not drawn to scale. A) Condensed mesenchyme (dark gray) surrounding the epithelial ureteric bud (dark gray). B) Pretubular aggregates (light gray) inferior to the branches of the ureteric bud. On the left, the pretubular aggregate is in its preinduction mesenchymal state. On the right, an epithelial aggregate is shown, that has undergone the mesenchymal to epithelial transformation. C) On the left, a simple comma-shaped tubule is present. On the right, an S-shaped tubule is present, with the portions that will contribute to the glomerulus "G", proximal tubule "P" and distal tubule "D" noted. A capillary (white) has begun to invade the glomerular cleft. (The epithelial nature of the ureteric bud is omitted for simplicity.) D) A maturing nephron connected to a collecting duct is shown. The glomerulus (actually much smaller in relation to remainder of nephron, is shown. The glomerular basement membrane is in black. Podocytes are light gray. E) Overall structure of the kidney. The cortex is unshaded, outer medulla light light gray and inner medulla darker gray. The branched network of collecting ducts is in black, and nephrons are light gray. The vascular component is omitted.

mesenchyme begins to form pretubular aggregates adjacent and inferior to the tips of the branching ureteric bud. These pretubular aggregates then undergo a mesenchymal to epithelial transformation to become the renal vesicle, which gives rise to the segmented nephron. The condensed mesenchyme is also hypothesized to generate a population of stem cells, which go on to constitute a population of cells in the extreme periphery of the nephrogenic zone in the developing kidney.[2] During kidney development, a portion of these stem cells remain undifferentiated and proliferative, maintaining a constant supply of precursor cells until the generation of additional nephrons ceases. The epithelial component of the adult kidney is therefore derived both from these putative stem cells, which give rise to the individual nephron units, and the ureteric bud derivatives, which give rise to the collecting ducts and ureter. Thus, several distinct steps can be defined during the differentiation of kidney epithelial cells derived from the metanephric mesenchyme.

Development of the Nephron

The simple tubule formed by the pretubular aggregate becomes progressively more complex, undergoing a process of segmentation to form the mature nephron. The glomerulus develops at the proximal end of each nephron. As the nephron undergoes segmentation, endothelial cells invade a cleft at its proximal end, to begin forming the glomerulus[3] (see Fig. 1). The glomerulus forms from the proximal end of the S-shaped body during the segmental differentiation of the nephron. A layer of columnar epithelial cells attached to a basement membrane forms one wall of a cleft within the S-shaped body, the other wall being the presumptive proximal tubule. A small group of capillaries and presumptive mesangial cells invades this cleft, causing a bulging out of this layer of epithelial cells.[3] As this simple capillary structure begins to branch into a more complex structure that ultimately involves 6-8 capillary loops, the epithelial cells lose most of their lateral cell attachments and migrate around the capillary loops, and extending foot process projections that form a scaffolding to support the capillary loops within the glomerulus. These interdigitated foot processes are in contact with each other by means of the slit-diaphragm, which is an adherens junction-like structure. The slit-diaphragm is the only point at which mature podocytes remain in intimate contact with each other. The extent to which foot processes are formed by migratory extensions from cell bodies, as opposed to selective loss of adhesion along the podocyte basal membrane, is not well understood.[4] However, one probable scenario for the assembly of these interdigitated foot processes is suggested by recent evidence that the slit-diaphragm is indeed an adherens junction-like structure.[5] In this scenario, the interdigitation occurs as the podocytes lose their cell-cell attachments, remaining only attached at the slit-diaphragm. In this sense, the interdigitation would not be the result of migratory events through which two separated cells establish contact, but rather the remodeling of a cell-cell junction. This issue is emphasized because podocyte biology, and particular the establishment of the slit-diaphragm complex is a major issue in kidney biology and disease, and is likely to importantly involve integrins and other adhesion molecules, as will be indicated later in this chapter.

The tubular portion of the nephron becomes segmented in a proximal-distal order, into the proximal convoluted tubule, descending and ascending loops of Henle, and distal convoluted tubule. The latter connects to the collecting ducts, which are derived from the ureteric bud derivatives, and not from the original mesenchymal component of the metanephric rudiment.

Branching Morphogenesis: Development of the Collecting System

The collecting system, made up of bundles of collecting ducts, are the tubules through which the filtrate produced by the nephrons is conducted out of the kidney and to the ureter and then the bladder. Water and salt absorption and excretion also occur in collecting ducts, under different regulatory mechanisms than are active in the tubular portions of the nephron. The collecting ducts are all derived from the original ureteric bud. So, whereas each nephron is an individual unit separately induced and originating from a distinct pretubular aggregate, the collecting ducts are the product of branching morphogenesis from a single epithelial bud. There is, however, considerable remodeling involved in forming collecting ducts from branches of ureteric bud.[6] The branching pattern is highly patterned, with the first several rounds of branching being somewhat symmetrical, followed by additional rounds of dichotomous branching, in which a main trunk of the collecting duct continues to extend towards the periphery of the developing kidney, while smaller buds branch as they induce new nephrons. There is also extensive remodeling to obtain the final structure of the collecting system. Originally, the ureteric bud derivatives are branching within a surrounding mesenchyme. Ultimately, they form a funnel shaped structure in which a cone-shaped grouping of ducts or papilla sits within a funnel or calyce that drains into the ureter. The mouse kidney has a single papilla and calyce, whereas a human kidney has 8-10 papillae, each of which drains into a minor calyce, with several minor calyces draining into a smaller number of major calyces.

Integrin Expression in the Developing Kidney

A survey of integrin expression in the different component tissues of the developing human kidney was first published by Korhonen and colleagues in 1990.[7] Their results and those of subsequent studies are summarized below. It is worth mentioning that the availability of new anti-integrin antibodies periodically results in revision of our knowledge of integrin distribution in the kidney, as they may allow detection of additional integrins, particulary in nonhuman tissues in which integrin expression could not previously be examined in detail.

Early Kidney

Integrin expression in the earliest stages of kidney development has not been well described. The ureteric bud derivatives express $\alpha3\beta1$ and $\alpha6\beta1$ integrins. Pretubular aggregates, having undergone the conversion into epithelium and formed renal vesicles, also begin to express $\alpha6\beta1$ integrin. $\alpha8\beta1$ integrin is a more recently described integrin found in the mesenchyme, particularly where it meets the ureteric bud.

Early Nephrons

In the comma and early S-shaped body, $\alpha6\beta1$ continues to be expressed, and is probably the most-widley expressed integrin in the epithelial components of the developing kidney. As the S-shaped body matures and displays greater proximal-distal segmentation, there begins to be more apparent compartmentalization of the $\alpha6\beta1$ expression, with expression in the glomerular compartment decreasing relative to expression in the portions that will give rise to the proximal and distal tubules. As expression of $\alpha6\beta1$ decreases in glomerular precursor cells, expression of $\alpha3\beta1$ becomes abundant at this end of the S-shaped body.

Glomerulus

$\alpha3\beta1$ integrin is the major integrin expressed by podocytes. It initially has a nonpolarized distribution, but as podocytes lose their cell-cell attachments and change from a conventional columnar epithelium to their previously described highly differentiated appearance, expression of $\alpha3\beta1$ integrin is confined to a basal distribution along the glomerular basement membrane. Mesangial cells are the second of three major cell types in the glomerulus, the third being endothelial cells. The developmental expression of integrins in the endothelial and mesangial component of the glomerulus has not been examined in detail. Mesangial cells, which are situated adjacent to endothelial cells on the vascular side of the glomerular basement membrane, express $\alpha1\beta1$, $\alpha2\beta1$, $\alpha5\beta1$, $\alpha8\beta1$ and αv-containing integrins.[7-9] Endothelial cells within the glomerulus express $\alpha5\beta1$ and αv-containing integrins.[7,9,10]

Proximal and Distal Tubules

$\alpha6\beta1$ is the major integrin expressed by proximal tubules, and this continues from the segementation of the S-shaped body through adult life. Both $\alpha2\beta1$ and $\alpha6\beta1$ are expressed in distal tubules. Whether there are more distinctive patterns of expression as the nephron becomes even more segmented into specialized sections has not been well-described.

Integrin Function

Evidence from Gene Targeting Experiments

The integrin knockout yielding the earliest kidney phenotype is that of α8 integrin. α8β1 integrin is a receptor for ostepontin and nephronectin.[11,12] In most α8 -/- embryos, the ureteric bud ceases to grow upon contact with the metanephric mesenchyme, resulting in renal agenesis.[13] In those α8 -/- mice that overcome this early block, a relatively normal kidney develops, with some reduction in renal mass and subtle additional changes in histology.[8] The renal agenesis phenotype suggests that α8β1 integrin transduces an important signal between the metanephric mesenchyme and the ureteric bud. The other known signals between these tissues are GDNF, a neurotrophin made by the mesenchyme that binds the c-Ret receptor on the ureteric bud to stimulate its growth and branching,[14-17] and an unidentified signal from the ureteric bud that stimulates the induction of the mesenchyme. In contrast to the α8 integrin mutant phenotype, in which the ureteric bud stops upon contact with the metanephric mesenchyme, in the absence of GDNF, the ureteric bud does not emerge at all from the Wolffian duct. Together, these phenotypes indicate that several signals between the mesenchyme and urteric bud are involved in stimulating growth and differentiation of the early kidney, including a diffusible signal required for emergence and continued growth of the bud, and presumably nondiffusible signals that are required once the bud makes contact with the mesenchyme. Further work will determine whether the major role of the α8β1 integrin interaction with nephronectin is to mechanically mediate contact between two cell types, or to biochemically regulate signaling pathways as a consequence of the epithelial:mesenchymal interaction, or both.

Integrin gene knockouts in mice have been striking as much for what they have not told us, with regard to kidney development, as for what they have told us. On the one hand, the α3 integrin knockout provided two major phenotypes, the failure to form a normal papilla, and inability to form podocyte foot processes.[18] On the other hand, in the absence of either α2β1 or α6β1 integrins, no abnormalities in kidney development have been identified.[19,20] Double mutant embryos, unable to express either α3β1 or α6β1 integrins, showed no additional kidney defects beyond those observed in the absence of α3β1 integrin,[21] although they could only be examined as far as embryonic day 16, and a detailed examination was not possible. Double mutants of α2 and other integrin genes have not yet been examined, nor have α2/α3/α6 triple mutants been derived. Nevertheless, these observations raise critical questions about the role of integrins in epithelial development. For example, α6β1 is the major, if not sole, integrin expressed by proximal tubule epithelial cells,[7] yet proximal tubules appeared unaffected in α6β1 integrin-deficient kidneys.[20] Collecting ducts express α2β1, α3β1 and α6β1,[7] so it is possible that α2β1 provided sufficient integrin function in the absence of both α3β1 and α6β1, though the normal kidney phenotype of α2β1 integrin-deficient mice[19] indicates that there is unlikely to be any specific function of that integrin (or, for that matter, α6β1 integrin). Only α3β1 integrin appears to have some functions unable to be provided by other integrins. What is to be made of these observations? It is possible that other basement membrane receptors, such as dystroglycan,[22] provide sufficient basement membrane attachment in the absence of integrins. It is also possible that there is a high degree of redundant function among the laminin-biding integrins, and that in some instances, there may be compensatory expression of integrins not normally expressed in particular locations, in the absence of the one usually expressed. Furthermore, it must be acknowledged that simply because an integrin is expressed does not mean it has an essential function in that cell type or location. Finally, it may be that some integrins have post-natal functions that would not be evident in studies of knockout mice that die during the perinatal period, although the normal development and survival of α2 integrin mutant mice tends to rule out nonredundant functions in the kidney for α2β1 integrin. Because both the α3 and α6 knockout mice die during the perinatal period, such that it has not been possible to study the functional consequences of integrin deficiency once the mouse becomes dependent on the kidney to help maintain physiologic homeostasis. Thus, the observations of distinct integrin repertoires among the different tubule types is intriguing in that it raises the possibility that the different integrins may have different physiologic or structural functions in mature epithelial cells, but at present there is little information that would allow further speculation on this matter.

The role of α3β1 integrin in podocyte biology has also been an area of great interest. As previously mentioned, podocytes are unable to form mature foot processes in the absence of α3β1 integrin. α3β1 integrin is presumed to mainly function as a receptor for the glomerular basement membrane (GBM), a thick basement membrane formed by the podocytes and the endothelial cells lining the capillary loops of the glomerulus. Maturation of the GBM occurs during kidney development, such that the laminin and type IV collagen subunits present at the beginning of glomerulogenesis are replaced with different subunits.[23] Specifically, (α1,α2) type IV collagen is replaced with (α3,α4,α5) type IV collagen, and (α1,β1,γ1) laminin-1 is replaced with (α5,β2,γ1) laminin-11. α3β1 integrin binds laminin-11 with higher affinity than it binds laminin-1,[24,25] suggesting that the laminin switch may result in α3β1 transducing signals that affect podocyte maturation. Whether the absence of a laminin-11 induced signal through α3β1 integrin is responsible for eliciting the formation of mature podocyte foot processes is an area of active investigation. Glomerular capillaries also failed to develop normally in α3β1 integrin deficient kidneys.[18] Capillary loops were fewer in number and those that were present were abnormally dilated, compared to normal glomerular capillaries. A more dramatic, though presumably related phenotype was observed in kidneys from mice carrying a targeted mutation in the gene encoding the α5 laminin subunit.[26] In α5 laminin-deficient glomeruli, which are unable to undergo the normal developmental switch from laminin-1 to laminin-11, glomeruli were completely unable to vascularize. Together these vascular defects suggest that the interaction of α3β1 integrin and laminin-11 may result in the activation of a set of signals crucial for vascular extension into the developing glomerulus, and establishment of the normal glomerular capillary loops.

Intriguingly, several other targeted mutations in mice that result in a failure to form or maintain podocyte foot processes are of genes that encode proteins associated with the slit-diaphragm,[27] the structure found between interdigitated foot processes that is, in effect, the cell-cell junction between adjacent foot processes. Similar to cell-cell junctions between more typical epithelial cells, the slit-diaphragm contains a member of the cadherin family, P-cadherin,[5] a protocadherin-FAT,[28] and ZO-1,[29] a protein commonly found in tight junctions. Nephrin, a specialized junctional protein that is a crucial component of the protein barrier between the circulation and Bowman's space within the glomerulus is also present at the slit-diaphragm.[30-32] These results raise the possibility that α3β1 integrin's function at the podocyte foot process may also be to maintain the structure of the slit-diaphragm, in addition to, or possibly instead of, its supposed role as a receptor the basement membrane. Indeed, in the absence of α3β1 integrin, podocytes do not dissociate from the basement membrane, and if anything, appear to have increased attachment to it in the absence of foot processes.[18] This attachment to the basement membrane may be mediated by dystroglycan. The possibility that α3β1 is participating in maintaining the structure of a cell-cell junction is supported by several lines of evidence. First, a long-standing observation in epithelial cell biology is that general blockade of integrin function in epithelial cells is that cells in vitro tend to round up, though they do not always detach.[33] This can variously be interpreted as a partial loss of cell-matrix adhesion, but also as a loss of cell-cell adhesion. The latter interpretation is supported by recent studies of immortalized kidney collecting duct epithelial cells in vitro, demonstrating that α3β1 integrin is a component of the adherens junction, and that an interaction between α3β1 and the tetrapsanin CD151, is required to maintain strong cadherin-mediated cell-cell adhesions.[34] Also of interest in the study, the gene expression of PTPμ, a tyrosine phosphatase known to be associated with E-cadherin, was also dependent on the α3β1:CD151 interaction.[34] These observations suggest that α3β1 integrin on podocyte foot processes may also have a role in stimulating cell-cell adhesion mediated at the slit-diaphragm, and that loss of foot processes in the absence of α3β1 integrin may relate as much to a deficit of cell-cell adhesion, as in cell-matrix adhesion.

Evidence from in Vitro Studies

Integrins have been studied extensively in a variety of epithelial cell lines, both in monolayer, and in three-dimensional systems that allow the formation of rudimentary branched tubules. The latter studies have been used as an in vitro model for branching morphogenesis. Some studies have also employed isolated ureteric buds placed in a three dimensional matrix of matrigel and type I collagen. Blockade of $\alpha3\beta1$ and $\alpha6\beta1$ integrins, by a combination of blocking antibodies and targeted mutations, resulted in reduced branching in these systems, a greater effect than observed on branching in $\alpha3/\alpha6$ integrin double mutant embryos.[35,36] While it is not unusual for there to be discrepancies between in vitro and in vivo experiments, it may be worthwhile to consider why this may be the case. One possibility is that the addition of a blocking antibody may cause an acute perturbation that the cell or tissue cannot circumvent, whereas the absence of a protein from the beginning of a cell or tissue's existence may allow if greater time to recruit or mobilize compensatory proteins or signaling pathways, Secondly, in vitro systems may be more dependent on specific ligands or receptors present in the experimental system, whereas in vivo environments may provide a greater diversity of ligands, such as basement membrane components, that allow cells to use a greater number of receptors to effect a biological process. Regardless, it will be of great interest in the future to determine which of the many signal transduction pathways known to be activated by integrin:matrix interactions are used by integrins to regulate branching morphogenesis, and whether this is done in direct coordination with signals transduced by growth factor receptors.

Summary

There is much left to learn about how integrins function in urogenital development. The questions remaining are broad ones: is the major role of integrins to serve as a mechanical link between the extracellular matrix and the cytoskeleton, or are they more importantly a component of a signal transduction module that elicits new cell behaviors as a consequence of an interaction with a component of the matrix. Or, do they fulfill both functions at the same time. The phenotypes described in this chapter are clear demonstrations of this puzzle. For example, fewer collecting duct epithelial cells are present in a3b1 integrin-deficient kidneys, but those that are present appear fairly normal. In contrast, all podocytes in these kidneys are very abnormal. Do these phenotypes reflect a structural or behavioral abnormality? In considering these questions, it is important to keep in mind that developmental processes must be considered very different from cell biological processes studied in tissue culture. Whereas mutations may cause a constant abnormality in uniform groups of cells in culture, development is a process in which groups of cells interact with other groups of cells. Thus, developmental phenotypes, including those if integrins, must be considered in terms of how interactions of one group of cells with their adjacent extracellular matrix affects the way they interact with neighboring groups of cells. To what extent the extensive information developed thus far about how integrins function in vitro will be relevant to their function in development is one of the most exciting aspects of integrin biology today.

References

1. Saxen L, Sariola H. Early organogenesis of the kidney. Pediatr Nephrol 1987; 1(3):385-92.
2. Bard JBL, McConnell JE, Davies JA. Towards a genetic basis for kidney development. Mechanisms of Development 1994; 48:3-11.
3. Potter EL. Development of the human glomerulus. Archives of Pathology 1965; 80:241-255.
4. Kreidberg JA. Podocyte differentiation and glomerulogenesis. J Am Soc Nephrol 2003; 14(3):806-14.
5. Reiser J et al. The glomerular slit diaphragm is a modified adherens junction. J Am Soc Nephrol 2000; 11(1):1-8.
6. Osathanondh V, Potter EL. Development of the human kidney as shown by microdissection II. Renal pelvis, calyces, and papillae. Archive of Pathology 1963; 76:53-65.
7. Korhonen M et al. The $\alpha1$-$\alpha6$ subunits of integrins are characteristically expressed in distinct segments of developing and adult human nephron. J Cell Biol 1990; 111:1245-1254.
8. Haas CS et al. Glomerular and renal vascular structural changes in alpha8 integrin-deficient mice. J Am Soc Nephrol 2003; 14(9):2288-96.

9. Hafdi Z et al. Distribution of alphavbeta3, alphavbeta5 integrins and the integrin associated protein—IAP (CD47) in human glomerular diseases. Cell Adhes Commun 2000; 7(6):441-51.
10. Roy-Chaudhury P et al. Importance of the tubulointerstitium in human glomerulonephritis. II. Distribution of integrin chains beta 1, alpha 1 to 6 and alpha V. Kidney International 1997; 52(1):103-10.
11. Brandenberger R et al. Identification and characterization of a novel extracellular matrix protein nephronectin that is associated with integrin alpha8beta1 in the embryonic kidney. J Cell Biol 2001; 154(2):447-58.
12. Denda S, Reichardt LF, Muller U. Identification of osteopontin as a novel ligand for the integrin alpha8 beta1 and potential roles for this integrin-ligand interaction in kidney morphogenesis. Molecular Biology of the Cell 1998; 9(6):1425-35.
13. Muller U et al. Integrin alpha8beta1 is critically important for epithelial-mesenchymal interactions during kidney morphogenesis. Cell 1997; 88(5):603-13.
14. Durbec P et al. GDNF signalling through the Ret receptor tyrosine kinase [see comments]. Nature 1996; 381(6585):789-93.
15. Jing S et al. GDNF-induced activation of the ret protein tyrosine kinase is mediated by GDNFR-alpha, a novel receptor for GDNF. Cell 1996; 85(7):1113-24.
16. Moore MW et al. Renal and neuronal abnormalities in mice lacking GDNF. Nature 1996; 382(6586):76-9.
17. Sanchez MP et al. Renal aegenesis and absence of enteric ganglions in mice lacking GDNF. Nature 1996; 382:70-74.
18. Kreidberg JA et al. Alpha 3 beta 1 integrin has a crucial role in kidney and lung organogenesis. Development 1996; 122(11):3537-47.
19. Chen J et al. The alpha(2) integrin subunit-deficient mouse: A multifaceted phenotype including defects of branching morphogenesis and hemostasis. Am J Pathol 2002; 161(1):337-44.
20. Georges-Labouesse E et al. Absence of integrin alpha 6 leads to epidermolysis bullosa and neonatal death in mice. Nature Genetics 1996; 13(3):370-3.
21. De Arcangelis A et al. Synergistic activities of α3 and α6 integrins are required during apical ectodermal ridge formation and organogenesis in the mouse. Development 1999; 126:3957-3968.
22. Henry MD, Campbell KP. A role for dystroglycan in basement membrane assembly. Cell 1998; 95(6):859-70.
23. Miner JH. Renal basement membrane components. Kidney International 1999; 56(6):2016-24.
24. Delwel GO et al. Distinct and overlapping ligand specificities of the alpha 3A beta 1 and alpha 6A beta 1 integrins: Recognition of laminin isoforms. Molecular Biology of the Cell 1994; 5(2):203-15.
25. Kikkawa Y, Sanzen N, Sekiguchi K. Isolation and characterization of laminin-10/11 secreted by human lung carcinoma cells. laminin-10/11 mediates cell adhesion through integrin alpha3 beta1. Journal of Biological Chemistry 1998; 273(25):15854-9.
26. Miner JH, Li C. Defective glomerulogenesis in the absence of laminin alpha5 demonstrates a developmental role for the kidney glomerular basement membrane. Developmental Biology (Orlando) 2000; 217(2):278-89.
27. Tryggvason K, Wartiovaara J. Molecular basis of glomerular permselectivity. Curr Opin Nephrol Hypertens 2001; 10(4):543-9.
28. Inoue T et al. FAT is a component of glomerular slit diaphragms. Kidney Int 2001; 59(3):1003-12.
29. Schnabel E, Anderson JM, Farquhar MG. The tight junction protein ZO-1 is concentrated along slit diaphragms of the glomerular epithelium. J Cell Biol 1990; 111(3):1255-63.
30. Holthofer H et al. Nephrin localizes at the podocyte filtration slit area and is characteristically spliced in the human kidney. Am J Pathol 1999; 155(5):1681-7.
31. Holzman LB et al. Nephrin localizes to the slit pore of the glomerular epithelial cell. Kidney Int 1999; 56(4):1481-91.
32. Ruotsalainen V et al. Role of nephrin in cell junction formation in human nephrogenesis. Am J Pathol 2000; 157(6):1905-16.
33. Larjava H et al. Novel function for beta 1 integrins in keratinocyte cell-cell interactions. Journal of Cell Biology 1990; 110(3):803-15.
34. Chattopadhyay N et al. alpha3beta1 integrin-CD151, a component of the cadherin-catenin complex, regulates PTPmu expression and cell-cell adhesion. J Cell Biol 2003; 163(6):1351-62.
35. Chen D et al. Differential expression of collagen- and laminin-binding integrins mediates ureteric bud and inner medullary collecting duct cell tubulogenesis. Am J Physiol Renal Physiol 2004; 287(4):F602-11.
36. Zent R et al. Involvement of laminin binding integrins and laminin-5 in branching morphogenesis of the ureteric bud during kidney development. Dev Biol 2001; 238(2):289-302.

CHAPTER 12

Integrin Regulation of Mammary Gland Development

Matthew J. Naylor and Charles H. Streuli*

Summary

The interaction between a cell and its microenvironment is critical in governing cellular function. Integrins, the major family of extracellular matrix receptors, link the matrix with the actin cytoskeleton and, together with providing a structural association between the inside and outside of the cells, they initiate a variety of downstream signalling pathways. Integrins control of a range of cellular functions in the mammary gland, including proliferation, differentiation, survival, migration, and ultimately tissue morphogenesis. The regulation of these cell phenotypes is essential to the normal development and functionality of the mammary gland, but can be subverted in the process of breast carcinogenesis.

Introduction

The mammary gland consists of a secretory luminal epithelium that is surrounded by a contractile myoepithelium, connecting to a ductal network that terminates at the nipple. This mammary epithelium is ensheathed by a laminin rich extracellular matrix (basement membrane) and embedded in a connective stroma, the mammary fat-pad (Fig. 1).

Development of the mammary gland is unique as it occurs primarily after birth, and in distinct stages that are closely linked to the reproductive status of the animal. Initially, the small ductal network present in the neonate gradually extends into the subtending mammary fat-pad until puberty, when rapid ductal elongation and bifurcation fill the fat-pad in response to ovarian steroid hormones. During each oestrous cycle and throughout early pregnancy, side branching of the ducts occurs. In a full pregnancy, mammary epithelium undergoes proliferation followed by differentiation, and alveolar buds progress into alveoli and later lobuloalveoli. These lobuloalveoli are the milk producing structures of the mammary gland, and represent the functional endpoint of mammopoiesis. Following weaning, involution occurs, and apoptosis of epithelia, together with extracellular matrix remodelling, results in the return to a mammary architecture resembling that of the nulliparous gland. This process of controlled proliferation, differentiation and apoptosis can occur repeatedly throughout the lifespan of the animal (Fig. 1).

Integrins are cell surface receptors for extracellular matrix (ECM) proteins. They are formed from α and β hetrodimers, and each combination has a unique specificity for a variety of ECM proteins such as laminins, collagens, vitronectin, and fibronectin.[1] Integrins provide a mechano-sensitive structural link between the cell's actin cytoskeleton and extracellular environment, participating in both inside-out (i.e., regulation from within the cell, of integrin receptor activity and specificity for different ECM proteins) and outside-in signalling (i.e., transduction of ECM signals into the cell).[2] Through initiation of their own downstream

*Corresponding Author: Charles H. Streuli—Faculty of Life Sciences, University of Manchester, Manchester, M13 9PT, U.K. Email: cstreuli@manchester.ac.uk

Integrins and Development, edited by Erik H.J. Danen. ©2006 Landes Bioscience.

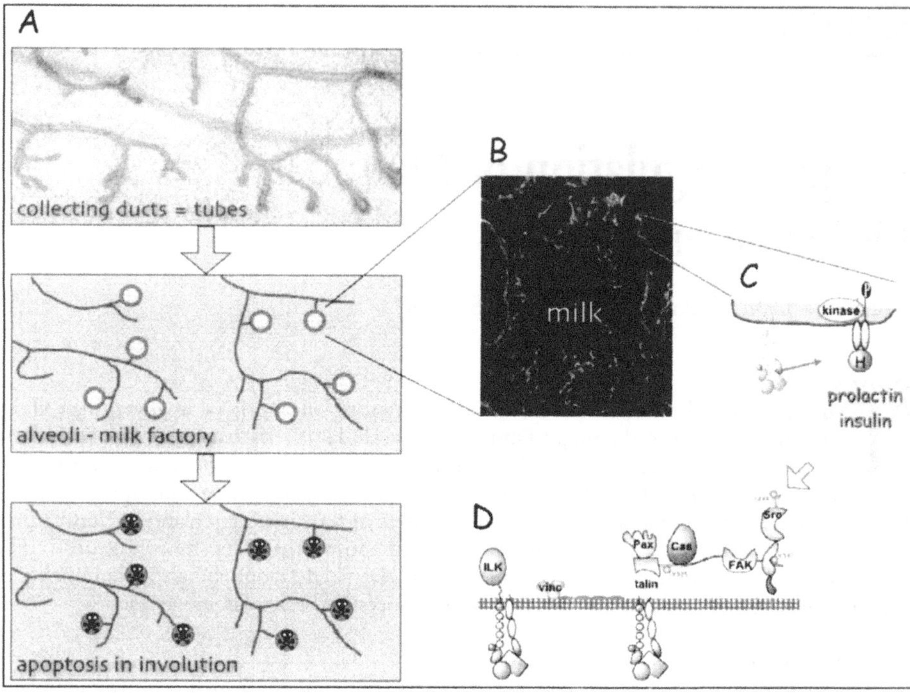

Figure 1. Diagrammatic illustration of mammary gland development. A) The top panel shows a whole-mount of nonpregnant mammary epithelium, with the epithelial ducts embedded in the stromal compartment of the tissue, containing fibroblasts and adipocytes. Blood vessels, neurones, and immune cells also inhabit the mammary stroma. The ducts themselves consist of a tube of polarised luminal epithelial cells surrounded by a layer of basal, myoepithelial cells, outside of which lies the basement membrane. The middle panel shows, in diagrammatic form, the emergence of lactational alveoli from specific points along the ducts. Following nursing, the secretory function of the mammary gland is no longer required, and cells within the alveoli undergo apoptosis (lower panel). B) Section through a lactating alveolus, showing the luminal epithelial cells surrounding the lumen. They are joined to each other via adherens junctions (β-catenin staining in green), and to the basement membrane (laminin staining in red) via integrins. The luminal cells in alveoli (but not ducts) contact the basement membrane directly. C) Diagrammatic illustration of the basal half of a luminal cell with integrin (left) and hormone receptor (right) embedded within the plasma membrane (blue line). The evidence from the authors' laboratory indicates that integrins have a role in regulating signalling downstream of both prolactin, insulin, and IGF receptors in mammary gland. D) Detailed drawing of some components within the integrin-containing adhesion complex. These components are all associated with mammary epithelial adhesion complexes, and research is under way to identify their roles in mammary development and function. Cas: $p130^{Cas}$; $pp125^{FAK}$: focal adhesion kinase; ILK: integrin-linked kinase; Pax: paxillin; Src: $pp60^{Src}$; vinc: vinculin.

signalling pathways, and via cross-talk with growth factor receptor signalling pathways, integrins have a critical role in a range of developmental processes, controlling cell functions as diverse as migration, proliferation, survival, and differentiation.[1]

The interaction between mammary epithelium and the ECM determines both the development and functionality of the mammary gland and will form the basis of this review (Fig. 1). The majority of the work in this field has been performed in tissue culture, however emerging studies are examining the role of integrins in mammary development in vivo. Here we focus on major functions of integrins identified to date in mammary biology, including cell phenotypes essential for normal development, as well as processes associated with tumourigenesis.

Distribution of Integrins in Mammary Epithelium

Integrins are present at the site of cell-ECM interaction of myoepithelial cells and on the basal and lateral surfaces of luminal alveolar cells. Immunohistochemical studies have identified α1, α2, α3, α5, α6, αv, β1 and β4 integrin subunits in myoepithelial cells and α2, α3, α6, β1 and β4 in luminal epithelial cells. The expression of these integrins is developmentally regulated. β1 and β4 integrin expression increases at pregnancy, and β1 is further upregulated in lactation, while β4 integrin declines.[3-6] α2 integrin levels are elevated in epithelial cells through pregnancy, but reduced in lactation.[5-8] α3 integrin expression increases in the course of postnatal development in myoepithelial cells,[4] while α5 integrin is upregulated for ductal side branching and declines in pregnancy and lactation.[3,9] α6 integrin increases during pregnancy in myoepithelial cells.[5,6]

While not exhaustive, the expression patterns identified in these studies indicate that hetrodimers containing the α2, α3, and α5 subunits are more likely involved in nulliparous development, while α6β1 and α6β4 may regulate aspects of development during pregnancy. Interestingly, β1 integrin expression peaks in lactation, suggestion a further role for this subunit during epithelial differentiation.

Integrin-Mediated Control of Mammary Epithelial Cell Function

The cell-ECM interactions that are mediated primarily by integrins have been studied extensively in cultured mammary cells. Since the functions attributed to integrins in culture parallel similar functions in vivo, they provide evidence for a regulatory role of integrins in mammary gland development.

Morphogenesis

Mammary epithelial cells grown on collagen form a monolayer, while those plated on reconstituted basement membrane form polarised spherical acinar structures that can undergo differentiation.[10] A laminin-derived signal is critical for polarisation as primary human luminal epithelial cells cultured on collagen form aggregates with reversed polarity, while culture on laminin results in correct luminal cell polarity.[11] In vivo polarity may also be influenced by cell-cell interactions with neighbours, since desmosomal cadherins determine the spatial juxtapositioning of luminal cells on the inside of ducts and myoepithelial cells on the outside.[12] The regulation of cellular position and context has an obvious role in the development and maintenance of mammary gland homeostasis and may also directly regulate various cell fates (these are discussed in greater detail later). For example, mammary epithelial cell lines are protected from apoptosis in the presence of purified basement membrane components when a 3-dimensional architecture is established,[13] indicating that suppression of apoptosis is influenced by tissue structure. While inhibition of proliferation and maintenance of differentiation have also been linked to cell polarity and tissue architecture, demonstration of cell differentiation in the complete absence of cellular polarity, when single cells are cultured within a laminin-rich basement membrane (BM) gel,[14] indicates that a specific integrin-ECM signal rather than polarity per se is required for mammary epithelial cell differentiation.

Survival

Primary mammary epithelial cells plated on a BM survive, while those on stromal collagen-type-I eventually undergo apoptosis, even in the presence of soluble survival factors.[15,16] This indicates the requirement of a specific ECM signal for cell survival, and cross-talk with the insulin/IGF signalling pathway, which is essential for cell survival, maybe the mechanism by which integrins prevent apoptosis. Indeed, while the proximal phosphorylation of the insulin and IGF type 1 receptors are substratum independent, the efficient activation of downstream effectors (e.g., Insulin Receptor Substrate, and PI3K signalling components) requires adhesion to the BM[17,18] (Lee Y-J and Streuli CH, unpublished data).

The complete detachment of cells from substratum results in apoptosis termed anoikis, which follows as a result of unligating integrins. In mammary epithelial cells, anoikis is dependent on the cytosol-to-mitochondria translocation of the pro-apoptotic protein Bax.[19] This translocation occurs with similar kinetics to a conventional signalling pathway (i.e., 15 mins), although subsequent events such as cytochrome c release take place 1-2 hours later.[20,21] Cell death can be rescued by replating cells onto the ECM within an hour, before the clustering of Bax within the mitochondrial membrane and release of cytochrome c.[20] The control of Bax localisation and thus apoptosis, is mediated through focal adhesion kinase, an immediate downstream effector of integrins,[19] and the intermediate steps involve the PI3K-PKB, but not Raf-Erk, pathways (Wang P, Pu Q-Q, Streuli CH et al, unpublished data).

Integrin-dependent translocation of Bax represents a novel function for integrins, controlling the subcellular shuttling of proteins from cytosol to mitochondria.

In addition to Bax, other members of the pro-apoptotic Bcl-2 family such as Bim may also be involved in the regulation of anoikis. Bim expression is induced following adhesion loss, which is associated with decreased β1 integrin and Erk activation;[22,23] furthermore decreased expression of Bim inhibits apoptosis in some cell lines. However, the involvement of Bim may be associated with apoptosis in response to loss of EGF signals, rather than anoikis per se.[22,23]

Adhesion-mediated regulation of mammary cell survival has important consequences for the developmental positioning of epithelia and their homeostasis. Since these cells normally survive in contact with BM, but not on the stromal collagens, there is a strong pressure on epithelia to remain associated with the BM – and therefore with the rest of the epithelium within the tissue. Conversely, cells migrating into stromal areas of the tissue undergo apoptosis and are removed. This has serious implications for cancer because it provides an innate mechanism for safeguarding against invasion during malignancy. Unfortunately, this protection fails when integrin survival pathways are usurped by mutations, so that malignant epithelia become able to survive in inappropriate ECM environments and at secondary metastatic sites.

Proliferation

A role for integrins in the regulation of mammary cell proliferation has been identified both from the culture of mammary tumour cells, as well as from in vivo studies. Human mammary epithelial cells grown on BM form acini and undergo growth arrest.[24] In contrast, breast cancer cells fail to undergo growth arrest under similar conditions, unless they are treated with a β1 integrin function-blocking antibody.[25] In mouse mammary gland in vivo, when β1 integrins are compromised through the expression of a dominant negative transgene, decreased cell proliferation occurs during pregnancy.[26] Moreover, a significant reduction in mammary proliferation occurs when the β1 integrin gene is deleted by homologous recombination.[27] Integrin regulation of cell cycle has been described in many systems and is associated with regulation of the EGFR, Shc, Akt and MAPK pathways.[28-30] In mammary epithelia, the β1 integrin deletion studies have demonstrated that it is required to maintain low levels of the cell cycle regulator, p21, and interestingly, the signal is mediated through the proximal integrin signaling protein, focal adhesion kinase.[27]

Differentiation

Several studies have indicated that a specific ECM-mediated signal is required for the differentiation of mammary epithelia. Lactationally-competent cells, isolated from pregnant mice and plated on a collagen substratum, rapidly loose their ability to synthesise the milk protein, β-casein, in response to the lactogenic hormone prolactin,[31,32] while reconstituted BM (principally comprised of laminin) enables differentiation.[14,33] The requirement of laminin as the ECM ligand mediating this signal was further demonstrated by the ability of purified laminin to drive both the transcriptional activation of the milk protein genes and milk protein synthesis, and by the inhibition of differentiation using function-blocking antibodies to laminin.[34-36]

β1 integrin is the ECM receptor mediating this interaction because function-blocking antibodies directed against this subunit also inhibit lactational differentiation in culture.[34] Moreover, deletion of the β1 integrin gene by Cremediated recombination of alleles surrounded by

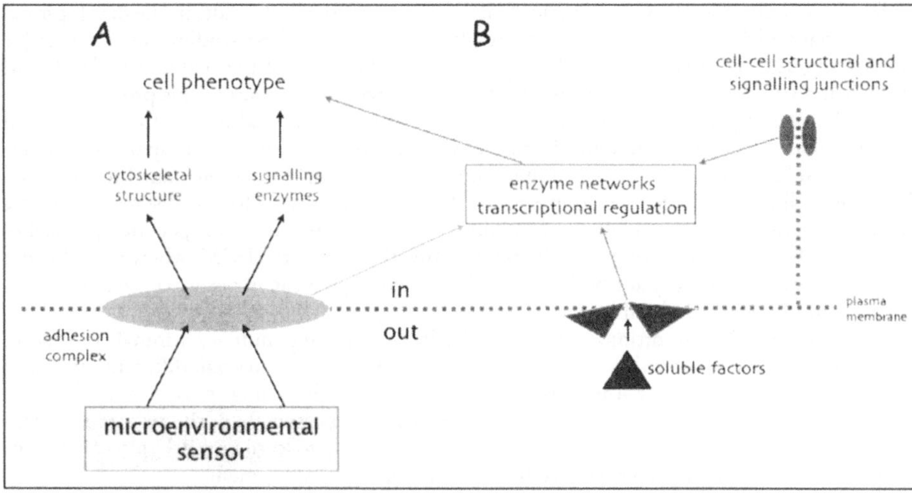

Figure 2. The holistic nature of signal transduction. A) The integrin-containing adhesion complex is a microenvironmental sensor that detects positional signals within the ECM, and influences cell phenotype through the cytoskeleton and signalling enzymes. B) Signals generated by the adhesion complex work together with those from conventional soluble factors, as well as cell-cell adhesions, to establish enzyme networks. In turn, these control cell phenotype directly, or indirectly by impacting on the activity of transcription factors.

LoxP sites, indicates that it is essential for differentiation and milk protein synthesis, and the absence of β1 integrin leads to a severe lactation phenotype.*

The reason that integrin is required for differentiation is that it provides a permissive signal for the prolactin receptor signalling pathway. How this is achieved is not understood. Components of the prolactin signalling pathway fail to undergo phosphorylation unless the mammary cells are in contact with laminin,[35] a process that maybe regulated by tyrosine phosphatases.[37] Additionally, the expression of TGF-α and TGF-β1, inhibitors of the transcription of the milk proteins β casein and whey acidic protein, is inhibited by cell contact with laminin.[38,39]

Identifying the mechanism of the crosstalk between integrin and cytokine signalling pathway is a major challenge to the field, but of key importance to our understanding of mammary differentiation and integrin mediated signalling. This crosstalk underscores the emerging concept that different types of extracellular signals control cell phenotype by setting up networks of intracellular signals, which depend upon each other and act in harmony (Fig. 2). If specific components of the network are absent, then the signals are not sufficient to generate a response, and through studies such as those described, it is now apparent that integrin-mediated signals are fundamental in establishing these networks.

Integrins in Mammary Gland Development

The role of integrins in mammary gland development in vivo remains poorly characterised, despite several studies that have used either function-blocking antibodies or a dominant-negative transgene to investigate mammopoiesis. Studies with integrin-inhibiting antibodies have revealed that development of the mammary ductal system during puberty is dependent on β1 integrin and laminin, but the α6 subunit does not appear to be involved.[40] In concordance with these studies a decrease in ductal branching also occurs in the mammary glands of α2β1

*Naylor MJ, Hynes NE, Streuli CH et al. Ablation of beta1 integrin in mammary epithelium reveals a key role for integrin in glandular morphogenesis and differentiation. Manuscript under revision.

integrin knockout mice,[41] while genetic ablation of either α6 or α3 results in the development of a normal and fully functional mammary gland.[42] Together these studies indicate that β1 integrin together with laminin regulates pubertal development of the mammary gland, and that α3 and α6 integrin are either dispensable or play redundant roles in this process.

Results that apparently conflict with this come from conditional deletion of β1 integrin from luminal epithelium where no ductal defects were detected during nulliparous development.[43] Similarly, expression of a dominant-negative β1 integrin transgene failed to show any defects in nulliparous ductal development.[26,30] However, the reason for the discrepancy is due to the inherent differences in the methodologies used because the latter genetic approaches only target the luminal cells of ducts which don't directly contact the ECM, whereas the former studies (e.g., with inhibitory antibodies) exert their phenotype through perturbation of integrins in the basal, myoepithelial cell layer.

Recent work using conditional deletion of β1 integrin from mammary luminal epithelium is uncovering a key role for this ECM receptor in proliferation,[27] survival, differentiation and alveolar morphogenesis (our unpublished data), and the field is now poised to explore the function of other integrin subunits in distinct cellular compartments of the mammary gland using similar genetic analysis. For example, it will be interesting to learn how deletion of integrins specifically within the myoepithelial cells affects mammary development.

Integrins in Mammary Tumourigenesis

Integrins have been implicated in the initiation, survival, progression and invasiveness of most tumour types, including breast carcinoma (reviewed in ref. 45-47).

The data concerning the expression of specific integrin dimers with breast cancer are somewhat confusing, as very few studies have actually correlated patient outcome with survival. With the exception of α1β1, most of the dimers expressed during normal breast and mammary gland development are also present in invasive carcinoma (reviewed in ref. 45). The best studied of these are the α2β1, α6β1 and α6β4 integrins. There is an emerging interest in the αv integrins, which become expressed in malignancy.

α2β1 integrin is maintained in benign breast lesions such as fibroadenomas,[48] with intermediate expression found in moderately differentiated adenocarcinomas, but it is reduced or lost from poorly differentiated adenocarcinomas.[48-50] In a culture model, knockdown of the α2 subunit in a highly differentiated breast cancer cell line resulted in characteristics consistent with a poor differentiated phenotype;[51] restoring expression in a poorly differentiated cell line led to the converse.[52] Additionally, reexpression of α2 integrin markedly reduced the in vivo tumourigenicity of poorly differentiated cells.[52] These studies argue for a role of α2β1 integrin in the maintenance of a normal epithelial phenotype and thus suppression of breast cancer metastasis: since this subunit is lost from fully differentiated lactational cells (see earlier), it may have a role in establishing epithelial polarisation[53] though this has yet to be confirmed for mammary gland.

The role of α6β4 is also somewhat controversial in terms of predicted function in breast cancer, based on expression patterns.[45] Loss of β4 integrin correlates with more invasive mammary tumours types,[54] as has increased expression of the same integrin.[55] However in terms of patient prognosis, α6β4 expression is consistent with poor survival.[55] Further evidence to suggest that β4 integrin is linked to malignancy is that ectopic expression increases the invasive potential of the breast cancer cell lines, Mm5MT[56] as well as the β4-defecient MDA-MB-435 breast cancer cells.[57] One of the paradoxes with α6β4 integrin having a role in the invasiveness of breast cancer has been that it is a component of hemidesmosomes, which form strong adhesive structures to stabilise epithelial positioning, but which are not permissive for motility. However in invasive carcinoma cells, α6β4 integrin interacts with actin microfilaments, rather than intermediate filaments, to induce filopodia and lamellae.[58] α6β4 integrin also activates several downstream signalling pathways such as PI3K, protein kinase A, and the GTPases Rho and Rac, which have a demonstrated role in the cancer migration, invasion and survival (reviewed in ref. 46).

Similar to $\alpha6\beta4$, the role of $\alpha6\beta1$ based on expression patterns is also somewhat controversial. $\alpha6\beta1$ is reported to be reduced or lost from some poorly differentiated invasive breast carcinomas,[59-61] but retained in others, with expression patterns consistent with a role in tumour motility.[62-64] However, when patient outcome is also assessed, expression of the $\alpha6$ subunit correlates with poor patient survival, although this is without distinguishing between the roles of $\alpha6\beta1$ or $\alpha6\beta4$ integrins.[61] Inhibition of $\alpha6\beta1$ using a dominant negative approach in a human breast cancer cell line results in impaired migration and adhesion in culture, and reduced tumour size in vivo due to decreased proliferation and increased apoptosis.[65,66] Moreover, function-blocking antibodies to $\beta1$ integrin inhibit proliferation of breast cancer cells that would otherwise form disorganised clusters in 3D culture, and revert their morphology to hollow acini.[25,29] This may be because integrins are associated with EGF receptors in breast cancer cells (but not in nontransformed normal mammary epithelia[23]), and inhibiting integrins also blocks EGF-mediated growth signals.[28] Although not conclusive, these studies together suggest a role for $\alpha6\beta1$ integrin in tumour cell survival, proliferation and thus progression.

Further evidence pinpointing an essential role for $\beta1$ integrin in cancer initiation and progression comes from a mouse model of tumourigenesis.[43] $\beta1$ integrin was deleted from mammary epithelium using the Cre/LoxP system and these mice were crossed with transgenic mice expressing the polyomavirus middle T oncogene. $\beta1$ integrin-null mammary glands display a dramatic reduction in tumourigenesis, with hyperplasias only forming from cells retaining the $\beta1$ integrin gene. Excision of $\beta1$ integrin after tumour induction results in decreased proliferation both in culture and in vivo. In addition, over-expression of integrin-linked kinase, which is an adaptor protein associated with the integrin-containing adhesion complex, also results in the induction of mammary hyperplasias and tumours in transgenic mice.[67] Together these studies further demonstrate a potential role for $\beta1$ integrin and downstream signalling effectors in the induction and proliferation of mammary tumours.

There is increasing evidence implicating the αv integrins in tumour invasion and metastasis, and in the breast they become more highly expressed than in the normal epithelium.[59] Breast cancer cell lines synthesise the $\alpha v\beta1$, $\alpha v\beta3$, and $\alpha v\beta5$ heterodimers, with higher levels correlating with increased invasion.[68,69] These integrins are important because they are involved with cell migration over vitronectin and fibronectin, and thereby provide the motile machinery for invasion; they also have a role in suppressing apoptosis. In some tumours, αv integrins become upregulated specifically at the leading front of an invasive tumour,[70] and they may have similar roles in the migration of breast cancers towards the lymphangiogenis system. Since these integrin subunits are unusually expressed on malignant cells, novel therapies directed at inhibiting the adhesive function of αv integrins may be beneficial for obstructing migration of breast cancer cells, as is the case for melanoma.[44]

Conclusions

This review has focused on the requirement for integrins in normal mammary gland function, as well as in tumourigenesis. The available evidence supports an important role for integrins in the control of most aspects of mammary gland development, either directly or via crosstalk with other receptor-mediated signalling pathways. While substantial data indicate a function for specific cell-ECM interactions in processes implicit in mammary gland development, such as survival, proliferation and differentiation, the precise mechanism or even the exact context in which integrin receptors are required for this regulation remains unknown. Studying integrins in mammary development and carcinogenesis has often provided conflicting results, which is not always helped by the use of cell lines adapted to grow under artificial culture conditions, but new genetic evidence is reinforcing the concept that integrins have a key part to play in both these processes. Although the paucity of genetic data thus far has primarily been due to the embryonic lethality resulting from ablation of the different integrin genes, the advent of new technologies, such as the Cre/LoxP system, is circumventing this problem,[71-74] and is being used to considerable advantage in the mammary gland.[27,43] Such studies now need to be expanded to include all integrin subunits as well as their downstream adaptor proteins and signalling components. This field is entering an exciting stage, as the conditional deletion of integrin genes from the epithelial, myoepithelial, and stromal compartments of the mammary gland will provide definitive data on their contribution to both mammopoiesis and carcinogenesis.

Acknowledgements

Matthew Naylor holds a National Health and Medical Research Council of Australia, Peter Doherty Fellowship (application I.D. number 303710). The Streuli lab is funded by a program grant from the Wellcome Trust.

References

1. Hynes RO. Integrins: Bidirectional, allosteric signaling machines. Cell 2002; 110:673-687.
2. Giancotti FG, Ruoslahti E. Integrin Signaling. Science 1999; 285:1028-32.
3. Huang RY, Ip MM. Differential expression of integrin mRNAs and proteins during normal rat mammary gland development and in carcinogenesis. Cell Tissue Res 2001; 303:69-80.
4. Deugnier MA, Moiseyeva EP, Thiery JP et al. Myoepithelial cell differentiation in the developing mammary gland: Progressive acquisition of smooth muscle phenotype. Dev Dyn 1995; 204:107-17.
5. Suzuki R, Atherton AJ, O'Hare MJ et al. Proliferation and differentiation in the human breast during pregnancy. Differentiation 2000; 66:106-15.
6. Suzuki RN, Entwistle A, Atherton AJ et al. The expression patterns of integrin subunits on human breast tissues obtained during pregnancy. Cell Biol Int 2002; 26:593-8.
7. Keely PJ, Wu JE, Santoro SA. The spatial and temporal expression of the alpha 2 beta 1 integrin and its ligands, collagen I, collagen IV, and laminin, suggest important roles in mouse mammary morphogenesis. Differentiation 1995; 59:1-13.
8. Prince JM, Klinowska TC, Marshman E et al. Cell-matrix interactions during development and apoptosis of the mouse mammary gland in vivo. Dev Dyn 2002; 223:497-516.
9. Woodward TL, Mienaltowski AS, Modi RR et al. Fibronectin and the alpha(5)beta(1) integrin are under developmental and ovarian steroid regulation in the normal mouse mammary gland. Endocrinology 2001; 142:3214-22.
10. Aggeler J, Ward J, Blackie LM et al. Cytodifferentiation of mouse mammary epithelial cells cultured on a reconstituted basement membrane reveals striking similarities to development in vivo. J Cell Sci 1991; 99(Pt 2):407-17.
11. Gudjonsson T, Ronnov-Jessen L, Villadsen R et al. Normal and tumor-derived myoepithelial cells differ in their ability to interact with luminal breast epithelial cells for polarity and basement membrane deposition. J Cell Sci 2002; 115:39-50.
12. Runswick SK, O'Hare MJ, Jones L et al. Desmosomal adhesion regulates epithelial morphogenesis and cell positioning. Nat Cell Biol 2001; 3:823-30.
13. Boudreau N, Werb Z, Bissell MJ. Suppression of apoptosis by basement membrane requires three-dimensional tissue organization and withdrawal from the cell cycle. Proc Natl Acad Sci USA 1996; 93:3509-13.
14. Streuli CH, Bailey N, Bissell MJ. Control of mammary epithelial differentiation: Basement membrane induces tissue-specific gene expression in the absence of cell-cell interaction and morphological polarity. J Cell Biol 1991; 115:1383-95.
15. Boudreau N, Sympson CJ, Werb Z et al. Suppression of ICE and apoptosis in mammary epithelial cells by extracellular matrix. Science 1995; 267:891-893.
16. Pullan S, Wilson J, Metcalfe A et al. Requirement of basement membrane for the suppression of programmed cell death in mammary epithelium. J Cell Sci 1996; 109(Pt 3):631-42.
17. Farrelly N, Lee YJ, Oliver J et al. Extracellular matrix regulates apoptosis in mammary epithelium through a control on insulin signaling. J Cell Biol 1999; 144:1337-48.
18. Lee YJ, Streuli CH. Extracellular matrix selectively modulates the response of mammary epithelial cells to different soluble signaling ligands. J Biol Chem 1999; 274:22401-8.
19. Gilmore AP, Metcalfe AD, Romer LH et al. Integrin-mediated survival signals regulate the apoptotic function of Bax through its conformation and subcellular localization. J Cell Biol 2000; 149:431-46.
20. Valentijn AJ, Metcalfe AD, Kott J et al. Spatial and temporal changes in Bax subcellular localization during anoikis. J Cell Biol 2003; 162:599-612.
21. Wang P, Valentijn AJ, Gilmore AP et al. Early events in the anoikis program occur in the absence of caspase activation. J Biol Chem 2003; 278:19917-25.

22. Reginato MJ, Mills KR, Paulus JK et al. Integrins and EGFR coordinately regulate the pro-apoptotic protein Bim to prevent anoikis. Nat Cell Biol 2003; 5:733-40.
23. Wang P, Gilmore AP, Streuli CH. Bim is an apoptosis sensor that responds to loss of survival signals delivered by epidermal growth factor but not those provided by integrins. J Biol Chem 2004; 279:41280-5.
24. Petersen OW, Ronnov-Jessen L, Howlett AR et al. Interaction with basement membrane serves to rapidly distinguish growth and differentiation pattern of normal and malignant human breast epithelial cells. Proc Natl Acad Sci USA 1992; 89:9064-8.
25. Weaver VM, Petersen OW, Wang F et al. Reversion of the malignant phenotype of human breast cells in three-dimensional culture and in vivo by integrin blocking antibodies. J Cell Biol 1997; 137:231-245.
26. Faraldo MM, Deugnier MA, Lukashev M et al. Perturbation of beta1-integrin function alters the development of murine mammary gland. EMBO J 1998; 17:2139-47.
27. Li N, Zhang Y, Naylor MJ et al. Beta1 integrins regulate mammary gland proliferation and maintain the integrity of mammary alveoli. EMBO J 2005; 24:1942-53.
28. Wang F, Weaver VM, Petersen OW et al. Reciprocal interactions between beta1-integrin and epidermal growth factor receptor in three-dimensional basement membrane breast cultures: A different perspective in epithelial biology. Proc Natl Acad Sci USA 1998; 95:14821-6.
29. Wang F, Hansen RK, Radisky D et al. Phenotypic reversion or death of cancer cells by altering signaling pathways in three-dimensional contexts. J Natl Cancer Inst 2002; 94:1494-503.
30. Faraldo MM, Deugnier MA, Thiery JP et al. Growth defects induced by perturbation of beta1-integrin function in the mammary gland epithelium result from a lack of MAPK activation via the Shc and Akt pathways. EMBO Reports 2001; 2:431-7.
31. Streuli CH, Bissell MJ. Expression of extracellular matrix components is regulated by substratum. J Cell Biol 1990; 110:1405-15.
32. Lee EY, Lee WH, Kaetzel CS et al. Interaction of mouse mammary epithelial cells with collagen substrata: Regulation of casein gene expression and secretion. Proc Natl Acad Sci USA 1985; 82:1419-23.
33. Barcellos-Hoff MH, Aggeler J, Ram TG et al. Functional differentiation and avleolar morphogenesis of primary mammary cultures on reconstituted basement membrane. Development 1989; 105:223-235.
34. Streuli CH, Schmidhauser C, Bailey N et al. Laminin mediates tissue-specific gene expression in mammary epithelia. J Cell Biol 1995; 129:591-603.
35. Streuli CH, Edwards GM, Delcommenne M et al. Stat5 as a target for regulation by extracellular matrix. J Biol Chem 1995; 270:21639-44.
36. Schmidhauser C, Bissell MJ, Myers CA et al. Extracellular-matrix and hormones transcriptionally regulate bovine beta-casein 5' sequences in stably transfected mouse mammary cells. Proc Natl Acad Sci USA 1990; 87:9118-9122.
37. Edwards GM, Wilford FH, Liu X et al. Regulation of mammary differentiation by extracellular matrix involves protein-tyrosine phosphatases. J Biol Chem 1998; 273:9495-500.
38. Lin CQ, Dempsey PJ, Coffey RJ et al. Extracellular matrix regulates whey acidic protein gene expression by suppression of TGF-alpha in mouse mammary epithelial cells: Studies in culture and in transgenic mice. J Cell Biol 1995; 129:1115-26.
39. Streuli CH, Schmidhauser C, Kobrin M et al. Extracellular matrix regulates expression of the TGF-beta 1 gene. J Cell Biol 1993; 120:253-60.
40. Klinowska TC, Soriano JV, Edwards GM et al. Laminin and beta1 integrins are crucial for normal mammary gland development in the mouse. Dev Biol 1999; 215:13-32.
41. Chen J, Diacovo TG, Grenache DG et al. The alpha(2) integrin subunit-deficient mouse: A multifaceted phenotype including defects of branching morphogenesis and hemostasis. Am J Pathol 2002; 161:337-44.
42. Klinowska TC, Alexander CM, Georges-Labouesse E et al. Epithelial development and differentiation in the mammary gland is not dependent on alpha 3 or alpha 6 integrin subunits. Dev Biol 2001; 233:449-67.
43. White DE, Kurpios NA, Zuo D et al. Targeted disruption of beta1-integrin in a transgenic mouse model of human breast cancer reveals an essential role in mammary tumor induction. Cancer Cell 2004; 6:159-70.
44. Mitjans F, Meyer T, Fittschen C et al. In vivo therapy of malignant melanoma by means of antagonists of alphav integrins. Int J Cancer 2000; 87:716-23.
45. Shaw LM. Integrin function in breast carcinoma progression. J Mammary Gland Biol Neoplasia 1999; 4:367-76.
46. Mercurio AM, Bachelder RE, Chung J et al. Integrin laminin receptors and breast carcinoma progression. J Mammary Gland Biol Neoplasia 2001; 6:299-309.

47. Taddei I, Faraldo MM, Teuliere J et al. Integrins in mammary gland development and differentiation of mammary epithelium. J Mammary Gland Biol Neoplasia 2003; 8:383-94.
48. Zutter MM, Mazoujian G, Santoro SA. Decreased expression of integrin adhesive protein receptors in adenocarcinoma of the breast. Am J Pathol 1990; 137:863-70.
49. Zutter MM, Krigman HR, Santoro SA. Altered integrin expression in adenocarcinoma of the breast. Analysis by in situ hybridization. Am J Pathol 1993; 142:1439-48.
50. Jones JL, Critchley DR, Walker RA. Alteration of stromal protein and integrin expression in breast—a marker of premalignant change? J Pathol 1992; 167:399-406.
51. Keely PJ, Fong AM, Zutter MM et al. Alteration of collagen-dependent adhesion, motility, and morphogenesis by the expression of antisense alpha 2 integrin mRNA in mammary cells. J Cell Sci 1995; 108(Pt 2):595-607.
52. Zutter MM, Santoro SA, Staatz WD et al. Reexpression of the alpha 2 beta 1 integrin abrogates the malignant phenotype of breast carcinoma cells. Proc Natl Acad Sci USA 1995; 92:7411-5.
53. Schwimmer R, Ojakian GK. The alpha 2 beta 1 integrin regulates collagen-mediated MDCK epithelial membrane remodeling and tubule formation. J Cell Sci 1995; 108(Pt 6):2487-98.
54. Rabinovitz I, Mercurio AM. The integrin alpha 6 beta 4 and the biology of carcinoma. Biochem Cell Biol 1996; 74:811-821.
55. Tagliabue E, Ghirelli C, Squicciarini P et al. Prognostic value of alpha 6 beta 4 integrin expression in breast carcinomas is affected by laminin production from tumor cells. Clin Cancer Res 1998; 4:407-10.
56. Sun H, Santoro SA, Zutter MM. Downstream events in mammary gland morphogenesis mediated by reexpression of the alpha2beta1 integrin: The role of the alpha6 and beta4 integrin subunits. Cancer Res 1998; 58:2224-33.
57. Shaw LM, Rabinovitz I, Wang HH et al. Activation of phosphoinositide 3-OH kinase by the alpha6beta4 integrin promotes carcinoma invasion. Cell 1997; 91:949-60.
58. Rabinovitz I, Mercurio AM. The integrin alpha6beta4 functions in carcinoma cell migration on laminin-1 by mediating the formation and stabilization of actin-containing motility structures. J Cell Biol 1997; 139:1873-84.
59. Koukoulis GK, Virtanen I, Korhonen M et al. Immunohistochemical localization of integrins in the normal, hyperplastic, and neoplastic breast. Correlations with their functions as receptors and cell adhesion molecules. Am J Pathol 1991; 139:787-99.
60. Gould VE, Koukoulis GK, Virtanen I. Extracellular matrix proteins and their receptors in the normal, hyperplastic and neoplastic breast. Cell Differ. Dev 1990; 32:409-16.
61. Friedrichs K, Ruiz P, Franke F et al. High expression level of alpha 6 integrin in human breast carcinoma is correlated with reduced survival. Cancer Res 1995; 55:901-6.
62. Jonjic N, Lucin K, Krstulija M et al. Expression of beta1 integrins on tumor cells of invassive ductal breast carcinoma. Pathol Res Practice 1993; 189.
63. Natali PG, Nicotra MR, Botti C et al. Changes in the expression of a6b4 integrin hetrodimer in primary and metastic breast cancer. Brit J Cancer 1992; 66:318-22.
64. Sager R, Anisowicz A, Neveu M et al. Identification by differential display of alpha6 integrin as a candidate tumour suppressor gene. FASEB J 1993; 7.
65. Wewer UM, Shaw LM, Albrechtsen R et al. The integrin alpha 6 beta 1 promotes the survival of metastatic human breast carcinoma cells in mice. Am J Pathol 1997; 151:1191-8.
66. Shaw LM, Chao C, Wewer UM et al. Function of the integrin alpha 6 beta 1 in metastatic breast carcinoma cells assessed by expression of a dominant-negative receptor. Cancer Res 1996; 56:959-63.
67. White DE, Cardiff RD, Dedhar S et al. Mammary epithelial-specific expression of the integrin-linked kinase (ILK) results in the induction of mammary gland hyperplasias and tumors in transgenic mice. Oncogene 2001; 20:7064-72.
68. Wong NC, Mueller BM, Barbas CF et al. Alphav integrins mediate adhesion and migration of breast carcinoma cell lines. Clin Exp Metastasis 1998; 16:50-61.
69. Meyer T, Marshall JF, Hart IR. Expression of alphav integrins and vitronectin receptor identity in breast cancer cells. Br J Cancer 1998; 77:530-6.
70. Thomas GJ, Lewis MP, Whawell SA et al. Expression of the alphavbeta6 integrin promotes migration and invasion in squamous carcinoma cells. J Invest Dermatol 2001; 117:67-73.
71. Aszodi A, Hunziker EB, Brakebusch C et al. Beta1 integrins regulate condrocyte rotation, G1 progression, and cytokinesis. Genes Dev 2003; 17:2465-79.
72. Blaess S, Graus-Porta D, Belvindrah R et al. Beta1-integrins are critical for cerebellar granule cell precursor proliferation. J Neurosci 2004; 24:3402-12.
73. Graus-Porta D, Blaess S, Senften M et al. Beta1-class integrins regulate the development of laminae and folia in the cerebral and cerebellar cortex. Neuron 2001; 31:367-79.
74. Schwander M, Leu M, Stumm M et al. Beta1 integrins regulate myoblast fusion and sarcomere assembly. Dev Cell 2003; 4:673-85.

Chapter 13

Cell Adhesion in Nervous System Development:

Integrin Functions in Glial Cells

Claudia S. Barros and Ulrich Müller*

Abstract

Integrins are a family of cell surface receptors that are expressed in many if not all tissues. Integrins mediate interactions with different ligands including extracellular matrix glycoproteins, cell surface counter-receptors, proteases and pathogens. The study of integrins has provided compelling evidence that they are instrumental for the development of the nervous system influencing many aspects of cell behavior such as cell proliferation, survival, axonal outgrowth and synapse formation. Recent findings provide strong evidence that integrins not only regulate the development and function of neuronal cells, but also the behavior of different glial subtypes. We will summarize here genetic, biochemical and cell biological experiments that demonstrate the importance of integrin-mediated adhesion and signaling in glial cells.

Introduction

The establishment of the three-dimensional cytoarchitecture of the nervous system is dependent on interactions between receptors on neurons and glial cells and molecules presented within the extracellular matrix (ECM) and by neighboring cells. Integrins are one class of receptors that mediate interactions with secreted ECM glycoproteins and membrane-anchored counter-receptors. Integrins are heterodimeric molecules consisting of an α and a β subunit (Fig. 1). Genes for 8 different β subunits and 18 different α subunits have been identified in the mammalian genome. They can form at least 24 different heterodimers.[1-3] Original identified as adhesion promoting molecules, integrins were thought to have a rather static role, providing a mechanical link between extracellular ligands and the cell interior. Studies carried out mainly with cells in culture have demonstrated that integrins also modulate the activity of a variety of signal transduction pathways. The integrin cytoplasmic domains have no intrinsic catalytic activity, but can interact with cytoplasmic proteins. The activation of signaling pathways is therefore likely achieved through the integrin-dependent recruitment and assembly of signaling complexes at the cytoplasmic leaflet of the plasma membrane. Integrins have also been shown to cooperate with other transmembrane molecules such as receptor tyrosine kinases in the modulation of signaling pathways. While many integrin-activated signaling pathways affect the dynamics of the actin cytoskeleton, regulating cell morphology, adhesion and migration, other pathways affect the activity of transcription factors, cell-cycle regulators and proteins that control cell survival and death.[2,4-7]

*Corresponding Author: Ulrich Müller—The Scripps Research Institute, Department of Cell Biology, 10550 N. Torrey Pines Rd., La Jolla, California 92037 U.S.A.
Email: umueller@scripps.edu

Integrins and Development, edited by Erik H.J. Danen. ©2006 Landes Bioscience.

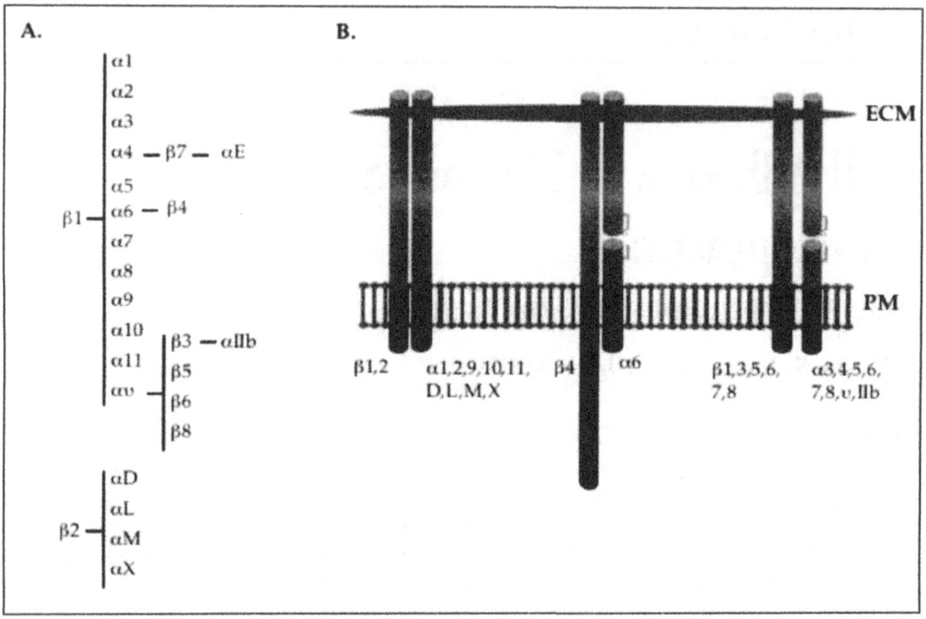

Figure 1. Mammalian integrin receptors. Integrins are cell surface heterodimers consisting of α and β subunits, each encoded by its own gene. A) The different mammalian αβ heterodimers are indicated. B) Each integrin subunit has a single transmembrane domain that spans the plasma membrane (PM), a large extracellular domain and a short cytoplasmic domain; the only exception is the integrin β4 subunit, which has a very large cytoplasmic domain. Several α subunits are proteolytically cleaved into two fragments, which are connected by a disulfide bond (-S-).

Genetic studies in model organisms including *D. melanogaster*, *C. elegans*, and *M. musculus* have demonstrated that integrins play essential roles in many tissues and organs. Studies on the function of integrins in the nervous system have provided evidence that they regulate the development and function of both neurons and glial cells. We will highlight here progress in two particularly informative areas. First, we will summarize current knowledge of the function of integrin/ECM interactions in Schwann cells and oligodendrocytes, the myelinating glial cells of the peripheral nervous system (PNS) and central nervous system (CNS), respectively. Second, we will describe integrin functions in radial glial cells, which serve both as neural stem cells and guideposts for migrating neurons. Other aspects of integrin function, such as their role in cell migration, axon outgrowth and in the development and function of synapses has been reviewed elsewhere[8-11] and will not be considered here.

Integrins, ECM Glycoproteins and Schwann Cells

Laminin Functions during Myelination of Peripheral Nerve

Myelination of peripheral nerve is carried out by a specialized glial subtype, the Schwann cells. Schwann cells migrate into developing axonal bundles and establish sheath-like extensions that infiltrate the axonal bundles. During this process, a 1:1 ratio is established between axons and Schwann cells. Subsequently, the sheath-like extensions of the Schwann cells enwrap the axons leading to the formation of a multilamellar myelin sheath (Fig. 2).[12] In vitro studies have provided evidence that myelination is dependent on contact of Schwann cells with both axons and ECM glycoproteins that surround the developing nerve.[13,14] The importance of the

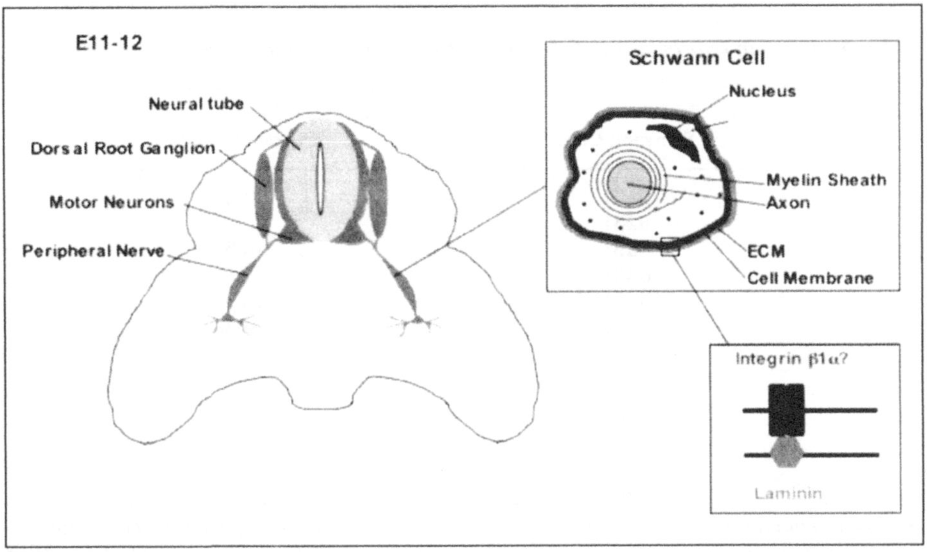

Figure 2. Integrins and myelination of peripheral nerve. A diagram of a cross section of an E11-12 embryo is shown on the left, and an enlargement of a cross section through peripheral nerve on the top right. The axon is surrounded by a myelin sheath that is formed by cytoplasmic extension of a Schwann cell. The Schwann cell also is in contact with LN, and interactions with LN are mediated by integrins containing the β1 subunit and α subunit(s) that still need to be defined.

ECM molecule Laminin (LN) in peripheral nerve has been confirmed by genetic findings. LNs are multimeric proteins consisting of α, β, and γ chains. In peripheral nerve, several LN isoforms are expressed, but the LN-2 isoform consisting of the α2, β1, and γ1 chain is particularly prominent.[15,16] Mutations in the *Lama2* gene (encoding the LN α2 chain) or mice and humans lead to congenital muscular dystrophy and peripheral neuropathy.[13-14] The symptoms include defects in myelination, suggesting that LN-2 affects Schwann cell development or function (Table 2). Mice that were genetically engineered to lack expression of the LN γ1 chain in peripheral nerve also show myelination defects (Table 2). The Schwann cells in these mutant mice proliferate normally, populate peripheral nerve and do not show excessive apoptosis. However, they are unable to differentiate and to synthesize myelin proteins.[17]

Integrins and Myelination of Peripheral Nerve

Schwann cells express several ECM receptors that are expected to mediate effects of ECM components on their development (Table 1). These include LN receptors such as integrins (α1β1, α2β1, α3β1, α6β1, α6β4, and α7β1) and the dystrophyin glycoprotein complex. The integrin α6β1 is expressed prior to birth in Schwann cell precursors and immature promyelinating Schwann cells (E12.5 – E15.5).[18] Dystroglycan and the integrin α6β4 appear perinatally, just prior to myelination (E18.5).[18] α2β1 and α3β1 are expressed at very low levels throughout Schwann cell development,[18-20] while α7β1 is synthesized by Schwann cells in mature but not fetal nerve.[21] In addition, the integrin α1β1 appears to be expressed in mature nerve, but only in non-myelin forming Schwann cells.[22] Finally, αυβ3 can be detected in Schwann cells in vitro and αυβ8 is present in adult nerve.[20] Experiments with function blocking antibodies and cells in culture have provided evidence that integrins containing the β1 subunit regulate myelination.[23] Genetic studies support this view. To inactivate β1 integrins in Schwann cells, mice with a floxed allele of the integrin β1 subunit gene (*Itgb1-flox*) were crossed with mice expressing CRE recombinase under control of the Schwann cell specific promoter of

Table 1. Integrin expression in Schwann cells and oligodendrocytes

Glial Integrins Schwann Cells	 Oligodendrocytes
α1β1	α6β1
α2β1	
α3β1	αυβ1
α4β1	αυβ3
α5β1	αυβ5
α6β1	αυβ8
α7β1	
α6β4	
αυβ3	
αυβ8	

Schwann cells and oligodendrocytes, the myelinating glial cells of the peripheral and central nervous system, express different sets of integrins as indicated. Some of the integrins are expressed in developmentally regulated patterns as described in detail in the text.

Table 2. Genetic analysis of the function of integrins and ECM glycoproteins

Cell Type	Integrin in Vivo Phenotype	ECM Ligand in Vivo Phenotype
Schwann Cells	β1: impaired axonal sorting; myelination defects α1, α7: no phenotype α1, α3, α6, α7, αυ, β3, β8: not reported α4: cell survival defects α5: cell proliferation defects	Ln-α2: myelination defects Ln-γ1: cell differentiation and myelination defects
Oligodendrocytes	β1, β3, β5, αυ: not reported α6: cell survival defects in brain stem and spinal cord	Ln-α2: myelination defects in forebrain and optic nerve Tn-C-precursor proliferation; migration defects along the optic nerve
Radial Glia	β1: differentiation and basement membrane anchorage defects	LN, entactin/nidogen, perlecan: basement membrane anchorage defects Tn-C: cell proliferation defects

The table summarizes our current knowledge of the functions of integrins and ECM glycoproteins in Schwann cells, oligodendrocytes and radial glial cells, as determined by genetic studies in mice.

the P0 gene. Deletion of the *Itgb1* gene leads to a marked delay in myelination. Schwann cells fail to subdivide axon bundles, do not progress past the premyelinating stage, and frequently fail to establish stable contact with the basement membrane. Some axons are not ensheathed, but the myelinated axons that form, although with a delay, seem normal (Table 2). These findings suggest that β1 integrins are essential for early stages during axonal ensheathment.[24] It is at present unclear which αβ1 heterodimers are the functional receptors in Schwann cells since no myelination defects have been observed in mice with targeted inactivation of the integrin α1, α2 and α7 subunits.[25-28] Mice lacking the α3 and α6 subunits die at birth,[29,30] preventing an analysis of the final stages of Schwann cell development and myelination to further define the function of individual integrin αβ1 heterodimers in Schwann cells. It will be important to generate mice that carry alleles of integrin α subunit genes suitable for tissue specific gene inactivation.

Signaling Pathways Downstream of Integrins

What are the mechanisms by which integrins and LN-2 regulate the development of Schwann cells? The differentiation of Schwann cells is perturbed in the absence of LN-2 in peripheral nerve, leading to defects in the expression of myelin proteins.[17] In cultured cells, adhesion of Schwann cells to the basal lamina is required for myelin-specific gene expression and myelination.[31-33] Interestingly, the proper organization of the actin cytoskeleton in Schwann cells is not only important to regulate cell shape changes but also to regulate gene expression.[34] While integrins are known regulators of actin assembly and cell morphology, no defects in gene expression have so far been reported for β1-deficient Schwann cells, but further quantitative studies are necessary to evaluate this possibility. The cytoplasmic signaling effectors that are activated by LN-2 and integrins also need to be determined. Schwann cells express known integrin downstream effectors such as focal adhesion kinase (FAK), paxillin and Fyn. These three proteins appear to form a complex with β1 integrins in differentiating Schwann cells that are co-cultured with neurons.[35,36] The tumor suppressor merlin/schwannomin binds to paxillin and may be part of an integrin-dependent protein complex.[37,38] Yet, the function of this protein complex in Schwann cells remains to be established. Up to now, no data are available showing a function for paxillin in myelination. Fyn has been implicated in myelination, but so far only in oligodendrocytes in the CNS and not in Schwann cells in the PNS.[39-43] Mutations in the gene encoding merlin/schwannomin lead to Neurofibromatosis type 1, an autosomal dominant disorder characterized by tumors, predominantly Schwannomas.[44] Some of the disease-causing mutations affect the protein domain of Schwannomin that interacts with paxillin, and the mutations reduce complex formation.[37] Schwannomin may link cell surface receptors such as integrins to the subplasmalemmal actin cortex. The association with the integrin complex at the cell surface might be a regulated process since it is only observed in cells grown at subconfluency.[37] However, Schwann cell-specific inactivation of the gene encoding merlin/schwannomin does not cause myelination defects, but leads to hyperplasia and the development of tumors.[45] These findings suggest that merlin/schwannomin may not act in a LN/integrin dependent signaling pathway. Intriguingly, recent findings have provided evidence that the integrins α4β1 and α5β1 that bind to fibronectin (FN), but not to LN, affect Schwann cells in distinct ways. α5β1 regulates Schwann cell proliferation, while α4β1 regulates Schwann cell survival (Table 2).[46] Further studies will be necessary to test whether schwannomin/merlin and these FN receptors act in a common pathway.

Dystroglycan and Peripheral Nerve Myelination

Genetic studies have provide convincing evidence that dystroglycan, which also binds to ECM components such as LN, is important in Schwann cell development. Originally, dystroglycan was identified as a component of the dystrophin glycoprotein complex in muscle. Mutations in genes encoding components of this complex lead to muscular dystrophy, but dystroglycan is also part of additional protein complexes in several cell types including Schwann cells.[47,48] Schwann cells that were genetically modified to lack dystroglycan initiate myelination and axonal sorting, but the myelin sheath folds excessively. Localization of Na^+ channels to nodes of Ranvier, the periodic gaps in the myelin sheath that allow for fast conduction of nerve impulses, is also defective.[49] The formation of nodes of Ranvier requires contact-dependent signals from Schwann cells.[50] Neurofascin and Nr-CAM have been implicated in mediating interactions of axons with Schwann cells at nodes of Ranvier.[51] Neurofascin and the β-subunit of Na^+ channels contain Ig-domains in their extracellular domain that mediate cis-interactions between both molecules.[52] Interestingly, dystroglycan localizes in Schwann cells at nodes of Ranvier and has an extracellular Ig-domain.[53,54] Thus, dystroglycan is a good candidate to mediate trans-interactions between Schwann cells and molecules in the nodal axolemma, such as neurofascin and Na^+ channels. In conclusion, current findings suggest that dystroglycan affects the development of Schwann cells at later stages than β1 integrins, and that some of its effects may be mediated by other ligands than LN. It would be interesting to inactivate dystroglycan and β1 integrins simultaneously, to define their distinct and potentially redundant functions in Schwann cells.

Integrins, ECM Glycoproteins and Oligodendrocytes

Oligodendrocytes are the myelinating glial cells of the CNS. Oligodendrocyte precursors arise in restricted areas of the CNS and disperse by migration to invade axonal tracts and initiate myelination.[55,56] Several studies suggest that ECM molecules and their integrin receptors control oligodendrocyte development and cooperate with soluble factors to promote oligodendrocyte proliferation and survival.

ECM Glycoproteins and Oligodendrocyte Function

ECM molecules such as FN, LN, vitronectin (VN), thrombospondin (TSP) and tenascin-C (TN-C) are expressed in appropriate patterns to be contacted by oligodendrocytes and their precursors. FN is expressed transiently in axonal white matter tracts during early development.[57-59] LN is present in the early developing CNS and in reduced amounts at later stages.[60-63] VN is expressed in axonal tracts during development and in the adult.[64] TSP and TN-C are also present in axonal tracts at different developmental stages.[64-69] Genetic studies support a role for TN-C and LN in myelination in the CNS (Table 2). In TN-C-deficient mice, oligodendrocyte precursor proliferation in the CNS is perturbed and the migration of precursors along the optic nerve is reduced.[70] Mice that are deficient for the LN-α2 subunit present defects in CNS myelination. These defects are apparently restricted to axonal tracts in the forebrain and optic nerve, while myelin in the spinal cord seems unaffected.[71] Together, these results suggest that ECM molecules have region specific effects on myelination.

Integrin Expression and Function in Oligodendrocytes

Expression of several integrin subunits has been observed within axonal tracts in vivo,[72] but it has been most carefully analyzed in oligodendrocytes in vitro (Table 1 and Fig. 3). Receptors for LN (α6β1), FN (αυβ1, αυβ3), VN (αυβ1, αυβ3, αυβ5, αυβ8), TSP (αυβ3), and TN-C (αυβ3) are expressed in distinct patterns. α6β1 and αυβ8 are expressed throughout development. αυβ1 is highly expressed on migrating oligodendrocyte precursors and is downregulated coincidentally with a decrease in their migration. αυβ3 is transiently upregulated and subsequently downregulated at the onset of final differentiation. Finally, αυβ5 is upregulated as oligodendrocytes differentiate and remains elevated at subsequent stages.[73,74] Studies with cells in culture suggest that different integrins have distinct functions in oligodendrocytes. αυβ1 promotes migration of oligodendrocyte precursors on purified ECM substrates,[75,76] while the integrin αυβ3 regulates proliferation of the precursors as well as of the immortalized oliogodendrocyte cell line CG-4.[70,77,78] αυβ5 appears to regulate differentiation, while the integrin α6β1 is required for cell survival and myelin sheath formation.[63,77,79-82] Finally, the function of αυβ8 in oligodendrocytes is unclear.

Genetic studies in mice have confirmed some of the in vitro findings. In mice lacking the integrin α6 subunit, the number of myelin-forming oligodendrocytes is decreases in parallel with increased apoptosis,[63] suggesting a function in cell survival (Table 2). α6β1 may be required for target-derived survival signaling, consistent with the finding that function-blocking

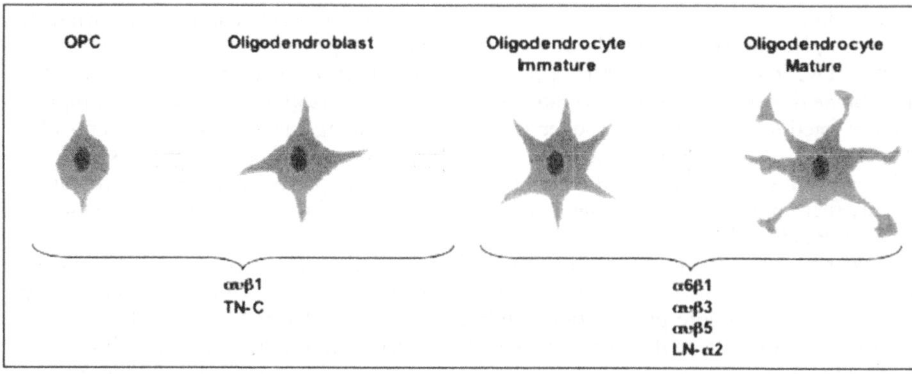

Figure 3. Integrins and oligodendrocytes. Oligodendrocytes differentiate in sequential stages. Proliferating oligodendrocyte precursors (OPCs) differentiate into oligodendroblasts, then into immature oligodendrocytes and finally into mature oligodendrocytes. Expression of integrins is differentially regulated during oligodendrocyte development as indicated.

antibodies to α6β1 reduce survival in co-culture of neurons and oligodendrocytes.[82] As outlined above, mutations that affect the LN-2 ligand for α6β1 lead to region specific defects in myelination,[71] and some of the effects could be caused by defects in survival, although myelin sheath formation could also be affected.[71,79] So far, no defects in myelination have been described for β3 and β5-deficient mice,[83-85] but more detailed studies will be necessary. Mice with a targeted inactivation of the genes encoding the αv and β8 subunits show intercerebral hemorrhage,[86-88] making the analysis of oligodendrocyte development difficult. The analysis of αv functions in oligodendrocytes will be facilitated by the availability of mice that carry a floxed allele of the gene encoding the integrin αv subunit.[88]

Integrin and Growth Factor Dependent Signaling Events

What are the cellular signal transduction pathways that are activated by integrins during oligodendrocyte differentiation? Oligodendrocyte proliferation and differentiation is regulated by contact with axonal targets and soluble factors, such as PDGF and neuregulins (NRGs).[89,90] At earlier stages of development, these soluble factors act as mitogens and inhibit differentiation, while later they become crucial in promoting oligodendrocyte survival.[91-96] Recent studies have led to the concept that integrins regulate the response of oligodendrocytes to secreted signaling molecules.[63] These findings could provide an explanation why PDGF and NRG can act both as mitogens and survival factors.

Initially, it has been demonstrated that ligands for the integrins αvβ3 such as VN affect the response of cultured oligodendrocyte precursors to PDGF. In the absence of VN, very high concentrations of PDGF are required to stimulate proliferation. In the presence of VN, PDGF effectively stimulates proliferation at concentrations that are in the physiologically relevant range.[78] The proliferative response induced by PDGF and VN is dependent on the non-receptor tyrosine kinase Lyn. PDGFαR, integrin αvβ3 and Lyn can be co-immunoprecipitated from cells, suggesting that they form a protein complex at the cell membrane. PKC and PI3K are also required for the proliferative response.[56,63,78,97,98] Studies with pharmacological inhibitors suggest that PKC is activated by the PDGF receptor. PKC then acts on the integrin heterodimer shifting it into a conformational state that allows high affinity ligand binding (so called "inside-out" signaling). Subsequent to ligand binding by the integrin, PI3K is activated and regulates downstream effectors to induce proliferation.[78]

In contrast to αvβ3, the integrin α6β1 cooperates with the PDGFαR receptor to regulate oligodendrocyte survival.[63] Survival signaling is dependent on PI3K and the non-receptor tyrosine kinase Fyn. Fyn forms a protein complex with the PDGFαR and the integrin α6β1. Intriguingly, the α6β1-PDGFαR-Fyn complex that promotes survival, but not the αυβ3-PDGFαR-Lyn complex that promotes proliferation, is localized to a specialized membrane domain known as lipid raft. Localization of the α6β1-PDGFαR-Fyn complex to the raft fraction is required for cooperative survival signaling, and increasing amounts of the PDGFαR appear to shift to the raft fraction when oligodendrocytes progress to a more differentiated state. Treatment of cells with Mn^{2+}, a known inducer of integrin ligand binding activity that mimics "inside-out signaling", is sufficient to induce integrin raft localization and survival signaling.[56,97,98]

Taken together, the findings suggest that PDGFαR signaling is contextual. The integrin αυβ3 and the PDGFαR receptor colocalize outside the raft fraction in oligodendrocyte precursors and promote proliferation, while the integrin α6β1 and the PDGFαR receptor colocalize in lipid rafts in differentiated oligodendrocytes and promote survival. This model is compelling, but there are several unresolved issues. For example, raft localization of α6β1 and PDGFαR is induced not only by LN but also by the control adhesive substrate Poly-D-Lysine (PDL). In contrast to LN, PDL does not cooperate with PDGFαR in survival signaling, suggesting that LN may have additional effects not mimicked by PDL.[56,97,98] In addition, a direct demonstration of the function of "inside-out signaling" and integrin-activation in survival signaling is still missing. In an attempt to address this issue, the integrin α6β1 was activated with Mn^{2+}, but surprisingly, the effect of Mn^{2+} on survival signaling could not be reversed by function blocking antibodies against the β1 integrin subunit.[98] To exclude nonspecific effects of Mn^{2+}, it would be important to activate integrins in distinct ways, for example by expressing constitutively active integrins. Alternatively, it would be interesting to engineer oligodendrocyte precursors that express integrin mutant forms that are locked in the inactive state.

Finally, similar in vitro studies suggest that integrins use distinct mechanisms to regulate the response of oligodendrocytes to NRGs. NRGs have been shown to regulate oligodendrocyte survival. When oligodendrocytes are plated onto the LN-2 ligand for the integrin α6β1, lower amounts of NRG are required to promote survival than in the absence of LN-2. Interestingly, while survival signaling via PDGF and via PDGF/α6β1 is PI3K dependent, NRG/α6β1-dependent survival is MAPK dependent.[63] It therefore appears that depending on the growth factor, different signaling pathways are activated to regulate cell survival.

Integrins, ECM Glycoproteins and Radial Glial Cells

Integrins and the Formation of the Radial Glia Scaffold

Glial cells with radial processes are located throughout many parts of the CNS (Fig. 4). They include the radial glial cells in the developing cerebral cortex (Fig. 4) and spinal cord, the Bergmann glia cells of the cerebellum, and the Müller glia cells of the retina. Some of these cells have long been implicated as guideposts for migrating neurons.[99-104] Converging genetic evidence from several laboratories provides strong evidence that β1 integrins are instrumental for the formation of the process of radial glial cells, Bergmann glial cells, and Müller glia cells. Initially, mice lacking the integrin α6β1 or double mutants lacking α3β1 and α6β1 have been shown to have defects in the formation of cortical layers, as well as in the assembly/maintenance of the meningeal basement membrane.[105,106] To address the mechanisms that cause the observed defects, and to clarify whether integrins are required in the meningeal cells that surround the brain or in neural tissue, the integrin β1 subunit and therefore all β1 integrins (including the integrin α3β1 and α6β1) have been inactivated in radial glial cells using CRE/LOX mediated gene inactivation. Expression of β1 integrins in the meningeal cell layer was not affected. The β1-deficient mice show similar defects in the formation of cortical layers as α3β1/α6β1-deficient mice.[107] Subsequent studies have shown that defects in the formation of cortical layers are caused in large parts by defects in radial glial fibers. In the absence of β1 integrins, radial glial

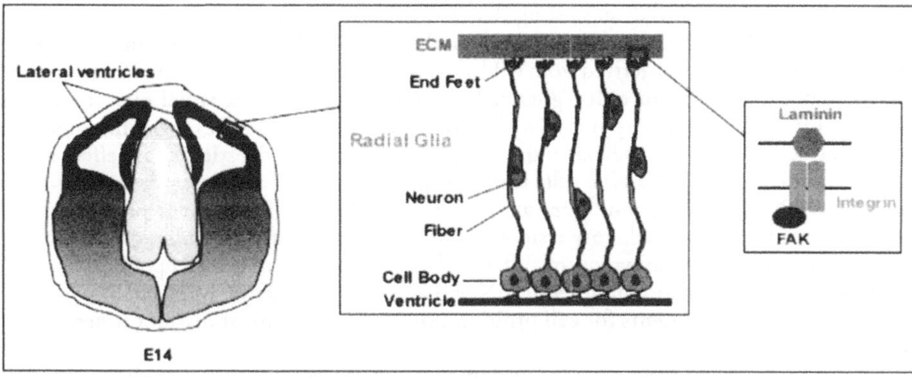

Figure 4. Integrins and radial glial cells. A crossection through an embryonic brain (E14) is shown, highlighting the developing cerebral cortex. Radial glial cells contain their cell bodies in the ventricular zone and extend long processes toward the brain surface where they are anchored with endfeet at the meningeal basement membrane (ECM). Genetic studies provide strong evidence that β1 integrins in radial glial cells mediate interactions with ECM components such as LN, and that FAK is required for β1 integrin function in radial glial cells.

processes develop, but are shorter, disorganized and loose their normal anchorage at the meningeal basement membrane (Table 2). This also leads to defects in the assembly and/or maintenance of the basement membrane.[107,108] Intriguingly, mice with mutations that affect components of the meningeal basement membrane such as perlecan, nidogen/entactin, and LN have similar defects to the mice with a CNS restricted β1 knock-out (Table 2).[109-112] Targeted inactivation of the integrin downstream effector focal adhesion kinase (FAK) in radial glial cells also leads to comparable defects.[113] Taken together, the findings suggest that an ECM-integrin-FAK signaling pathway is essential for the formation and maintenance of the radial glial network in the cerebral cortex (Fig. 4). Similar defects have been observed in β1-deficient mice in radial glial processes in the hippocampus, cerebellum and retina,[107,108,114] suggesting that β1 integrins have a central function in organizing radial glial processes throughout the CNS. Finally, defects in the interaction of radial glial fibers with the meningeal basement membrane have been observed in mice with a CNS restricted knock-out of the gene encoding dystroglycan, suggesting that β1 integrins and dystroglycan may cooperate to regulate radial glia differentiation and attachment.[115]

Integrins, Radial Glia and Cell Proliferation in the CNS

Recent findings have shown that some radial glial cells in the CNS not only serve as guideposts for migrating cells, but also as precursors for neuronal cells and glial cells. Radial glial cells therefore fulfill a dual function in the generation of neuronal and glial subtypes and in guiding neuronal migration.[100-102,116,117] This raises the question to what extent integrins and their ECM ligand may influence not only the morphological differentiation of these cells but also their proliferation and differentiation. Several lines of evidence have linked ECM glycoproteins to the regulation of cell proliferation in the CNS. ECM glycoproteins such as FN, TN-C, LN-1 and the LN-α2 subunit are expressed in the ventricular neuroepithelium.[58,59,118-122] Using cells in culture, it has been shown that LN and FGF act synergistically to regulate the proliferation of neuroepithelial cells.[123] Recent findings have further demonstrated that FN also affects proliferation.[124] While the in vivo significance of these findings still needs to be demonstrated, studies with cells in culture and genetic studies in mice suggest a role for TN-C in the regulation of cell proliferation in the CNS. In knock-out mice lacking TN-C, cell proliferation in cortical areas of the CNS is reduced, suggesting that TN-C may have a stimulatory role during cell proliferation (Table 2).[70] In support of this hypothesis, FGF stimulates cell proliferation more effectively in neurospheres derived from wild-type when compared to TN-C-deficient animals. TN-C may directly act on radial glial cells as staining for RC2, a radial glial marker, is decreased in TN-C-deficient animals.[120] However, the function of ECM components in regulating cell proliferation is likely more complex. For example, the stimulatory effect of FN on cell proliferation of wild-type cells is inhibited by TN-C,[124] suggesting that under some conditions TN-C can have inhibitory and not stimulatory functions. It is at present unclear how these seemingly contradictory results can be explained. As one possible explanation, the precise amounts and combination of different ECM molecules may be important. In vivo, local variation in the expression levels of FN and TN-C (and other ECM glycoproteins) may create distinct environments for cell proliferation/differentiation at distinct times during development.

Several experiments have provided information on the receptors that mediate effects of ECM glycoproteins on cell proliferation. Studies with LN fragments demonstrate that synergistic activation of cell proliferation by LN and FGF is observed with a proteolytic LN fragment that contains the integrin binding site.[123] Surprisingly, while proliferating cells in neurospheres express the LN-binding integrin α6β1, function blocking antibodies against the α6 subunit do not affect neurosphere proliferation.[124] It is possible that effects were not observed because other LN receptors, including dystroglycan (which also binds the LN-fragments that stimulate proliferation in vitro),[48] may mediate effects on cell proliferation. In contrast, effects of FN on cell proliferation appear to depend on integrins. Neurospheres express the FN receptors α5β1, αυβ1, αυβ5 and αυβ8. RGD peptides that block interactions of these receptors with FN inhibit the effect of FN on proliferation. A partial block is observed with β1 integrin antibodies, suggesting that the effect is mediated by α5β1 and/or αυβ1. Neuroepithelial cells that are FACS sorted according to high β1 expression levels have a higher propensity to generate neurospheres than cells with lower β1 expression levels.[118,124] However, there are some conflicting results. While antibodies to β1 integrins can inhibit the effect of ECM components on proliferation, no such block is observed when the β1 integrin gene is inactivated in the cells using CRE/LOX-mediated gene inactivation.[118] The reason for this discrepancy is unknown. It is possible that genetic perturbations are more severe than antibody perturbation, leading to the activation of pathways that compensate for the loss of β1 integrins. Further biochemical, cell biological and genetical experiments are necessary to resolve this issue.

Conclusions

Studies from many laboratories have provided compelling evidence that integrin/ECM interactions control the development and function of glial cells, including the differentiation of radial glial cells, oligodendrocytes and Schwann cells. However, the picture is at present still incomplete. For example, genetic studies support a role for β1 integrins in radial glial cells and Schwann cells, but a genetic confirmation for the functions of β1 and αυ integrins in oligodendrocytes is largely missing. Likewise, the signaling pathways that operate downstream of integrins are just starting to be identified. Finally, the functions of integrins in glial cells of the adult nervous system has not been reported. The availability of refined tools for the genetic perturbation of integrin expression will undoubtfully shed new light on these unresolved issues.

Acknowledgements

Work in the authors laboratory was supported by funding from the NIH (U.M., RO1-NS046456) and the Christopher Reeve Paralysis Foundation (C.S.B.).

References

1. Bokel C, Brown NH. Integrins in development: moving on, responding to, and sticking to the extracellular matrix. Dev Cell 2002; 3(3):311-21.
2. Hynes RO. Integrins: bidirectional, allosteric signaling machines. Cell 2002; 110(6):673-87.
3. Müller U. Integrins and Extracellular Matrix in Animal Models. Handb. Exp. Pharm. Behrens J, Nelson WJ, eds. Vol. 165. New York: Springer, 2004:217-241.
4. Brakebusch C, Fassler R. The integrin-actin connection, an eternal love affair. EMBO J 2003; 22(10):2324-33.
5. Giancotti FG, Ruoslahti E., Integrin signaling. Science 1999; 285(5430):1028-32.
6. Hynes RO. Integrins: versatility, modulation, and signaling in cell adhesion. Cell 1992; 69(1):11-25.
7. Schwartz MA, Ginsberg MH. Networks and crosstalk: integrin signalling spreads. Nat Cell Biol 2002; 4(4):E65-8.
8. Clegg DO et al. Integrins in the development, function and dysfunction of the nervous system. Front Biosci 2003; 8:d723-50.
9. Clegg DO. Novel roles for integrins in the nervous system. Mol Cell Biol Res Commun 2000; 3(1):1-7.
10. Milner R, Campbell IL. The integrin family of cell adhesion molecules has multiple functions within the CNS. J Neurosci Res 2002; 69(3):286-91.
11. Nakamoto T, Kain KH, Ginsberg MH. Neurobiology: New connections between integrins and axon guidance. Curr Biol 2004; 14(3):R121-3.
12. Mirsky R, Jessen KR. The neurobiology of Schwann cells. Brain Pathol 1999; 9(2):293-311.
13. Previtali SC et al. Role of integrins in the peripheral nervous system. Prog Neurobiol 2001 64(1):35-49.
14. Scherer SS. Myelination: some receptors required. J Cell Biol 2002; 156(1):13-5.
15. Sanes JR et al. Molecular heterogeneity of basal laminae: isoforms of laminin and collagen IV at the neuromuscular junction and elsewhere. J Cell Biol 1990; 111(4):1685-99.
16. Patton BL et al. Distribution and function of laminins in the neuromuscular system of developing, adult, and mutant mice. J Cell Biol 1997; 139(6):1507-21.
17. Chen ZL, Strickland S. Laminin gamma1 is critical for Schwann cell differentiation, axon myelination, and regeneration in the peripheral nerve. J Cell Biol 2003; 163(4):889-99.
18. Previtali SC et al. Expression of laminin receptors in schwann cell differentiation: evidence for distinct roles. J Neurosci 2003; 23(13):5520-30.
19. Hsiao LL et al. Expression of laminin isoforms by peripheral nerve-derived connective tissue cells in culture. Comparison with epitope distribution in normal human nerve and neural tumors in vivo. Lab Invest 1993; 68(1):100-8.
20. Milner R et al. Division of labor of Schwann cell integrins during migration on peripheral nerve extracellular matrix ligands. Dev Biol 1997; 185(2):215-28.
21. Previtali SC et al. Schwann cells synthesize alpha7beta1 integrin which is dispensable for peripheral nerve development and myelination. Mol Cell Neurosci 2003; 23(2):210-8.
22. Stewart HJ et al. Expression and regulation of alpha1beta1 integrin in Schwann cells. J Neurobiol 1997; 33(7):914-28.
23. Fernandez-Valle C et al. Anti-beta 1 integrin antibody inhibits Schwann cell myelination. J Neurobiol 1994; 25(10):1207-26.
24. Feltri ML et al. Conditional disruption of beta 1 integrin in Schwann cells impedes interactions with axons. J Cell Biol 2002; 156(1):199-209.
25. Gardner H et al. Deletion of integrin alpha 1 by homologous recombination permits normal murine development but gives rise to a specific deficit in cell adhesion. Dev Biol 1996; 175(2):301-13.
26. Chen J et al. The alpha(2) integrin subunit-deficient mouse: a multifaceted phenotype including defects of branching morphogenesis and hemostasis. Am J Pathol 2002; 161(1):337-44.
27. Holtkotter O et al. Integrin alpha 2-deficient mice develop normally, are fertile, but display partially defective platelet interaction with collagen. J Biol Chem 2002; 277(13):10789-94.
28. Mayer U et al. Absence of integrin alpha 7 causes a novel form of muscular dystrophy. Nat Genet 1997; 17(3):318-23.
29. Kreidberg JA et al. Alpha 3 beta 1 integrin has a crucial role in kidney and lung organogenesis. Development 1996; 122(11):3537-47.
30. Georges-Labouesse E et al. Absence of integrin alpha 6 leads to epidermolysis bullosa and neonatal death in mice. Nat Genet 1996; 13(3):370-3.
31. Carey DJ, Todd MS, Rafferty CM. Schwann cell myelination: induction by exogenous basement membrane-like extracellular matrix. J Cell Biol 1986; 102(6):2254-63.

32. Eldridge CF, Bunge MB, Bunge RP. Differentiation of axon-related Schwann cells in vitro: II. Control of myelin formation by basal lamina. J Neurosci 1989; 9(2):625-38.
33. Fernandez-Valle C et al. Expression of the protein zero myelin gene in axon-related Schwann cells is linked to basal lamina formation. Development 1993; 119(3):867-80.
34. Fernandez-Valle C et al. Actin plays a role in both changes in cell shape and gene-expression associated with Schwann cell myelination. J Neurosci 1997; 17(1):241-50.
35. Chen LM, Bailey D, Fernandez-Valle C. Association of beta 1 integrin with focal adhesion kinase and paxillin in differentiating Schwann cells. J Neurosci 2000; 20(10):3776-84.
36. Fernandez-Valle, Wood PM, Bunge MB. Localization of focal adhesion kinase in differentiating Schwann cell/neuron cultures. Microsc Res Tech 1998; 41(5):416-30.
37. Fernandez-Valle C et al. Paxillin binds schwannomin and regulates its density-dependent localization and effect on cell morphology. Nat Genet 2002; 31(4):354-62.
38. Obremski VJ, Hall AM, Fernandez-Valle C. Merlin, the neurofibromatosis type 2 gene product, and beta1 integrin associate in isolated and differentiating Schwann cells. J Neurobiol 1998; 37(4):487-501.
39. Sperber BR et al. A unique role for Fyn in CNS myelination. J Neurosci 2001; 21(6):2039-47.
40. Sperber BR, McMorris FA. Fyn tyrosine kinase regulates oligodendroglial cell development but is not required for morphological differentiation of oligodendrocytes. J Neurosci Res 2001; 63(4):303-12.
41. Biffiger K et al. Severe hypomyelination of the murine CNS in the absence of myelin-associated glycoprotein and fyn tyrosine kinase. J Neurosci 2000; 20(19):7430-7.
42. Seiwa C et al. Fyn tyrosine kinase participates in the compact myelin sheath formation in the central nervous system. Neurosci Res 2000; 37(1):21-31.
43. Umemori H et al. Initial events of myelination involve Fyn tyrosine kinase signalling. Nature 1994; 367(6463):572-6.
44. Ramesh V. Merlin and the ERM proteins in Schwann cells, neurons and growth cones. Nat Rev Neurosci 2004; 5(6):462-70.
45. Giovannini M et al. Conditional biallelic Nf2 mutation in the mouse promotes manifestations of human neurofibromatosis type 2. Genes Dev 2000; 14(13):1617-30.
46. Haack H, Hynes RO. Integrin receptors are required for cell survival and proliferation during development of the peripheral glial lineage. Dev Biol 2001; 233(1):38-55.
47. Michele DE, Campbell KP. Dystrophin-glycoprotein complex: post-translational processing and dystroglycan function. J Biol Chem 2003; 278(18):15457-60.
48. Henry MD, Campbell KP. Dystroglycan inside and out. Curr Opin Cell Biol 1999; 11(5):602-7.
49. Saito F et al. Unique role of dystroglycan in peripheral nerve myelination, nodal structure, and sodium channel stabilization. Neuron 2003; 38(5):747-58.
50. Ching W et al. Clustering of neuronal sodium channels requires contact with myelinating Schwann cells. J Neurocytol 1999; 28(4-5):295-301.
51. Peles E, Salzer JL. Molecular domains of myelinated axons. Curr Opin Neurobiol 2000; 10(5):558-65.
52. Ratcliffe CF et al. Sodium channel beta1 and beta3 subunits associate with neurofascin through their extracellular immunoglobulin-like domain. J Cell Biol 2001; 154(2):427-34.
53. Saito F et al. Characterization of the transmembrane molecular architecture of the dystroglycan complex in schwann cells. J Biol Chem 1999; 274(12):8240-6.
54. Bozic D, Engel J, Brancaccio A. Sequence analysis suggests the presence of an IG-like domain in the N-terminal region of alpha-dystroglycan which was crystallized after mutation of a protease susceptible site (Arg168→His). Matrix Biol 1998; 17(7):495-500.
55. Pfeiffer SE, Warrington AE, Bansal R. The oligodendrocyte and its many cellular processes. Trends Cell Biol 1993; 3(6):191-7.
56. Baron W, Colognato H, Ffrench-Constant C. Integrin-growth factor interactions as regulators of oligodendroglial development and function. Glia 2005; 49(4):467-79.
57. Pearlman AL, Sheppard AM. Extracellular matrix in early cortical development. Prog Brain Res 1996; 108:117-34.
58. Sheppard AM et al. Neuronal production of fibronectin in the cerebral cortex during migration and layer formation is unique to specific cortical domains. Dev Biol 1995; 172(2):504-18.
59. Chun JJ, Shatz CJ. A fibronectin-like molecule is present in the developing cat cerebral cortex and is correlated with subplate neurons. J Cell Biol 1988; 106(3):857-72.
60. McLoon SC et al. Transient expression of laminin in the optic nerve of the developing rat. J Neurosci 1988; 8(6):1981-90.
61. Morissette N, Carbonetto S. Laminin alpha 2 chain (M chain) is found within the pathway of avian and murine retinal projections. J Neurosci 1995; 15(12):8067-82.

62. Powell SK et al. Laminin-like proteins are differentially regulated during cerebellar development and stimulate granule cell neurite outgrowth in vitro. J Neurosci Res 1998; 54(2):233-47.
63. Colognato H et al. CNS integrins switch growth factor signalling to promote target-dependent survival. Nat Cell Biol 2002; 4(11):833-41.
64. Neugebauer KM et al. Vitronectin and thrombospondin promote retinal neurite outgrowth: developmental regulation and role of integrins. Neuron 1991; 6(3):345-58.
65. O'Shea KS, Rheinheimer JS, Dixit VM. Deposition and role of thrombospondin in the histogenesis of the cerebellar cortex. J Cell Biol 1990; 110(4):1275-83.
66. Bartsch S et al. Expression of tenascin in the developing and adult cerebellar cortex. J Neurosci 1992; 12(3):736-49.
67. Bartsch U et al. Tenascin demarcates the boundary between the myelinated and nonmyelinated part of retinal ganglion cell axons in the developing and adult mouse. J Neurosci 1994; 14(8):4756-68.
68. Garwood J et al. The extracellular matrix glycoprotein Tenascin-C is expressed by oligodendrocyte precursor cells and required for the regulation of maturation rate, survival and responsiveness to platelet-derived growth factor. Eur J Neurosci 2004; 20(10):2524-40.
69. Laywell ED et al. Enhanced expression of the developmentally regulated extracellular matrix molecule tenascin following adult brain injury. Proc Natl Acad Sci USA 1992; 89(7):2634-8.
70. Garcion E, Faissner A, Ffrench-Constant C. Knockout mice reveal a contribution of the extracellular matrix molecule tenascin-C to neural precursor proliferation and migration. Development 2001; 128(13):2485-96.
71. Chun SJ et al. Integrin-linked kinase is required for laminin-2-induced oligodendrocyte cell spreading and CNS myelination. J Cell Biol 2003; 163(2):397-408.
72. Pinkstaff JK et al. Integrin subunit gene expression is regionally differentiated in adult brain. J Neurosci 1999; 19(5):1541-56.
73. Milner R, Ffrench-Constant C. A developmental analysis of oligodendroglial integrins in primary cells: changes in alpha v-associated beta subunits during differentiation. Development 1994; 120(12):3497-506.
74. Milner R et al. Expression of alpha vbeta3 and alpha vbeta8 integrins during oligodendrocyte precursor differentiation in the presence and absence of axons. Glia 1997; 21(4):350-60.
75. Tiwari-Woodruff SK et al. OSP/claudin-11 forms a complex with a novel member of the tetraspanin super family and beta1 integrin and regulates proliferation and migration of oligodendrocytes. J Cell Biol 2001; 153(2):295-305.
76. Milner R et al. A role in migration for the alpha V beta 1 integrin expressed on oligodendrocyte precursors. J Neurosci 1996; 16(22):7240-52.
77. Blaschuk KL, Frost EE, Ffrench-Constant C. The regulation of proliferation and differentiation in oligodendrocyte progenitor cells by alphaV integrins. Development 2000; 127(9):1961-9.
78. Baron W, Shattil SJ, Ffrench-Constant C. The oligodendrocyte precursor mitogen PDGF stimulates proliferation by activation of alpha(v)beta3 integrins. EMBO J 2002; 21(8):1957-66.
79. Relvas JB et al. Expression of dominant-negative and chimeric subunits reveals an essential role for beta1 integrin during myelination. Curr Biol 2001; 11(13):1039-43.
80. Buttery PC et al. Mapping regions of the beta1 integrin cytoplasmic domain involved in migration and survival in primary oligodendrocyte precursors using cell-permeable homeopeptides. Biochem Biophys Res Commun 1999; 259(1):121-7.
81. Corley SM et al. Astrocytes attenuate oligodendrocyte death in vitro through an alpha(6) integrin-laminin-dependent mechanism. Glia 2001; 36(3):281-94.
82. Frost EE et al. Integrins mediate a neuronal survival signal for oligodendrocytes. Curr Biol 1999; 9(21):1251-4.
83. Huang X et al. Normal development, wound healing, and adenovirus susceptibility in beta5-deficient mice. Mol Cell Biol 2000; 20(3):755-9.
84. Hodivala-Dilke KM et al. Beta3-integrin-deficient mice are a model for Glanzmann thrombasthenia showing placental defects and reduced survival. J Clin Invest 1999; 103(2):229-38.
85. Reynolds LE et al. Enhanced pathological angiogenesis in mice lacking beta3 integrin or beta3 and beta5 integrins. Nat Med 2002; 8(1):27-34.
86. Zhu J et al. beta8 integrins are required for vascular morphogenesis in mouse embryos. Development 2002; 129(12):2891-903.
87. McCarty JH et al. Defective associations between blood vessels and brain parenchyma lead to cerebral hemorrhage in mice lacking alphav integrins. Mol Cell Biol 2002; 22(21):7667-77.
88. McCarty JH et al. Selective ablation of alphav integrins in the central nervous system leads to cerebral hemorrhage, seizures, axonal degeneration and premature death. Development 2005; 132(1):165-76.

89. Barres BA, Raff MC. Control of oligodendrocyte number in the developing rat optic nerve. Neuron 1994; 12(5):935-42.
90. Buonanno A, Fischbach GD. Neuregulin and ErbB receptor signaling pathways in the nervous system. Curr Opin Neurobiol 2001; 11(3):287-96.
91. Noble M et al. Platelet-derived growth factor promotes division and motility and inhibits premature differentiation of the oligodendrocyte/type-2 astrocyte progenitor cell. Nature 1988; 333(6173):560-2.
92. Barres BA et al. Cell death and control of cell survival in the oligodendrocyte lineage. Cell 1992; 70(1):31-46.
93. Canoll PD et al. GGF/neuregulin is a neuronal signal that promotes the proliferation and survival and inhibits the differentiation of oligodendrocyte progenitors. Neuron 1996; 17(2):229-43.
94. Canoll PD et al. GGF/neuregulin induces a phenotypic reversion of oligodendrocytes. Mol Cell Neurosci 1999; 13(2):79-94.
95. Calver AR et al. Oligodendrocyte population dynamics and the role of PDGF in vivo. Neuron 1998; 20(5):869-82.
96. Calaora V et al. Neuregulin signaling regulates neural precursor growth and the generation of oligodendrocytes in vitro. J Neurosci 2001; 21(13):4740-51.
97. Baron W et al. Regulation of integrin growth factor interactions in oligodendrocytes by lipid raft microdomains. Curr Biol 2003; 13(2):151-5.
98. Decker L, Baron W, Ffrench-Constant C. Lipid rafts: microenvironments for integrin-growth factor interactions in neural development. Biochem Soc Trans 2004; 32(Pt3):426-30.
99. Hatten ME. Central nervous system neuronal migration. Annu Rev Neurosci 1999; 22:511-39.
100. RakicDevelopmental and evolutionary adaptations of cortical radial glia. Cereb Cortex 2003; 13(6):541-9.
101. Fishell G, Kriegstein AR. Neurons from radial glia: the consequences of asymmetric inheritance. Curr Opin Neurobiol 2003; 13(1):34-41.
102. Nadarajah B, Parnavelas JG. Modes of neuronal migration in the developing cerebral cortex. Nat Rev Neurosci 2002; 3(6):423-32.
103. Willbold E, Layer PG. Muller glia cells and their possible roles during retina differentiation in vivo and in vitro. Histol Histopathol 1998; 13(2):531-52.
104. Muller T, Kettenmann H. Physiology of Bergmann glial cells. Int Rev Neurobiol 1995; 38:341-59.
105. De Arcangelis A et al. Synergistic activities of alpha3 and alpha6 integrins are required during apical ectodermal ridge formation and organogenesis in the mouse. Development 1999; 126(17):3957-68.
106. Georges-Labouesse E et al. Essential role of alpha 6 integrins in cortical and retinal lamination. Curr Biol 1998; 8(17):983-6.
107. Graus-Porta D et al. Beta1-class integrins regulate the development of laminae and folia in the cerebral and cerebellar cortex. Neuron 2001; 31(3):367-79.
108. Blaess S et al. Beta1-integrins are critical for cerebellar granule cell precursor proliferation. J Neurosci 2004; 24(13):3402-12.
109. Arikawa-Hirasawa E et al. Perlecan is essential for cartilage and cephalic development. Nat Genet 1999; 23(3):354-8.
110. Costell M et al. Perlecan maintains the integrity of cartilage and some basement membranes. J Cell Biol 1999; 147(5):1109-22.
111. Dong L et al. Neurologic defects and selective disruption of basement membranes in mice lacking entactin-1/nidogen-1. Lab Invest 2002; 82(12):1617-30.
112. Halfter W et al. A critical function of the pial basement membrane in cortical histogenesis. J Neurosci 2002; 22(14):6029-40.
113. Beggs H.E et al. FAK deficiency in cells contributing to the basal lamina results in cortical abnormalities resembling congenital muscular dystrophies. Neuron 2003; 40(3):501-14.
114. Foerster E et al. Reelin, Disabled 1, and b1-class integrins are required for the formation of the radial glial scaffold in the hippocampus. Proc Natl Acad Sci USA 2002; in press.
115. Moore SA et al. Deletion of brain dystroglycan recapitulates aspects of congenital muscular dystrophy. Nature 2002; 418(6896):422-5.
116. Campbell K, Gotz M. Radial glia: multi-purpose cells for vertebrate brain development. Trends Neurosci 2002; 25(5):235-8.
117. Hatten ME. New directions in neuronal migration. Science 2002; 297(5587):1660-3.
118. Campos LS et al. Beta1 integrins activate a MAPK signalling pathway in neural stem cells that contributes to their maintenance. Development 2004; 131(14):3433-44.
119. Gates MA et al. Cell and molecular analysis of the developing and adult mouse subventricular zone of the cerebral hemispheres. J Comp Neurol 1995; 361(2):249-66.

120. Garcion E et al. Generation of an environmental niche for neural stem cell development by the extracellular matrix molecule tenascin C. Development 2004; 131(14):3423-32.
121. Stewart GR, Pearlman AL. Fibronectin-like immunoreactivity in the developing cerebral cortex. J Neurosci 1987; 7(10):3325-33.
122. Stettler EM, Galileo DS. Radial glia produce and align the ligand fibronectin during neuronal migration in the developing chick brain. J Comp Neurol 2004; 468(3):441-51.
123. Drago J, Nurcombe V, Bartlett PF. Laminin through its long arm E8 fragment promotes the proliferation and differentiation of murine neuroepithelial cells in vitro. Exp Cell Res 1991; 192(1):256-65.
124. Jacques TS et al. Neural precursor cell chain migration and division are regulated through different beta1 integrins. Development 1998; 125(16):3167-77.

Chapter 14

Roles for Integrins and Associated Proteins in the Haematopoietic System

Alessandra Cambi, Suzanne F.G. van Helden and Carl G. Figdor*

Summary

The haematopoietic system is a highly dynamic system where cell adhesion and cell migration are tightly controlled. During both embryonic and adult haematopoiesis, the finely orchestrated action of several integrins contributes to the generation of all different mature blood cells. Within the immune system, integrins regulate cell-cell and cell-extracellular matrix interaction during homeostasis as well as inflammation. Accumulating evidence suggests that many of the integrin-mediated functions occur through physical interaction of integrins with other associated proteins. The goal of this chapter is to describe the role of integrins—particularly α4, β2, and β7—and integrin-interacting molecules in the haematopoietic and immune systems.

Introduction

The haematopoietic system consists of stem cells that self replicate and differentiate into specific progenitor cells to eventually generate all the different mature blood cells (Fig. 1).

In steady-state conditions, stem cells and progenitors reside in the bone marrow (BM) microenvironment, where various cytokines and extracellular matrix (ECM) components are secreted.[1] Stem cells as well as progenitor cells express several adhesion receptors (selectins, integrins, and sialomucins) that mediate specific cell-cell and cell-ECM contacts.[2,3] These contacts play an important role in the regulation of homing and retention of progenitors within the BM and their growth, survival and differentiation.[4-6] However, so far no adhesion receptor has been identified that is exclusively expressed on stem cells. Likewise, no adhesive ligand, except haemonectin for the granulocyte lineage,[7] has been identified to be specifically present in the BM microenvironment.

The adhesive processes of leukocytes have been the subject of numerous studies. These bone-marrow derived cells with diverse form and function circulate in the blood in a resting state of low adhesiveness. They display a particular pattern of adhesion receptors that can change in a signal- and time-dependent manner.

While some of these receptors, among which β1 integrins, are present also in other cell types, the β2 and β7 integrins are exclusively expressed on the cell surface of leukocytes.[8]

The β2 integrins family comprehends four members, αLβ2 (CD11a/CD18; LFA-1), αMβ2 (CD11b/CD18; MAC-1), αXβ2 (CD11c/CD18; p150,95), and αDβ2 (CD11d/CD18).[9] Each of the β2 integrins recognizes one or more members of the intercellular adhesion molecule (ICAM) family.[9] In addition, αLβ2 binds to JAM-1, expressed at the tight junction of

*Corresponding Author: Carl G. Figdor—Department of Tumor Immunology, Nijmegen Centre for Molecular Life Sciences, Radboud University, Nijmegen, The Netherlands.
Email: c.figdor@ncmls.kun.nl

Integrins and Development, edited by Erik H.J. Danen. ©2006 Landes Bioscience.

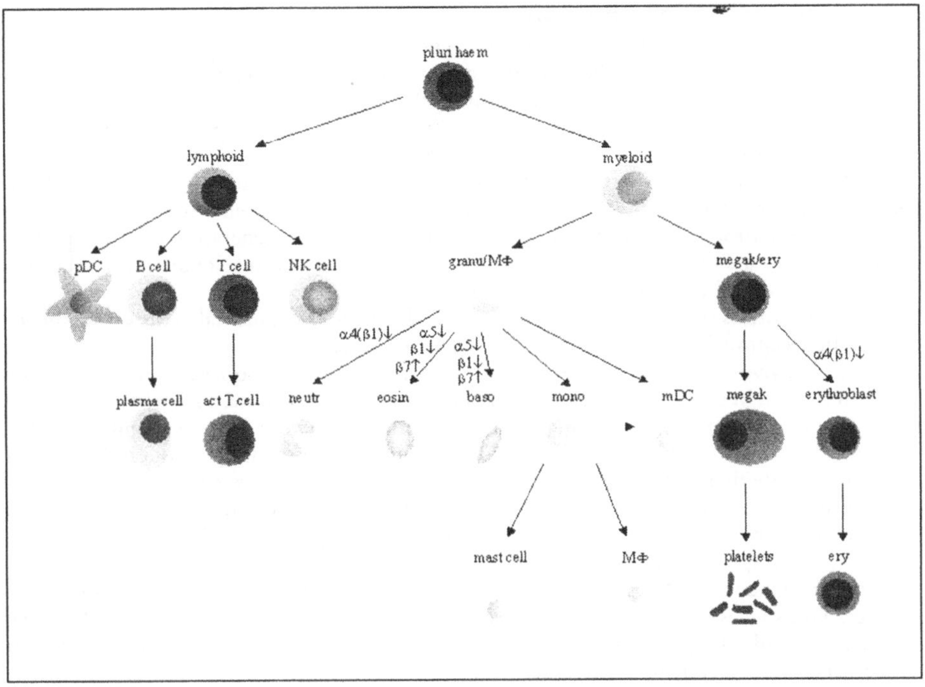

Figure 1. Haematopoietic lineage. Pluri haem: pluripotent haematopoietic stem cell; lymphoid: lymphoid progenitor; myeloid: myeloid progenitor; pDC: plasmacytoid dendritic cell; NK cell: natural killer cell; granu/MF: granulocyte/macrophage progenitor; megak/ery: megakaryocyte/erythrocyte progenitor; act T cell: activated T cell; neutr: neutrophil; eosin: eosinophil; baso: basophil; mono: monocyte; mDC: myeloid dendritic cell; megak: megakaryocyte; Mϕ: macrophage; ery: erythrocyte. ↓: down regulation; ↑: up regulation.

resting endothelium,[10] and αDβ2 binds to vascular cell adhesion molecule-1 (VCAM-1),[11] while αMβ2 and αβ2 can also interact with polysaccharides.[12,13] The RGD motif, a critical feature in many integrin ligands, is not required for ligands of the β2 integrins family. The most extensively studied member of this family is αLβ2. This integrin mediates leukocyte migration across the endothelium both during normal lymphocyte recirculation and in response to inflammatory signals.[14-17] αLβ2 has also been shown to be involved in the formation of the "immunological synapse", the contact site between T lymphocytes and antigen presenting cells.[18]

The β7 integrins are specifically expressed by distinct subsets of T lymphocytes and in the murine immune system they can be found also on subsets of skin-derived dendritic cells (DCs) in peripheral lymph nodes.[19] This family consists of only two members: α4β7, expressed on a subset of memory $CD4^+$ and $CD8^+$ T lymphocytes, which binds to the mucosal addressin cell adhesion molecule (MAdCAM-1),[20] and αEβ7, which interacts with epithelial-specific cadherin (E-cadherin).[21,22]

In this chapter, we shall focus on the above-mentioned families of integrins and their role in both the haematopoietic and the immune system. Finally, the importance of functional integrins will be discussed in the context of integrin-associated disorders.

Integrins in the Haematopoiesis

Development of haematopoietic stem cells and their differentiated progeny occurs during fetal as well as adult life and relies on interactions with the BM microenvironment that provides support and stimuli for proliferation and differentiation.

During embryo development, gradual colonization of all haematopoietic organs occurs,[23] and in mouse after embryonic day 10 definitive pluripotent progenitors are found in the fetal liver, thymus, bone marrow, and spleen. However, active haematopoiesis begins at different time points, depending on the development of each haematopoietic environment. In humans, after birth, erythropoiesis, myelopoiesis, B lymphopoiesis, and generation of T lymphocyte progenitors, occurs in the BM, while in mouse erythropoiesis and myelopoiesis can take place both in the BM and in the spleen.

Especially the use of knock-out or chimera mice has been instrumental in dissecting the role of specific receptors in embryonic and adult haematopoiesis.

Several in vivo studies have demonstrated the involvement of integrins in multiple steps of haematopoiesis. In particular, the expression of α4 integrins is regulated during haematopoiesis; these receptors are expressed on early multipotent progenitors but are downregulated during erythrocyte and neutrophil differentiation.[24] In addition, α4 integrins present on the progenitor cells mediate their interaction with BM stromal cells, which express VCAM-1.[25]

The analysis of α4-null mice has shown that while erythropoiesis in the yolk sac can occur in the absence of α4 integrins, definitive haematopoiesis in spleen and BM is severely compromised.[26] More specifically, α4-null erythroid progenitors can colonize fetal liver, BM, and spleen, where they can differentiate, although they produce only low numbers of mature erythrocytes. This suggests a defect in proper expansion of the erythroid populations in absence of α4 integrins and is in agreement with in vitro data showing an inhibition of erythroid development when anti- α4 integrin antibodies were used.[27]

Studies in α4-null mice have also shown that myeloid and lymphoid development depends on α4 integrins. Although terminal differentiation of T and B lymphocytes, monocytes, and granulocytes occurs during fetal development in the absence of α4 integrins, only a small percentage of α4-null progenitors can complete differentiation. This defect becomes more evident after birth, when the bone marrow starts playing a major role in haematopoietic development.[28]

A widely accepted model for the function of α4 integrins in the regulation of hematopoiesis is based on the assumption that within the stroma, three different compartments for progenitor cells exist: self-renewal, expansion, and maturation.[29] The balance among different signals (cytokines, chemokines, and cell-cell contacts) in each niche would determine the fate of the committed progenitors. α4 integrins could provide a retention signal for progenitors to remain in the expansion compartment. As a consequence, α4-deficient progenitors prematurely detach, lacking the expansion phase, and rapidly shift towards differentiation. This might explain the low numbers of progenitors found in α4-deficient hematopoietic organs and the low yield of mature cells. In this way, α4 integrins act as major regulators of the balance between proliferation and differentiation during hematopoiesis.[26] This hypothesis is further substantiated by studies regarding adult hematopoiesis in α4 conditional-knockout mice.[30] Here, a dramatic increase of circulating progenitors was observed, suggesting their early release from the BM. Moreover, these circulating α4-null progenitor cells were partially unable to home to the BM, in agreement with the observation that α4 integrin-function-blocking antibodies were able to selectively reduce BM homing. Since the α4 subunit can associate with either the β1 or the β7 subunit, α4-null hematopoietic cells lack both α4β1 and α4β7 integrins. Nevertheless, no defects in hematopoietic cell development were observed in β7-null mice,[31] suggesting a primary role for the α4β1 combination.

Although β7 integrins seem to play a minor role in haematopoiesis, it should be mentioned that up-regulation of their expression levels together with a parallel down-regulation of β1 and α5 integrins accompanies eosinophilic-basophilic lineage commitment (Fig. 1).[32]

Further supporting evidence of the dominant role of β1 integrins in haematopoiesis comes from numerous studies on the involvement of β1 integrins in fetal and adult haematopoiesis. The use of antibodies against β1 integrins has been shown to severely reduce the reentry of hematopoietic progenitors into the spleen and the BM, after human stem cell transplantation.[25] The absence of β1 integrins also causes an impaired migration of haematopoietic stem cells towards the fetal liver.[33]

However, α4 and β1 integrins are not the only adhesion receptors that regulate haematopoiesis. A synergistic contribution of β2 integrins was also reported. Although deficiencies in β2 integrins alone do not have a major impact on homing and repopulation kinetics, the addition of an α4 functional blockade to β2 deficiency results in over 90% reduction of the homing in contrast to the 40% inhibition observed with α4 blockade alone.[34]

Similar synergy between β1 and β2 integrins is observed in the mobilization of hematopoietic stem and progenitor cells from the BM to the blood. Function-blocking antibodies against β1 and β2 integrins, used separately or in combination, together with β2-null mice have shown that abrogation of β2 integrin function contributes significantly to cell mobilization when α4β1 integrin function is inhibited simultaneously.[35] However, it remains unclear at what stage this synergy occurs: at the stage of reversible adhesion or by enhancement of migration. A direct role for β2 integrins, particularly αLβ2 and αMβ2, in cytokine-induced mobilization was also established. Specifically, anti-αLβ2 blocking antibodies, but not anti-α4β1 antibodies, prevent mobilization of hematopoietic progenitor cells induced by IL-8.[36] In contrast, administration of antibodies against αLβ2 and αMβ2 synergistically enhances stem cell-mobilization induced by granulocyte colony-stimulating factor (G-CSF).[37] The exact mechanisms regulating retention in- and mobilization from the BM are largely unknown; however, recent studies indicate that a fine-tuned interplay between cytokines/chemokines, proteases, and adhesion molecules is fundamental. A clear example is the regulation of integrin-dependent adhesion by the chemokine CXCL12 during B lymphopoiesis.[38] In human BM, CXCL12, also known as stromal cell-derived factor 1, triggers a sustained α4β1-dependent adhesion response in progenitor B cells. This response decreases during B cell development in the BM and is completely absent in circulating fully mature B cells, which only exhibit transient CXCL12-enhanced adhesion.[38] These observations provide a developmental cell-stage specific mechanism where chemokines regulate hematopoiesis through sustained rather than short-lived activation of integrin adhesion.

In conclusion, during all different stages of haematopoiesis, cell-cell and cell-extracellular matrix interactions via adhesion receptors are tightly regulated both in time and space to guarantee the correct balance between retention and mobilization as well as proliferation and differentiation of stem cells and their progenitors.

Integrins in the Immune System

The human body is protected against pathogens such as viruses, bacteria, and parasites by the constantly alert immune system. This sophisticated defense system is guaranteed by a multitude of well-orchestrated interactions between different cell types, which derive from a specific progenitor cell generated from haematopoietic stem cells. Immune cells belong to either the innate or the adaptive immune system. Innate immunity is mediated by macrophages, natural killer (NK) cells, granulocytes, and monocytes that interact with pathogens in a non-specific manner. The adaptive immunity is based on T and B lymphocytes that specifically recognize a certain pathogenic antigen (Ag) and build up an immunological memory, which will provide a faster and enhanced reaction upon repeated encounter with the same Ag. At the interface between innate and adaptive immunity are the dendritic cells (DCs). These highly specialized cells take up and process Ag at the periphery and then migrate to the lymph nodes, where they present Ag-derived peptides to resting lymphocytes, thereby initiating the adaptive immune response.

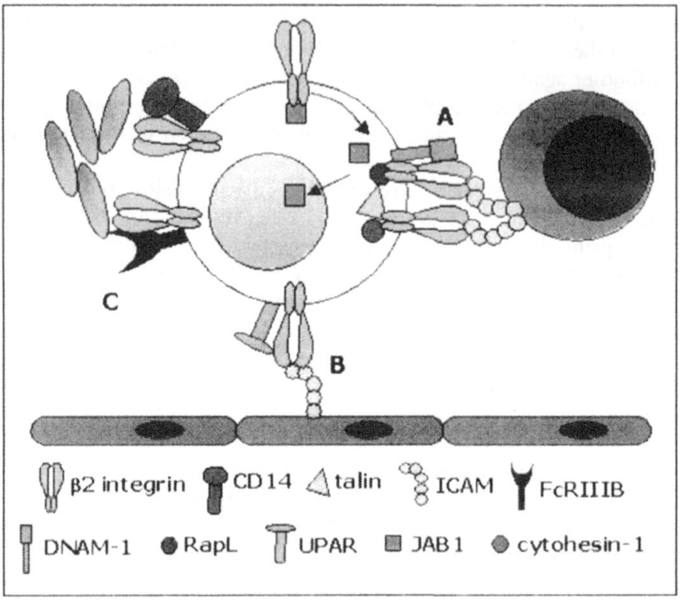

Figure 2. β2 integrins and interacting proteins in the immune system. Integrins are involved in cell-cell interactions (A), cell-endothelium interactions (B) and cell-pathogen interactions (C). Interacting proteins are involved in activating integrins, signal transduction and cytoskeletal rearrangement.

Integrins play important roles in various phases of the immune response by mediating cell-pathogen, cell-cell, and cell-extracellular matrix interactions. They are not merely 'sticky' receptors, but deliver 'outside-in' signals across the cell membrane by establishing *cis* associations with other receptors on the same cell, thus forming multi-molecular assemblies (Fig. 2).

Several integrin-associated proteins have been identified that can modulate integrin function in different ways: by altering the affinity of the integrin for the ligand by directly altering integrin conformation; by multimerization and recruitment of additional integrin receptors to form integrin clusters; or by the association of integrins with other molecules to form an active signaling complex that also acquires increased ligand binding potential.

Most integrin-associated proteins identified so far do not share any common structural relationship (see Table 1). Both GPI-anchored, trans-membrane or cytoplasmic proteins have been reported to associate with integrins.[39-41]

The association of integrins with GPI-linked proteins has raised the question whether integrin function could be modulated by the lipid microenvironment on the cell membrane. Indeed, there is increasing evidence that lipid microdomains enriched in glycosphingolipids and cholesterol, also known as *lipid rafts* (see Box 1) provide scaffolding platforms where different types of receptors—including integrins—having similar affinity for a certain lipid microenvironment, can meet and subsequently mediate specific signaling events.

In general, the involvement of integrins in membrane supramolecular complexes constituted by proteins as well as specific lipids is an emerging area of research particularly important in the immune system.

Antigen Recognition

A central element in immunity is the recognition of invading pathogens by leukocytes. The latter can be broadly divided into lymphocytes and phagocytes. Phagocytes (monocytes, macrophages, and granulocytes) form the 'first line' of defense and are equipped with a variety of

Table 1. Beta1, 2 and 7 interacting proteins in the haematopoietic system

Interacting Protein	Integrin	Function
Cytoplasmic proteins		
Cytoskeletal-associated proteins		
α-actinin[98,99]	β1, β2	actin-bundling protein
Talin[82,100-102]	β1, β2, β7	actin-binding protein, adaptor protein
Filamin[100,103,104]	β1, β2, β7	actin-binding protein
Myosin-X[84]	β1	actin-based motor protein
Adaptor and signaling proteins		
Cytohesin-1[61,62,105]	β2	ARF guanine nucleotide exchange factor
Cytohesin-3[106]	β2	ARF guanine nucleotide exchange factor
ILK[107]	β1	Integrin-linked kinase
FAK[108]	β1, β2	Focal adhesion associated kinase
RapL[66]	αL	Rap1 associating protein
Rack1[109]	β1, β2	WD protein, receptor for activated PKC
WAIT-1[90]	α4β7, αEβ7	WD protein
Paxillin[78,110]	α4β1	intracellular adaptor protein
RanBPM[111]	β1, β2	molecular scaffold protein
ICAP-1[112]	β1	integrin-cytoplasmic-domain associated protein-1
Nischarin[113]	α5β1	cytoplasmic protein
Nuclear/cytoplasmic proteins		
JAB1[60]	β2	Jun activating binding protein-1
FHL2(DRAL)[114]	β1, β2	LIM-only protein
Calcium binding proteins		
Calnexin[115]	β1	membrane-bound calcium-binding chaperone in ER
Calreticulin[116]	α	luminal ER calcium-binding protein
Peripheral proteins		
ProMMP-2[117]	β2	pro matrix metallo protease
ProMMP-9[117]	β2	pro matrix metallo protease
Glycosylphosphatidyl-inositol (GPI)-anchored proteins		
UPAR(CD87)[75,118]	αLβ2, αMβ2	urokinase-type plasminogen activator receptor
FcγRIIIB(CD16B)[119]	αMβ2	IgG Fc receptor
CD14[47,50,120]	αMβ2	LPS-binding protein
GPI-80[121]	β2	GPI-anchored protein
Transmembrane proteins		
Tetraspannins		
CD9[122]	α4β1, α5β1	membrane microdomain stabilizing proteins
CD53[123]	α4β1	
CD63[123]	α4β1, β2	
CD81[71,123]	α4β1	
CD82[123]	α4β1	
CD151/PETA-3[124]	α4β1, α5β1	
Transporters/channels		
CD98[125]	β1	amino acid transporter
16K H-ATPase[126]	β1	ATP-driven proton transporter in vesicles
Kv1.3[127]	β1	voltage-gated K^+channel
Other transmembrane proteins		
CD44[72,128]	α4β1	adhesion receptor
Caveolin-1[129]	β1	transmembrane adaptor protein
DNAM-1(CD226)[63]	αLβ2	transmembrane protein
EWI-2[80]	α4β1	transmembrane protein

Box 1. What are lipid rafts?

The plasma membrane has a nonrandom and asymmetrical distribution of lipid molecules within the bilayer. These lipids are mainly phospholipids, (glyco)-spingolipids, and cholesterol. Phospholipids tend to be loosely packed in bilayers, forming *liquid disordered* membrane. In contrast, sphingolipids have longer and more saturated alkyl chains than phospholipids and, in membranes, they show stronger lateral cohesion, generating tightly packed regions (*liquid ordered* domains). Cholesterol preferentially interacts with sphingolipids and occupies the space between the acyl chain, confering to these liquid ordered domains the capacity to float in the disordered phospholipid membrane. The combination of sphingolipids and cholesterol in small domains is responsible for the formation of lipid rafts microdomains.[130]

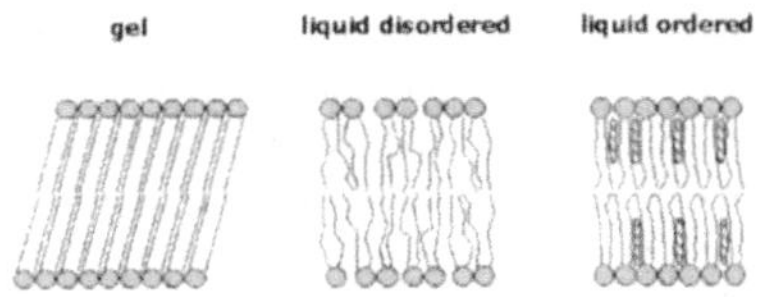

The size of rafts and their kinetics are still largely unknown, although several recent studies using different high-resolution microscopy techniques suggest that the basic raft units can be as small as 5-70 nm in diameter and the half-life can range from milliseconds to minutes.[131] Several membrane proteins have been shown to reside in lipid rafts, either permanently or temporarily, and this association can be facilitated by covalent lipid modification of the protein molecule. In fact, addition of lipid chain (N-myristoylation, palmitoylation, etc) to proteins increases their hydrophobicity and therefore their propensity to associate with cellular membranes.[132]
Lipid rafts on the plasma membrane can serve as platforms where different receptors with similar affinity for the lipid domains meet and orchestrate various signaling event.[130] Moreover, the existence of lipid domains also on the membrane of cell organelles seems to be responsible for appropriate protein/lipid sorting to specific cell compartments.[133] Several viruses or bacteria exploit lipid rafts on the plasma membrane to enter the host cell.[134]
Several recent reports have shown that lipid rafts can influence integrin-mediated cell adhesion and signaling. Membrane cholesterol is directly involved in the association of CD47 molecules with G proteins and the αVβ3 integrin: at low cholesterol levels, the signaling complex dissociates.[135] Moreover, cholesterol depletion strongly inhibits αLβ2-mediated adhesion of murine T cells[136] and abrogates rapid chemokine triggering of a4-integrin-dependent peripheral blood lymphocytes adhesion to endothelium.[137]
A better understanding of integrin function in the lipid microenvironment of plasma membranes will hopefully provide clear clues regarding integrin-mediated phenomena.

pathogen uptake receptors including Fc receptors (FcR), complement receptors, among which CR3 (the αMβ2 integrin also known as Mac-1) and CR4 (the αXβ2 integrin also known as p150/95), and more recently discovered carbohydrate-binding proteins (such as C-type lectins and scavenger receptors). As discussed above, cross-talk between associating FcR and αMβ2 or αXβ2 enhances their functional activity.[42]

Using Fluorescence Resonance Energy Transfer (FRET), which allows detection of interacting molecules within 10 nm distance, FcRIIIB was shown to interact with αMβ2 on neutrophils membrane.[43] In fact, since FcRIIIB is a GPI-anchored molecule without intracellular portion, it completely depends on lateral association with αMβ2 for signal transduction. This association is invaluable to effectively promote antibody-dependent phagocytosis of opsonized microorganisms.[43] Similarly, it has recently been shown that an interaction between FcγRIII and αMβ2 enhances binding to iC3b, a critical complement-3 fragment that opsonizes pathogens and apoptotic cells after complement activation, on monocytes.[44] This interaction is lost during monocyte differentiation into dendritic cells (DCs),[45] potent antigen-presenting cells, where binding of iC3b via αMβ2 is impaired.[44]

Another GPI-linked molecule that associates with αMβ2 is CD14. This protein binds to bacterial lipopolysaccharide and subsequently associates with αMβ2 contributing to the generation of proinflammatory cytokines. Interestingly, this association occurs only in cell that circulate in the blood and is lost upon adhesion on a substrate, emphasizing the dynamic nature of these lateral interactions.[43]

The integrin αMβ2 is involved in pathogen recognition by mediating adhesion of lymphocytes to the fungus *Candida albicans.*[46] This interaction occurs via cooperation between the I domain of the α subunit and the lectin-like domain of the β2 chain. The αMβ2-mediated adhesion of lymphocytes to *C.albicans* is a fundamental prerequisite to trigger the antifungal effect of lymphocytes and to inhibit hyphal growth of this pathogenic fungus.[46]

Several other microorganisms have also been shown to exploit this β2 integrin to gain access into the host cells. For example, interaction between αMβ2 and *Histoplasma capsulatum, Blastomyces dermatitidis,*[47] *Mycobacterium tuberculosis,*[48] and *Leishmania sp.*[49] have been documented.

Another member of the β2 integrin family, αLβ2, has been shown to mediate the attachment of bacteria to the host cell. More specifically, the periodontopathogenic organism, *Porphyromonas gingivalis,* possess fimbriae on its cell surface that mediate adherence to human monocytes. The bacterium uses these fimbriae to bind CD14, Toll-like receptor-2 and the αL subunit, thereby inducing production of proinflammatory cytokines, such as IL-6. The interaction does not seem to depend on the intact αLβ2 dimer, since antibodies against αL, but not β2, inhibited IL-6 production on monocytes after challenge with *P. gingivalis.*[50] The same integrin also contributes to cell-to-cell transmission of HIV-1 by interacting with virus-anchored host ICAM-1.[51,52] In this way, virus dissemination exploits interactions between host molecules that are incorporated within HIV-1 and their natural counter-receptors.

A further example of viruses exploiting integrins to gain entry to host cell is the human parvovirus B19 that requires activation of α5β1 for internalization.[53] The major cell surface receptor for this virus is the blood group P antigen, which only allows binding but not internalization. Replication of the pathogenic parvovirus B19 is restricted to erythroid progenitor cells. Mature red blood cells, which express high levels of P antigen but not the α5β1 integrin, bind parvovirus but do not allow viral entry. By contrast, primary human erythroid progenitor cells express high levels of both P antigen and α5β1, which facilitate β1 integrin-mediated entry of parvovirus B19.[53]

Antigen Presentation

A stable interaction between a T lymphocyte and an antigen-presenting cell (APC) is a prerequisite to effectively trigger the signaling cascade for T cell activation and effector function.[54]

The sole interaction between TCR and peptide/MHC complex is not strong enough to mediate this stable interaction. The αLβ2 integrin, highly expressed on T cells, and its counter receptor ICAM-1, present on the surface of APCs, are key mediators of this interaction.[14,55] More specifically, αLβ2, together with the integrin-associated cytoskeletal protein talin, forms an external ring surrounding a central area rich in TCR, thus participating to the establishment of the immunological synapse (IS) during antigen-dependent interactions with APCs.[56] After 1 hour of conjugation, the TCR disappears from the IS, while clustered αLβ2 is present up to 4 hours.[57] This suggests that prolonged localization of αLβ2 at the T cell-APC facilitates molecular engagements at the contact site necessary for complete T cell activation, which results in T cell proliferation and enhancement of IL-2 production. Besides stabilizing the T cell-APC contact, αLβ2 can provide costimulatory signals, other than CD28, that support and promote TCR-induced T cell activation.[58,59] As such, αLβ2 modulates IL-2 transcription level by interacting with Jun-activating binding protein 1 (JAB-1).[60] JAB-1 colocalizes with αLβ2 at the cell surface by associating to the integrin cytoplasmic tails. The activation of αLβ2 induces phosphorylation of the β2 integrin chain and subsequent release of JAB-1. The nuclear pool of JAB-1 increases and binding of c-Jun-containing activator protein-1 (AP-1)-complexes to their DNA consensus site is enhanced, resulting in activation of AP-1-driven transcription.[56] Along the same line, a second cytoplasmic protein, cytohesin-1 (Cyh-1), contributes to αLβ2-mediated costimulation in a JAB-1 independent manner.[61] Cyh-1, a guanine-nucleotide exchange factor originally identified as dynamic regulator of αLβ2 function in leukocyte arrest on endothelium,[62] is able to mediate Erk1/2 activation that is required for IL-2 production. The underlaying mechanism is based on physical interactions between the C-terminal PH domain of Cyh-1 and the β2 cytoplasmic tail of αLβ2, which results in enhanced clustering of the integrin at the cell surface and strengthening of ICAM-1 binding both in monocytic and lymphocytic cell lines.[62]

An additional partner of αLβ2 that contributes to costimulatory signals is the leukocyte adhesion molecule DNAM-1 (CD226).[63] Upon stimulation of peripheral blood T lymphocytes by anti-CD3, physical association between αLβ2 and DNAM-1 has been reported. Once associated, cross-linking of LFA-1 induces Fyn-mediated tyrosine phosphorylation of DNAM-1. This complex of αLβ2, DNAM-1 and Fyn is found at the IS both in $CD4^+$ and $CD8^+$ T cells. However, while in $CD4^+$ T cells this complex colocalizes with lipid rafts, on $CD8^+$ T cells the αLβ2-mediated costimulatory signal seems lipid raft independent.[63] The relevance of this different involvement of lipid rafts in the different T cell subsets however remains unclear.

Interestingly, when DNAM-1 is recruited into lipid rafts, it can associate with another αLβ2 interacting protein: the small GTPase Rap-1.[64] The constitutively active form of Rap-1 (Rap1V12) increases both affinity and avidity of LFA-1 for ICAM-1, thereby regulating adhesion strength between T cell and APC.[65] This occurs by association of Rap-1 with its effector RAPL.[62] Identified by yeast two-hybrid screening, RAPL was found to associate with the constitutively active Rap1V12, but not with inactive Rap-1, and to coimmunoprecipitate with αLβ2 in cells expressing Rap1V12.[66] Moreover, in vivo RAPL has been reported to colocalize with αLβ2 at membrane protrusions and at the IS, thereby perhaps controlling spatial regulation of αLβ2.[66]

The interaction of αLβ2 with several partners at the IS triggers multiple lymphocyte signaling pathways that eventually will lead to proper T cell activation and expansion.

Leukocyte Migration

Resting T lymphocytes continuously circulate between blood and lymph nodes, waiting for the right APC to provide the proper stimulus. This constant movement of naïve T cells ensures that the immune system continuously surveys for as many different intruders as possible. The ability of naïve T cells to enter lymph nodes or tissues relays on sets of adhesion molecules that are used in a specific time- and space-dependent manner.[67] Although the majority of integrins on unstimulated T cells exist in an inactive state, a small subset of integrins are in a high-affinity state and are particularly important for spontaneous adhesion to endothelium.[68] Under shear flow, the initial interaction of T cells with endothelium consists in rolling of the T cells on the endothelial surface. Besides selectins, also the integrin α4β1 contributes by establishing quick low-affinity interactions with its counter receptor VCAM-1.[69] While affinity modulation does not seem to play any role in regulating tethering and rolling, sub-second induction of α4β1 clustering at the leukocyte-substrate contact zone, upon chemokine stimulation, enhances leukocyte avidity to VCAM-1.[70] This α4β1-mediated transient rolling is a prerequisite for firm adhesion to and migration through the endothelial barrier for lymphocytes as well as monocytes. Recently, it has been shown that the lateral association of α4β1 with the tetraspanin CD81 is critical for rapid adhesion strengthening in monocytes and primary murine B cells interacting with VCAM-1 under shear flow.[71]

Most recent studies show that α4β1 can also mediate firm adhesion of human lymphoid cells onto endothelium by forming a bimolecular complex with CD44.[72] This receptor mediates rolling of activated T cells on hyaluronan on endothelial cells. By both coimmunoprecipitation and transfection studies, it was shown that CD44 specifically associates with α4β1 and not with αLβ2, and that its cytoplasmic tail is essential for this interaction. Lack of the cytoplasmic tail of CD44 abrogates the α4β1 function, inhibiting in vivo lymphocyte trafficking to sites of inflammation.[72]

Engagement of α4β1 with VCAM-1 increases binding between αLβ2 and ICAM-1.[73] More specifically, this α4β1-mediated enhancement of αLβ2 adhesiveness requires the presence of the urokinase receptor (uPAR; CD87) as an activation transducer on the leukocyte cell membrane.[74] This GPI-linked receptor can, besides its role in the fibrinolytic system, upregulate cell adhesion by forming a multimeric receptor complex where tyrosine kinases as well as β2 integrins localize.[75] In this way, the α4 integrin which mediates rolling also stimulates the β2 integrins to initiate firm adhesion and transmigration.

The αLβ2-ICAM-1 pair mediates migration of T cells across the endothelium by stabilizing the formation of a docking structure between T and endothelial cells.[76] Recently, the junctional adhesion molecule-1 (JAM-1) was shown to be a ligand of the integrin αLβ2 during T cell transmigration.[77]

T cells must not only cross the endothelium but must also migrate through the perivascular basement membrane into the tissue or lymph node.

The integrin α4β1 promotes cell migration and antagonizes cell spreading. In particular, the association of the cytoplasmic protein paxillin to the cytoplasmic tail of α4 is essential for α4β1-mediated inhibition of cell spreading.[78] In fact, α4-paxillin interaction promotes focal adhesion kinase (FAK) phosphorylation, which in turn is essential for integrin-dependent cell migration and disassembly of focal adhesions.[78,79]

It has been already mentioned that α4β1-CD81 complexes strengthen the adhesion to VCAM-1 under shear flow. Recently, another cell-surface partner of α4β1, the 4 immunoglobulin domains trans-membrane protein EWI-2, has been documented to regulate α4β1-mediated inhibition of cell spreading rather than interaction with VCAM-1 under shear flow.[80] EWI-2 sequesters α4β1-CD81 complexes into larger α4β1-CD81-EWI-2 complexes, thereby limiting the availability of α4β1 to spread and ruffle on VCAM-1.[80]

Cells move by coordinating the generation of protrusions (lamellipodia) at the front of the cell, which lead to new attachments, and the detachment of previous adhesions at the rear of the cell. The integrin αLβ2 has been also shown to mediate T cell migration on ICAM-1 by influencing the acto-myosin cytoskeleton dynamic organization. This occurs by a compartmentalized regulation of myosin light chain kinase (MLCK) and Rho kinase (ROCK), at the leading edge and at the trailing edge, respectively.[81] The coordination between these two kinases mediates the forward movement of the migrating T cells.

Interestingly, not only do integrins influence the cytoskeletal organization, but the cytoskeleton also modulates integrin activity. The specific binding of the cytoskeletal protein talin to integrin β subunit cytoplasmic tail leads to conformational rearrangements of integrin extracellular domains, thereby increasing the affinity for the ligand.[82] FRET measurements between cyan fluorescent protein- and yellow fluorescent protein-fused αL and β2 provided in vivo evidence for α–β tail associations in the inactive form that are disrupted or rearranged after talin binding and subsequent integrin activation.[83] The unconventional myosin-X (Myo10) is an actin-based motor protein that was recently shown by yeast two-hybrid to interact with integrin β subunit cytoplasmic tail.[84] Knock-down of Myo10 by short interfering RNA impaired integrin function in cell adhesion, whereas overexpression of Myo10 stimulated filopodia formation and elongation in an integrin-dependent fashion. Moreover, Myo10 is also able to relocalize β1 integrins at the tips of the filopodia.[84] Both talin and Myo10 bind to the β chain cytoplasmic tail NPXY motif, which is conserved in six of eight human β integrin cytoplasmic domains (including β1, β2, and β7), and the interaction is mediated by the FERM domain present on both talin and Myo10.[82,84] A tight control of these interactions is fundamental to the regulation of integrin-mediated adhesion and migration.

Although the mechanism regulating integrin-mediated migration can be common for almost all integrins, the integrin repertoire, responsible for different ligand specificity, expressed on the cell surface by each immune cell specifically determines the migration routes towards tissues and/or lymph nodes. In this way, integrins collaborate with other families of receptors involved in homing, such as selectins and chemokine receptors. Recent in vivo studies showed an impaired Th2, but not Th1, effector homing to lungs in allergen-challenged β2-null mice.[85] Moreover, in the spleen, αLβ2 and α4β1 integrins function together to promote both T and B cell transit from the marginal zone into white pulp cords.[86]

β7 integrins are differentially expressed on two subsets of memory T lymphocytes, $\alpha E\beta 7^{HI}\alpha 4\beta 7^{LO}$ and $\alpha E\beta 7^{LO}\alpha 4\beta 7^{HI}$, and regulate their trafficking to mucosal sites.[87,a] In particular, α4β7 is expressed on a distinct subset of memory CD4 and CD8 T cells and directs their trafficking to intestinal sites of inflammation, while αEβ7 is present on intraepithelial lymphocytes, where it mediates their retention between intestinal epithelial cells, and on $CD4^+CD25^+$ and $CD4^+CD25^-$ regulatory T cells (Treg).[87,88] The importance of β7 integrins in lymphocyte migration into gut-associated lymphoid organs is shown by the phenotype of L-selectin/β7 integrin double knock-out mice. In these animals, more than 95% lymphocytes were unable to attach to high endothelial venules of Peyer's patches and other lymphoid tissues.[89] Using a yeast interaction trap screen, a human WD repeat protein, termed WAIT-1, was isolated that interacts with the integrin β7 cytoplasmic tail.[90] WAIT-1 also binds to the cytoplasmic domains of α4 and αE but not to those of integrin β1, β2, and αL subunits. This association of WAIT-1 and β7 integrin was confirmed by coprecipitation, and the binding site for WAIT-1 was mapped to a short membrane-proximal region of the β7 cytoplasmic tail. How this interaction modulates the binding capacity of β7 integrins is at present unknown.

In conclusion, the multitude of integrin-mediated interactions illustrates the importance of these receptors for the immune system, not only to facilitate migration but also to assist in mediating antigen specific signals by stabilizing cell-cell interactions.

Integrin-Associated Disorders

One of the best characterized integrin-associated diseases is leukocyte adhesion deficiency (LAD), a rare autosomal-recessive immunodeficiency.[91] This syndrome results from a variety of mutations in the β2 chain that prevent normal formation of the integrin heterodimer and almost completely abolish cell surface expression.[91] This disease is associated with an inability of leukocytes to firmly adhere to blood vessels, with subsequent persistent leukocytosis, dramatically reduced accumulation of neutrophils and monocytes at extravascular sites, impaired tissue remodeling, and recurring life-threatening bacterial infection.[92]

Very recently, a variant of this syndrome, LAD-III, has been documented.[93] Although LAD-III shares some clinical features with LAD-I, these patients have only moderately reduced levels of β2 integrins on the surface of circulating leukocytes. This syndrome is characterized by the defective ability of integrins, like α4β1 and all β2, to undergo G-protein coupled receptor-mediated stimulation at endothelial contacts.[93] This results in normal leukocyte rolling but impaired firm adhesion in response to chemoattractants displayed on the endothelial surface. Therefore, in LAD-III patients, integrins are present at the cell surface but are unable to sense activating inside-out stimuli (an abnormal Rap-1 machinery seems to be the cause of this syndrome[94]), while they can normally respond to stimulatory antibodies or cations.[93]

Accumulating evidence indicates the contribution of abnormal expression and function of integrins to hematological malignancies.

Defective adhesion through α4β1 and α5β1 integrins on chronic myelogenous leukemia (CML) is in part responsible for the abnormal premature CML $CD34^+$ cells in the blood.[95] Impairment of integrin-mediated regulation of CML progenitor growth and subsequent increased ability of CML $CD34^+$ cells to survive without interacting with the BM would contribute to the massive expansion of CML progenitors.[95]

Adhesion defects regarding α4β1 and αLβ2 have been documented for malignant cells within B-lineage acute lymphoblastic leukemia.[96] In these patients, although the expression levels of the integrins were normal, no integrin-mediated adhesion could be detected, and on

[a] HI and LO indicate high and low expression level, respectively.

the majority of cases no activation with stimuli like phorbol 12-myristate 13-acetate (PMA) could be obtained for either αLβ2 or α4β1.[96] Whether this integrin unresponsiveness is due to mutations in the integrin molecule or in integrin-associated molecules is not known. The integrin αLβ2 and its counter receptor ICAM-1 have been also reported to be involved in enhanced survival of T-lineage acute lymphoblastic leukemia cell on BM stroma.[97]

The study of LAD disorders and tumor-associated integrin impairments can provide useful genetic and biochemical models to identify integrin binding partners that regulate both inside-out and outside-in activation mechanisms.

Conclusion

In this chapter we discussed the role of integrins in cells of the hematopoietic system. During the past 20 years a wealth of information has become available illustrating the importance of this family of proteins, not only for the proper development of the many cell types that arise from stem cells, but also for the proper functioning of the immune system. By now we know that integrin functions can be regulated at multiple levels and are indispensable for a properly functioning immune system. However, the precise signaling mechanisms are still unresolved, illustrating the complexity of integrin signaling. Novel techniques such as multiphoton laser microscopy and high resolution fluorescence microscopy will certainly reveal the dynamics of fluorescent protein tagged integrins unraveling the remaining secrets of this intriguing family of receptors.

References

1. Lemischka IR. Microenvironmental regulation of hematopoietic stem cells. Stem Cells 1997; 15(Suppl 1):63-68.
2. Simmons P, Levesque J, Zannettino A. Adhesion molecules in haemopoiesis. Baillieres Clin Haematol 1997; 10:485-492.
3. Yoder M, Williams D. Matrix molecule interactions with hematopoietic stem cells. Exp Hematol 1995; 23:961-967.
4. Papayannopoulou T, Craddock C. Homing and trafficking of hemopoietic progenitor cells. Acta Haematol 1997; 97:97-109.
5. Hurley RW, McCarthy JB, Verfaillie C. Direct adhesion to bone marrow stroma via fibronectin receptors inhibits hematopoietic progenitor proliferation. J Clin Invest 1995; 96:511-512.
6. Jiang Y, Prosper F, Verfaillie CM. Opposing effects of engagement of integrins and stimulation of cytokine receptors on cell cycle progression of normal human hematopoietic progenitors. Blood 2000; 95:846-854.
7. Campbell AD, Long MW, Wicha MS. Haemonectin, a bone marrow adhesion protein specific for cells of granulocyte lineage. Nature 1987; 329:744-746.
8. Hynes RO. Integrins: Versatility, modulation, and signaling in cell adhesion. Cell 1992; 69:11-25.
9. Harris ES, McIntyre TM, Prescott SM et al. The leukocyte integrins. J Biol Chem 2000; 275(31):23409-23412.
10. Ostermann G, Weber KS, Zernecke A et al. JAM-1 is a ligand of the β2 integrin LFA-1 involved in transendothelial migration of leukocytes. Nat Immunol 2002; 3:151-158.
11. van der Vieren M, Crowe DT, Hoekstra D et al. The leukocyte integrin alpha D beta 2 binds VCAM-1: Evidence for a binding interface between I domain and VCAM-1. J Immunol 1999; 163:1984-1990.
12. Wright SD, Levin SM, Jong MT et al. CR3 (CD11b/CD18) expresses one binding site for Arg-Gly-Asp-containing peptides and a second site for bacterial lipopolysaccharide. J Exp Med 1989; 169(1):175-183.
13. Ingalls RR, Golenbock DT. CD11c/CD18, a transmembrane signaling receptor for lipopolysaccharide. J Exp Med 1995; 181(4):1473-1479.
14. Van Kooyk Y, van de Wiel-van Kamenade P, Weder P et al. Enhancement of LFA-1-mediated cell adhesion by triggering through CD2 or CD3 on T lymphocytes. Nature 1989; 342:811-813.
15. Dustin ML, Springer TA. T-cell receptor cross-linking transiently stimulates adhesiveness through LFA-1. Nature 1989; 341:619-624.
16. Lub M, van Kooyk Y, Figdor CG. Ins and outs of LFA-1. Immunol Today 1995; 16:479-483.
17. Carman C, Jun CDV, Salas A et al. Endothelial cells proactively form microvilli-like membrane projections upon Intercellular Adhesion Molecule 1 engagement of leukocyte LFA-1. J Immunol 2003; 171:6135-6144.

18. Grakoui A, Bromley SK, Sumen C et al. The immunological synapse: A molecular machine controlling T cell activation. Science 1999; 285:221-227.
19. Pribila JT, Itano AA, Mueller KL et al. The alpha 1 beta 1 and alpha E beta 7 integrins define a subset of dendritic cells in peripheral lymph nodes with unique adhesive and antigen uptake properties. J Immunol 2004; 172(1):282-291.
20. Rott LS, Briskin MJ, Andrew DP et al. A fundamental subdivision of circulating lymphocytes defined by adhesion to mucosal addressin cell adhesion molecule-1: Comparison with vascular cell adhesion molecule-1 and correlation with β7 integrins and memory differentiation. J Immunol 1996; 156:3727-3736.
21. Cepek KL, Parker CM, Madara JL et al. Integrin αEβ7 mediates adhesion of T lymphocytes to epithelial cells. J Immunol 1993; 150:3459-3470.
22. Higgins JMG, Manlebrot DA, Shaw SK et al. Direct and regulated interaction of integrin αEβ7 with E-cadherin. J Cell Biol 1998; 140:197-210.
23. Delassus S, Cumano A. Circulation of hematopoietic progenitors in the mouse embryo. Immunity 1996; 4:97-106.
24. Hemler ME, Lobb RR. The leukocyte β1 integrins. Curr Opin Hematol 1995; 2:61-67.
25. Williams DA, Rios M, Stephens C et al. Fibronectin and VLA-4 in haematopoietic stem cell-microenvironment interactions. Nature 1991; 352:438-441.
26. Arroyo AG, Yang JT, Rayburn H et al. α4 integrins regulate the proliferation/differentiation balance of multilineage hematopoietic progenitors in vivo. Immunity 1999; 11:555-566.
27. Yanai N, Sekine C, Yagita H et al. Roles for integrin very late activation antigen-4 in stroma-dependent erythropoiesis. Blood 1994; 83:2844-2850.
28. Arroyo AG, Yang JT, Rayburn H et al. Differential requirements for α4 integrins during fetal and adult hematopoiesis. Cell 1996; 85:997-1008.
29. Weissman IL. Developmental switches in the immune system. Cell 1994; 76:207-218.
30. Scott LM, Priestley GV, Papayannopoulou T. Deletion of α4 integrins from adult hematopoietic cells reveals roles in homeostasis, regeneration, and homing. Mol Cell Biol 2003; 23(24):9349-9360.
31. Wagner N, Lohler J, Kunkel EJ et al. Critical role for β7 integrins in formation of the gut-associated lymphoid tissue. Nature 1996; 382:366-370.
32. Lundhal J, Sehmi R, Moshfegh A et al. Distinct phenotypic adhesion molecule expression on human cord blood progenitors during early eosinophilic commitment: Upregulation of β7 integrins. Scand J Immunol 2002; 56:161-167.
33. Hirsch E, Iglesias A, Potocnik AJ et al. Impaired migration but not differentiation of haematopoietic stem cells in the absence of b1 integrins. Nature 1996; 380:171-175.
34. Papayannopoulou T, Priestley GV, Nakamoto B et al. Molecular pathways in bone marrow homing: Dominant role of α4β1 over β2-integrins and selectins. Blood 2001; 98(8):2403-2411.
35. Papayannopoulou T, Priestley GV, Nakamoto B et al. Synergistic mobilization of hemopoietic progenitor cells using concurrent β1 and β2 integrin blockade or β2-deficient mice. Blood 2001; 97(5):1282-1288.
36. Pruijt JFM, van Kooyk Y, Figdor CG et al. Anti-LFA-1 blocking antibodies prevent mobilization of hematopoietic progenitor cells induced by interleukin-8. Blood 1998; 91(11):4099-4105.
37. Velders GA, Pruijt JFM, Verzaal P et al. Enhancement of G-CSF-induced stem cell mobilization by antibodies against the β2 integrins LFA-1 and Mac-1. Blood 2002; 100(1):327-333.
38. Glodek AM, Honczarenko M, Le Y et al. Sustained activation of cell adhesion is a differentially regulated process in B lymphopoiesis. J Exp Med 2003; 197(4):461-473.
39. Porter JC, Hogg N. Integrins take partners: Cross-talk between integrins and other membrane receptors. Trends Cell Biol 1998; 8:390-396.
40. Hemler ME. Integrin associated proteins. Curr Opin Cell Biol 1998; 10:578-585.
41. Brown EJ. Integrin-associated proteins. Curr Opin Cell Biol 2002; 14:603-607.
42. Ortiz-Stern A, Rosales C. Cross-talk between Fc receptors and integrins. Immunol Lett 2003; 90:137-143.
43. Todd RF, Petty HR. β2(CD11/CD18) integrins can serve as signaling partners for other leukocyte receptors. J Lab Clin Med 1997; 129:492-498.
44. Preynat-Seauve O, Villiers CL, Jourdan G et al. An interaction between CD16 and CR3 enhances iC3b binding to CR3 but is lost during differentiation of monocytes into dendritic cells. Eur J Immunol 2004; 34:147-155.
45. Steinman RM. The dendritic cell system and its role in immunogenicity. Annu Rev Immunol 1991; 9:271-296.
46. Forsyth CB, Mathews HL. Lymphocyte adhesion to Candida albicans. Infect Immun 2002; 70(2):517-527.

47. Newman SL, Chaturvedi S, Klein BS. The WI-1 antigen of Blastomyces dermatitidis yeasts mediates binding to human macrophage CD11b/CD18 (CR3) and CD14. J Immunol 1995; 154:753-761.
48. Cywes C, Godenir NL, Hoppe HC et al. Nonopsonic binding of Mycobacterium tuberculosis to human complement receptor type 3 expressed in Chinese hamster ovary cells. Infect Immun 1996; 64(12):5373-5383.
49. Russell DG, Wright SD. Complement receptor type 3 (CR3) binds to an Arg-Gly-Asp-containing region of the major surface glycoprotein, gp63, of Leishmania promastigotes. J Exp Med 1988; 168(1):279-292.
50. Ogawa T, Asai Y, Hashimoto M et al. Bacterial fimbriae activate human peripheral blood monocytes utilizing TLR2, CD14 and CD11a/CD18 as cellular receptors. Eur J Immunol 2002; 32:2543-2550.
51. Bounou S, Giguère J, Cantin R et al. The importance of virus-associated host ICAM-1 in human immunodeficiency virus type 1 dissemination depends on the cellular context. FASEB J 2004; 18(11):1294-1296.
52. Hioe CE, Chien Jr PC, Lu C et al. LFA-1 expression on target cells promotes human immunodeficiency virus type 1 infection and transmission. J Virol 2001; 75(2):1077-1082.
53. Weigel-Kelley KA, Yoder MC, Srivastava A. Alpha5beta1 integrin as a cellular coreceptor for human parvovirus B19: Requirement of functional activation of beta1 integrin for viral entry. Blood 2003; 102(12):3927-3933.
54. Miller MJ, Wei SH, Parker I et al. Two-photon imaging of lymphocyte motility and antigen response in intact lymph node. Science 2002; 296:1869-1873.
55. Sims TN, Dustin ML. The immunological synapse: Integrins take the stage. Immunol Rev 2002; 186:100-117.
56. Monks CRF, Freiberg BA, Kupfer H et al. Three-dimensional segregation of supramolecular activation clusters in T cells. Nature 1998; 395:82-86.
57. Lee KH, Holdorf AD, Dustin ML et al. T cell receptor signaling precedes immunological synapse formation. Science 2002; 295:1539-1542.
58. van Seventer GA, Shimizu Y, Horgan KJ et al. The LFA-1 ligand ICAM-1 provides an important costimulatory signal for T cell receptor-mediated activation of resting T cells. J Immunol 1990; 144:4579-4586.
59. Abraham C, Miller J. Molecular mechanisms of IL-2 gene regulation following costimulation through LFA-1. J Immunol 2001; 167:5193-5201.
60. Bianchi E, Denti S, Granata A et al. Integrin LFA-1 interacts with the transcriptional coactivator JAB1 to modulate AP-1 activity. Nature 2000; 404(6778):617-621.
61. Perez O, Mitchell D, Jager GC et al. Leukocyte functional antigen 1 lowers T cell activation thresholds and signaling through cytohesin-1 and Jun-activating binding protein 1. Nat Immunol 2003; 4(11):1083-1092.
62. Weber KSC, Weber C, Ostermann G et al. Cytohesin-1 is a dynamic regulator of distinct LFA-1 functions in leukocyte arrest and transmigration triggered by chemokines. Curr Biol 2001; 11:1969-1974.
63. Shibuya K, Shirakawa J, Kameyama T et al. CD226 (DNAM-1) is involved in lymphocyte function-associated antigen 1 costimulatory signal for naïve T cell differentiation and proliferation. J Exp Med 2003; 198(12):1829-1839.
64. Ralston KJ, Hird SL, Zhang X et al. The LFA-1-associated molecule PTA-1 (CD226) on T cells forms a dynamic molecular complex with protein 4.1G and human discs large. J Biol Chem 2004; 279(32):33816-33828.
65. Sebzda E, Bracke M, Tugal T et al. Rap1A positively regulates T cells via integrin activation rather than inhibiting lymphocyte signaling. Nat Immunol 2002; 3:251-258.
66. Katagiri K, Maeda A, Shimonaka M et al. RAPL, a Rap1-binding molecule that mediates Rap1-induced adhesion through spatial regulation of LFA-1. Nat Immunol 2003; 4(8):741-748.
67. Butcher EC, Williams M, Youngman K et al. Lymphocyte trafficking and regional immunity. Adv Immunol 1999; 72:209-253.
68. Chen C, Mobley JL, Dwir O et al. High affinity very late antigen-4 subsets expressed on T cells are mandatory for chemokine-independent adhesion strengthening but not for rolling on VCAM-1 in shear flow. J Immunol 1999; 162(2):1084-1095.
69. Berlin C, Bargatze RF, Campbell JJ et al. α4 integrins mediate lymphocyte attachment and rolling under physiologic flow. Cell 1995; 80:413-422.
70. Grabovsky V, Feigelson S, Chen C et al. Subsecond induction of alpha4 integrin clustering by immobilized chemokines stimulates leukocyte tethering and rolling on endothelial vascular cell adhesion molecule 1 under flow conditions. J Exp Med 2000; 192(4):495-506.

71. Feigelson SW, Grabovsky V, Shamri R et al. The CD81 tetraspanin facilitates instantaneous leukocyte VLA-4 adhesion strengthening to vascular cell adhesion molecule 1 (VCAM-1) under shear flow. J Biol Chem 2003; 278:51203-51212.
72. Nandi A, Estess P, Siegelman M. Bimolecular complex between rolling and firm adhesion receptors required for cell arrest; CD44 association with VLA-4 in T cell extravasation. Immunity 2004; 20(4):455-465.
73. Chan JR, Hyduk SJ, Cybulsky MI. Alpha 4 beta 1 integrin/VCAM-1 interaction activates alpha L beta 2 integrin-mediated adhesion to ICAM-1 in human T cells. J Immunol 2000; 164(2):746-753.
74. May AE, Neumann FJ, Schömig A et al. VLA-4 (a4b1) engagement defines a novel activation pathway for β2 integrin-dependent leukocyte adhesion involving the urokinase receptor. Blood 2000; 96(2):506-513.
75. Bohuslav J, Horeisí V, Hansmann C et al. Urokinase plasminogen activator receptor, β2-integrins, and Src-kinases within a single receptor complex of human monocytes. J Exp Med 1995; 181:1381-1390.
76. Barreiro O, Yanez-Mo M, Serrador JM et al. Dynamic interaction of VCAM-1 and ICAM-1 with moesin and ezrin in a novel endothelial docking structure for adherent leukocytes. J Cell Biol 2002; 157:1233-1245.
77. Ostermann G, Weber KSC, Zernecke A et al. JAM-1 is a ligand of the β2 integrin LFA-1 involved in transendothelial migration of leukocytes. Nat Immunol 2002; 3:151-158.
78. Liu S, Thomas SM, Woodside DG et al. Binding of paxillin to a4 integrins modifies integrin-dependent biological responses. Nature 1999; 402:676-681.
79. Rose DM, Liu S, Woodside DG et al. Paxillin binding to the a4 integrin subunit stimulates LFA-1 (integrin aLb2)-dependent T cell migration by augmenting the activation of focal adhesion kinase/proline-rich tyrosine kinase-2. J Immunol 2003; 170:5912-5918.
80. Kolesnikova TV, Stipp CS, Rao RM et al. EWI-2 modulates lymphocyte integrin α4β1 functions. Blood 2004; 103(8):3013-3019.
81. Smith A, Bracke M, Leitinger B et al. LFA-1-induced T cell migration on ICAM-1 involves regulation of MLCK-mediated attachment and ROCK-dependent detachment. J Cell Sci 2003; 116(Pt 15):3123-3133.
82. Tadokoro S, Shattil SJ, Eto K et al. Talin binding to integrin β tails: A final common step in integrin activation. Science 2003; 302:103-106.
83. Kim M, Carman CV, Springer TA. Bidirectional transmembrane signaling by cytoplasmic domain separation in integrins. Science 2003; 301(5640):1720-1725.
84. Zhang H, Berg JS, Li Z et al. Myosin-X provides a motor-based link between integrins and the cytoskeleton. Nature Cell Biol 2004; 6(6):523-531.
85. Lee SH, Prince JE, Rais M et al. Differential requirement for CD18 in T-helper effector homing. Nature Med 2003; 9(10):1281-1286.
86. Lo CG, Lu TT, Cyster JG. Integrin-dependence of lymphocyte entry into the splenic white pulp. J Exp Med 2003; 197(3):353-361.
87. Rott LS, Briskin MJ, Andrew DP et al. A fundamental subdivision of circulating lymphocytes defined by adhesion to mucosal addressin cell adhesion molecule-1 –comparison with vascular cell adhesion molecule-1 and correlation with β7 integrins and memory differentiation. J Immunol 1996; 156:3727-3736.
88. Lehmann J, Huehn J, de la Rosa M et al. Expression of the integrin αEβ7 identifies unique subsets of CD25+ as well as CD25-regulatory T cells. Proc Natl Acad Sci USA 2002; 99(20):13031-13036.
89. Steeber DA, Tang ML, Zhang XQ et al. Efficient lymphocyte migration across high endothelial venules of mouse Peyer's patches requires overlapping expression of L-selectin and beta7 integrin. J Immunol 1998; 161(12):6638-6647.
90. Rietzler M, Bittner M, Kolanus W et al. The human WD repeat protein WAIT-1 specifically interacts with the cytoplasmic tails of beta7-integrins. J Biol Chem 1998; 273(42):27459-27466.
91. Anderson DC, Springer TA. Leukocyte adhesion deficiency: An inherited defect in the Mac-1, LFA-1, and p150,95 glycoproteins. Annu Rev Med 1987; 38:175-194.
92. Bunting M, Harris ES, McIntyre TM et al. Leukocyte adhesion deficiency syndrome: Adhesion and tethering defects involving β2 integrins and selectin ligands. Curr Opin Hematol 2002; 9:30-35.
93. Alon R, Etzioni A. LAD-III, a novel group of leukocyte integrin activation deficiencies. Trends Immunol 2003; 24(10):561-566.
94. Kinashi T, Aker M, Sokolovsky-Elsenberg M et al. LAD-III, a leukocyte adhesion deficiency syndrome associated with defective Rap1 activation and impaired stabilization of integrin bonds. Blood 2004; 103(3):1033-1036.

95. Bhatia R, Munthe HA, Verfaillie CM. Role of abnormal integrin-cytoskeletal interactions in impaired beta1 integrin function in chronic myelogenous leukemia hematopoietic progenitors. Exp Hematol 1999; 27(9):1384-1396.
96. Geijtenbeek TB, van Kooyk Y, van Vliet SJ et al. High frequency of adhesion defects in B-lineage acute lymphoblastic leukemia. Blood 1999; 94(2):754-764.
97. Winter SS, Sweatman JJ, Lawrence MB et al. Enhanced T-lineage acute lymphoblastic leukaemia cell survival on bone marrow stroma requires involvement of LFA-1 and ICAM-1. Br J Haematol 2001; 115(4):862-871.
98. Otey CA, Pavelko FM, Burridge K. An interaction between alpha actinin and the β1 integrin subunit in vitro. J Cell Biol 1990; 111:721-729.
99. Pavelko FM, La Roche SM. Activation of human neutrophils induces an interaction between the integrin beta 2-subunit (CD18) and the actin binding protein alpha-actinin. J Immunol 1993; 151:3795-3807.
100. Pfaff M, Liu S, Erle DJ et al. Integrin β cytoplasmic domains differentially bind to cytoskeletal proteins. J Biol Chem 1998; 273:6104-6109.
101. Calderwood DA, Zent R, Grant R et al. The talin head domain binds to integrin beta subunit cytoplasmic tails and regulates integrin activation. J Biol Chem 1999; 274:28071-28074.
102. Horwitz A, Duggan K, Buck CA et al. Interaction of plasma membrane fibronectin receptor with talin-a transmembrane linkage. Nature 1986; 320:531-533.
103. Sharma CP, Ezzell RM, Arnaout MA. Direct interaction of filamin (ABP-280) with the beta 2-integrin subunit CD18. J Immunol 1995; 154:3461-3470.
104. Loo DT, Kanner SB, Aruffo A. Filamin binds to the cytoplasmic domain of the β1-integrin. Identification of amino acids responsible for this interaction. J Biol Chem 1998; 269:18311-18314.
105. Geiger C, Nagel W, Boehm T et al. Cytohesin-1 regulates beta-2 integrin-mediated adhesion through both ARF-GEF function and interaction with LFA-1. EMBO J 2000; 19:2525-2536.
106. Korthauer U, Nagel W, Davis EM et al. Anergic T lymphocytes selectively express an integrin regulatory protein of the cytohesin family. J Immunol 2000; 164:308-318.
107. Hannigan GE, Leung-Hagesteijn C, Fitz-Gibbon L et al. Regulation of cell adhesion and anchorage-dependent growth by a new beta 1-integrin-linked protein kinase. Nature 1996; 379:91-96.
108. Schaller MD, Otey CA, Hildebrand JD et al. Focal adhesion kinase and paxillin bind to peptides mimicking β integrin cytoplasmic domains. J Cell Biol 1995; 130:1181-1187.
109. Liliental J, Chang DD. Rack1, a receptor for activated protein kinase C, interacts with integrin beta subunit. J Biol Chem 1998; 273:2379-2383.
110. Hyduk SJ, Oh J, Xiao H et al. Paxillin selectively associates with constitutive and chemoattractant-induced high affinity {alpha}4{beta}1 integrins: Implications for integrin signaling. Blood 2004, [DOI 10.1182/blood-2003-12-4402].
111. Denti S, Sirri A, Cheli A et al. RanBPM is a phosphoprotein that associates with the plasma membrane and interacts with the integrin LFA-1. J Biol Chem 2004; 279:13027-13034.
112. Zhang XA, Hemler ME. Interaction of the integrin beta1 cytoplasmic domain with ICAP-1 protein. J Biol Chem 1999; 274:11-19.
113. Alahari SK, Lee JW, Juliano RL. Nischarin, a novel protein that interacts with the integrin alpha5 subunit and inhibits cell migration. J Cell Biol 2000; 151:1141-1154.
114. Wixler V, Geerts D, Laplantine E et al. The LIM-only protein DRAL/FHL2 binds to the cytoplasmic domain of several alpha and beta integrin chains and is recruited to adhesion complexes. J Biol Chem 2000; 275:33669-33678.
115. Lenter M, Vestweber D. The integrin chains beta 1 and alpha 6 associate with the chaperone calnexin prior to integrin assembly. J Biol Chem 1994; 269:12263-12268.
116. Rojiani MV, Finlay BB, Gray V et al. In vitro interaction of a polypeptide homologous to human Ro/SS-A antigen (calreticulin) with a highly conserved amino acid sequence in the cytoplasmic domain of integrin alpha subunits. Biochemistry 1991; 30:9859-9866.
117. Stefanidakis M, Bjorklund M, Ihanus E et al. Identification of a negatively charged peptide motif within the catalytic domain of progelatinases that mediates binding to leukocyte beta 2 integrins. J Biol Chem 2003; 278:34674-34684.
118. Xue W, Kindzelskii AL, Todd 3rd RF et al. Physical association of complement receptor type 3 and urokinase-type plasminogen activator receptor in neutrophil membranes. J Immunol 1994; 152:4630-4640.
119. Xia Y, Borland G, Huang J et al. Function of the lectin domain of Mac-1/complement receptor type 3 (CD11b/CD18) in regulating neutrophil adhesion. J Immunol 2002; 169:6417-6426.
120. Zarewych DM, Kindzelskii AL, Todd 3rd RF et al. LPS induces CD14 association with complement receptor type 3, which is reversed by neutrophil adhesion. J Immunol 1996; 156:430-433.

121. Sendo D, Takeda Y, Ishikawa H et al. Localization of GPI-80, a beta2-integrin-associated glycosylphosphatidyl-inositol anchored protein, on strongly CD14-positive human monocytes. Immunobiology 2003; 207:217-221.
122. Rubinstein E, Le Naour F, Billard M et al. CD9 antigen is an accessory subunit of the VLA integrin complexes. Eur J Immunol 1994; 24:3005-3013.
123. Mannion BA, Berditchevski F, Kraeft SK et al. Transmembrane-4 superfamily proteins CD81 (TAPA-1), CD82, CD63, and CD53 specifically associated with integrin alpha 4 beta 1 (CD49d/CD29). J Immunol 1996; 157:2039-2047.
124. Fitter S, Sincock PM, Jolliffe CN et al. The transmembrane 4 superfamily protein CD151 (PETA-3) associates with beta 1 and alpha IIb beta 3 integrins in haemopoietic cell lines and modulates cell-cell adhesion. Biochem J 1999; 338:61-70.
125. Miyamoto YJ, Mitchell JS, McIntyre BW. Physical association and functional interaction between beta1 integrin and CD98 on human T lymphocytes. Mol Immunol 2003; 39:739-751.
126. Skinner MA, Wildeman AG. beta(1) integrin binds the 16-kDa subunit of vacuolar H(+)-ATPase at a site important for human papillomavirus E5 and platelet-derived growth factor signaling. J Biol Chem 1999; 274:23119-23127.
127. Levite M, Cahalon L, Peretz A et al. Extracellular K(+) and opening of voltage-gated potassium channels activate T cell integrin function: Physical and functional association between Kv1.3 channels and beta1 integrins. J Exp Med 2000; 191:1167-1176.
128. Nandi A, Estess P, Siegelman M. Bimolecular complex between rolling and firm adhesion receptors required for cell arrest; CD44 association with VLA-4 in T cell extravasation. Immunity 2004; 20:455-465.
129. Wary KK, Mainiero F, Isakoff SJ et al. The adaptor protein Shc couples a class of integrins to the control of cell cycle progression. Cell 1996; 87:733-743.
130. Simons K, Toomre D. Lipid rafts and signal transduction. Nat Rev Mol Cell Biol 2000; 1:31-39.
131. Manes S, Ana Lacalle R, Gomez-Mouton C et al. From rafts to crafts: Membrane asymmetry in moving cells. Trends Immunol 2003; 24(6):320-326.
132. Bhatnagar RS, Gordon JI. Understanding covalent modifications of proteins by lipids: Where cell biology and biophysics mingle. Trends Cell Biol 1997; 7:14-20.
133. van Meer G, Sprong H. Membrane lipids and vesicular traffic. Curr Opin Cell Biol 2004; 16(4):373-378.
134. Manes S, del Real G, Martinez-AC. Pathogens: Raft hijackers. Nat Rev Immunol 2003; 3(7):557-568.
135. Green JM, Zhelesnyak A, Chung J et al. Role of cholesterol in formation and function of a signaling complex involving alphavbeta3, integrin-associated protein (CD47), and heterotrimeric G proteins. J Cell Biol 1999; 146(3):673-682.
136. Marwali MR, Rey-Ladino J, Dreolini L et al. Membrane cholesterol regulates LFA-1 function and lipid raft heterogeneity. Blood 2003; 102(1):215-222.
137. Shamri R, Grabovsky V, Feigelson SW et al. Chemokine stimulation of lymphocyte alpha 4 integrin avidity but not of leukocyte function-associated antigen-1 avidity to endothelial ligands under shear flow requires cholesterol membrane rafts. J Biol Chem 2002; 277(42):40027-40035.

Index

C

D

E

F

T

U

V

W

Z

For Product Safety Concerns and Information please contact our
EU representative GPSR@taylorandfrancis.com Taylor & Francis
Verlag GmbH, Kaufingerstraße 24, 80331 München, Germany